**Resounding Prais**
***Nicholas A. Christ***

# Blueprint

## The Evolutionary Origins of a Good Society

"A dazzlingly erudite synthesis of history, philosophy, anthropology, genetics, sociology, economics, epidemiology, statistics, and more. It uses everything from shipwrecks to the primatologist Jane Goodall to make its pro-kindness case, and it inadvertently shames you into realizing that while most of us, standing at the buffet of knowledge, content ourselves with a pork chop and rice pudding, Christakis pillages the carving station and the omelet station and the soup array and the make-your-own-sundae bar."

—Frank Bruni, *New York Times*

"Mr. Christakis's deep optimism (and considerable evidence) about the arc of human society bending toward good is uplifting. Along the way he delves fascinatingly into human cultures and customs, exploring, for instance, why monogamy and marriage have become so common (though not universal), and what friendship really means, from an evolutionary perspective."

— *The Economist*

"An encouraging, detailed, and persuasive antidote to misanthropy."

—David P. Barash, *Wall Street Journal*

"A multidisciplinary tour de force . . . It brims with good news about human nature."

—Andew Billen, *The Times* (London)

"Christakis sees us as genetically predisposed to be good to one another, even beyond our immediate group. *Blueprint* interweaves engaging examples of people, places, and events to offer hope that humans can form communities under even the most challenging circumstances, such as the small-scale societies that emerge after shipwrecks. Christakis proposes that a 'social suite' of patterns and processes predisposes us to work together to create a 'morally good society,' which enhances individual and group fitness."

—Augustin Fuentes, *Nature*

"*Blueprint* is a refreshing reminder that when people say we're all in this together, it's not just a platitude—it's evolution." —Bill Gates

"The diversity of our cultures and personal identities masks the fact that we are one. In this brilliant, beautiful, and sweeping book, Christakis shows how eight universal human tendencies have bound us together, and given us dominion over our planet, our lives, and our common fate. A masterful achievement that is surely the best and most original science book of the year." —Daniel Gilbert, author of *Stumbling On Happiness*

"Nicholas Christakis is a pioneer in bridging the conceptual chasm between the choices of individual people and the shaping of an entire society. In this timely and fascinating book, he shows how the better angels of our nature, rooted in our evolutionary past, can bring forth an enlightened and compassionate civilization." —Steven Pinker, author of *Enlightenment Now*

"In a book of great wisdom and unusual breadth, Christakis pulls together philosophy, history, anthropology, sociology, genetics, and evolutionary biology to make an extraordinarily optimistic argument: evolution has prewired us for goodness. At a moment when the dark history of the early twentieth century suddenly seems relevant again, it's a relief to be reminded of why so many efforts to reengineer human society have failed—and of why the better side of human nature often triumphs in the end."

—Anne Applebaum, author of the
Pulitzer Prize–winning *Gulag: A History*

"In this wisely optimistic book, Christakis explores the evolutionary imperative of forming bonds that are both cultural and genetic. His writing is colorful, personal, and often exuberant."

—Andrew Solomon, author of *Far from the Tree*

"Christakis brings to general readers his most famous theory: the genetic profile of both humans and animals dictates the types of societies that they create. Using a plethora of accessible examples that range from the social behavior of dolphins and chimpanzees to the tenets that link human behavior in a myriad of settings, from reality shows to arranged marriages, along with a generous look into the author's own past, Christakis reminds us that leadership, friendship, and group tendencies are all rooted in the most fundamental mechanism of our biological sorting: natural selection. A must-read for anyone interested in how we find ourselves wholly divided

into political, religious, and workplace silos, and where these separations may lead us."
—Hope Jahren, author of *Lab Girl*

"A remarkable achievement! Christakis explains, in the most lucid and accessible way imaginable, how our genetic and cultural heritages are deeply intertwined. The story of human nature is no fairy tale, but it nevertheless reveals our potential, and our proclivity, for good."
—Angela Duckworth, author of *Grit*

"In this brilliant and humane book, Christakis defends an optimistic view of humanity. Human nature is not solitary and brutish—we are social beings, capable of intimate ties and great kindness, blessed with extraordinary potential. *Blueprint* is clear, persuasive, and vitally important."
—Paul Bloom, author of *Against Empathy*

"Christakis has found that all human cultures converge on a consistent style of social network, and in *Blueprint* he explores the reasons why. The answer, he boldly argues, lies in our genes. Digging widely, Christakis shows that a gene-based account does not have to challenge the impact of culture, nor does it commit the analysis to reductionism or determinism. *Blueprint* stakes a powerful claim for a richer incorporation of biology into the social sciences."
—Richard Wrangham, author of *Catching Fire*

"In *Blueprint,* Christakis shows that goodness has a biological purpose. More than an ideal pushed upon us by moral and religious leaders, goodness is a survival tactic demanded by our very genes. Christakis's argument about our common humanity, made in such a powerful and vivid fashion, is an important one for these unstable times. He shows that kindness and love are not merely things we can do—but things we must do."
—Brandon Stanton, founder of Humans of New York

"Tribalism is all around us, but it does not have to be. After all, we are all human. In lively and engaging prose, Christakis shows what is possible, and what we must do."
—Eric Schmidt, former chairman of Google

"Mixing brilliant insights with vivid and memorable storytelling, *Blueprint* is both deeply scholarly and, at the same time, a genuine pleasure read."
—Greg Lukianoff, coauthor of *The Coddling of the American Mind*

"Come for the gripping stories about shipwrecks, communes, and Antarctic outposts. Stay for the sociology of networks. As social connectivity and the pace of change both increase in the twenty-first century, Christakis is the essential guide, and this is the essential book. A joy to read, and a warning about the challenges of creating new societies and institutions within which real human beings can flourish."

—Jonathan Haidt, coauthor of *The Coddling of the American Mind*

"A magnificent achievement. If you think you understand human nature, think again; Christakis will open your eyes and make you gasp. A special bonus: His book is inspiring and deeply optimistic. The perfect book for our time."

—Cass R. Sunstein, coauthor of *Nudge*

"As an historian, I probably tend to overemphasize the darker side of human nature—our remarkable capacity as a species for generating war and revolution, manias, and panics. As a physician and a social scientist, Christakis is here to tell me to lighten up. 'There is more that unites us than divides us,' he argues in this deeply erudite and engaging book, 'and society is basically good.' If, like me, you respond to that claim with skepticism, you have a treat in store. Christakis will change your view of the naked ape."

—Niall Ferguson, author of *The Square and the Tower*

"We live in a time rife with 'us' versus 'them' divisions based on class, religion, ethnicity, and politics. But in this majestic, important, and enjoyable book, Christakis rightfully reminds us that we also evolved to live together, cooperate, and thrive in complex, diverse social groups. Now more than ever we need to understand and tap into these deep and fundamental adaptations that help us live and work side by side, value each other, and pursue common cause."

—Daniel E. Lieberman, author of *The Story of the Human Body*

"In the media and online, we live with a daily barrage of the things that divide us—the differences among individuals, groups, and whole societies seem to define the ways we interact with one another. With a broad sweep of history and a deep knowledge of genetics and social science, Christakis takes us along a different path, one that is as important as it is timely. Whether in hunter-gatherer societies, small bands of people brought together by chance, or Silicon Valley corporations, our societies are linked by the common bonds of humanity. In *Blueprint,* Christakis shows how we are much more than divisiveness and division; we are programmed to build and thrive in societies based on cooperation, learning, and love."

—Neil Shubin, author of *Your Inner Fish*

"Christakis takes us on a spellbinding tour of how evolution brings people together, setting the stage for our modern world where online networks connect people in new and unprecedented ways. Our genes don't work in isolation; rather they equip our species with the capacity to join together and make great things. This powerful and fascinating book shows the fundamental good that lies within us, that connects us, and that helps us cooperate beyond the survival of the fittest."

—Marc Andreessen, cofounder and general partner of Andreessen Horowitz

"*Blueprint* is a brilliant and provocative tour de force that could not be more timely. I don't think I've learned this much from a book in a long time. Christakis is the rare author who can combine rigor and erudition with page-turning readability. Filled with fascinating studies, including experiments from his own lab, *Blueprint* ultimately offers reason for hope grounded in science for our difficult times.

—Amy Chua, author of *Battle Hymn of the Tiger Mother*

"In an era when borders close and the perceived differences among us drive the public narrative, Christakis travels across societies and continents to remind us that we share much more than what keeps us apart. *Blueprint* unveils the communities and the social networks that define our successes and failures, and it celebrates the universality of human experience. A powerful and gripping book."

—Albert-László Barabási, author of *The Formula*

"At a time when it seems that nothing can unite us, Christakis cuts through our divisions to reveal a rich and poignant look at our shared human nature. Christakis's trademark passion and broad scholarship are full-throttle as he lays out the ancient recipe for our shared humanity. Compelling, absorbing, and chock-full of delightful examples of what humans can do when they band together, *Blueprint* is a must-read."

—Coren Apicella, University of Pennsylvania

"A blueprint for constructing a good society arrives when we most need it. Christakis has outrageous optimism, rooted firmly in biological and social science, that we will prevail. With a voice that is joyous and uplifting, he teaches us about the core of our nature—this obligatory patterning of ourselves into units called society, with the building blocks being love, friendship, cooperation, and learning. What an enlightened house this blueprint will build if we, the occupants, heed his message about the possibilities that lie within us."

—Mahzarin R. Banaji, coauthor of *Blindspot*

"One of the world's leading social scientists is on the hunt for the biological bounds of human culture, for what we are capable of as a species, and for society's generic tendencies. In this eloquent, wide-ranging book, Christakis finds what turns out to be the good news about what it means to be human."

—Gary King, Harvard University

"*Blueprint* is a timely, powerful, and riveting demonstration of the inherent suite of sensibilities that drive our social life and cultural evolution. An authoritative integration of the social and evolutionary sciences, this engrossing work's great achievement is to definitively shift the focus of social inquiry from what differentiates us to our common humanity, and to show that, while we may be primed for conflict, we are also hardwired for love, friendship, and cooperation, inviting us, should we choose, toward a humane society."

—Orlando Patterson, author of *The Cultural Matrix*

"*Blueprint* is highly original and engrossing. Christakis is a fluent and lucid writer with an arresting personal voice. At the heart of the book is what he describes as 'the social suite'—a set of cultural universals that constitute the core and the blueprint of all societies. Integral to the universality of the social suite is his contention that these key features of all human societies are shaped by natural selection and encoded in our genes. Christakis calls into question a false dichotomy between cultural and genetic evolution. Rather, he regards the two as coexisting in ways that recurrently intersect and influence each other. He shows that the similarities that exist between the social attributes of human and animal societies bind humans together in a way that heightens our common humanity. *Blueprint* is a richly interdisciplinary, deeply documented, brilliant opus on how our long evolutionary history bends toward a good society."

—Renée C. Fox, professor emerita, University of Pennsylvania

"*Blueprint* is an exciting volume that constitutes a major scientific contribution of broad interest. It is a fascinating account of how genes and culture interact and how this knowledge provides the foundations for establishing a Good Society."
—Ernst Fehr, University of Zurich

"With *Blueprint,* a thoughtful discourse on our genetic disposition toward community, Christakis offers a compelling argument, replete with engaging stories, against the reductive notions of so many late-stage capitalists and libertarians. We are wired for society, for cooperation, for engagement, for collective action. Darwin still applies: the survival of the fittest may mean the survival of those among us who can see beyond themselves and work with others doing the same. And therein lies some real cause for optimism."
—David Simon, writer and producer
for *The Wire* and *The Deuce*

"*Blueprint* is an extraordinarily readable and entertaining book that is also one of the most profound among recently published books on evolution. It brings to bear a long history of research to show that cooperation and pro-social traits of humans are genetically based and are the result of evolution by natural selection. By doing this, Christakis corrects one the most frequent misperceptions about biological evolution, namely that inter-individual competition is a law of nature. I only wish this book had been published decades earlier."
—Gunter Wagner, Yale University

"It's rare for a physician to become a prominent social scientist. It's even rarer for him to write a book that opens your eyes to a fresh way of understanding the world. Nicholas Christakis zooms out, Yuval Noah Harari–style, to look at how evolution shapes civilizations. *Blueprint* is a remarkably broad, deep, and provocative exploration of how good societies may be shaped less by historical forces and more by natural selection."
—Adam Grant, author of *Originals*

Also by Nicholas A. Christakis

*Connected: The Surprising Power of Our Social Networks and How They Shape Our Lives*

*Death Foretold: Prophecy and Prognosis in Medical Care*

# BLUEPRINT

## THE EVOLUTIONARY ORIGINS OF A GOOD SOCIETY

NICHOLAS A. CHRISTAKIS

Little, Brown Spark
New York Boston London

Little, Brown Spark
Hachette Book Group
1290 Avenue of the Americas, New York, NY 10104
littlebrownspark.com

Originally published in hardcover by Little, Brown Spark, March 2019
First Little, Brown Spark trade paperback edition, March 2020

Little, Brown Spark is an imprint of Little, Brown and Company, a division of Hachette Book Group, Inc. The Little, Brown Spark name and logo are trademarks of Hachette Book Group, Inc.



ISBN 978-0-316-23003-2 (hc) / 978-0-316-23004-9 (trade pb) / 978-0-316-423915 (international pb)
LCCN 2018957420

Printing 2, 2025

LSC-H

Printed in the United States of America

*The world is better*
*the closer you are to*
*Erika*

# Contents

# Preface: Our Common Humanity

When I was a boy spending the summer in Greece in July of 1974, the military dictators unexpectedly fell from power. A former prime minister, Konstantinos Karamanlis, returned from exile to Syntagma (Constitution) Square in central Athens. Enormous crowds gathered in all the avenues approaching the square, and my mother, Eleni, took me and my brother Dimitri out into the city that night. In the preceding hours, the junta had sent scores of trucks with armed men and megaphones into the streets. "People of Athens," the soldiers blared, "this doesn't concern you. Stay inside."

My mother ignored the warnings. We got as close as a block from Syntagma Square, near the royal palace and the national zoo. She boosted us onto a huge stone wall topped with a wrought-iron fence that kept the animals on the other side from escaping. Dimitri and I stood with our backs pressed against the metal rails in the narrow bit of ledge that was available to us, and my mother stood below us wedged in among everyone else.

The crowd was packed body to sweaty body. When Karamanlis arrived in Athens in the middle of the night, the crowd pulsed with power. The masses began chanting slogans revealing their pent-up frustrations with years of dictatorial rule and foreign meddling: "Down with the torturers!" "Out with the Americans!"

Perhaps oddly for a man who has spent his adult life studying social phenomena, I have never liked crowds. As a child clinging to the fence, I remember feeling excited but mostly afraid. Even at age twelve, I knew that I was witnessing something unusual—certainly an event unlike anything I had ever seen—and that scared me.

The crowd grew louder and angrier. I remember not understanding why, if they were supposed to be celebrating, the people were so agitated. I looked down at my mother with complex feelings of pride and alarm because she—my beautiful, gentle mother—was getting into the spirit of things too. She was proud to be Greek and, like many of her fellow citizens, was rejoicing at the restoration of democracy. I also knew that she was intensely devoted to our education and wanted us to participate in and learn from this historic event. She was the sort of parent who took us to civil rights marches and anti-war demonstrations back in the United States and wanted us to see the world.

But I also felt fear because I could see in my mother's eyes that she was being swept away by a powerful force. I watched uneasily as she became more animated. I worried that she might forget us up on that stone ledge or that we might get separated as the crowds shifted. Suddenly, as the anti-American epithets grew louder, she pointed up at me and my brother and shouted, "Να *οι Αμερικανοί!*"—"*There* are the Americans!"

What could possibly have possessed her to do this? I grew up on a heavy diet of Greek mythology, and I wonder if, in that moment, the story of the child-killing mother Medea might have flashed before my eyes.

To this day, I don't know what my mother meant to accomplish by her outburst. She was a deeply sensible and loving caregiver who adopted children of diverse racial backgrounds in addition to having her own. Why would she draw reckless attention to her beloved sons' outsider status in the midst of a volatile crowd? Did she believe that such a gesture might work to cool the ardor of an unthinking mob? I can't ask her these questions because she died when I was twenty-five and she was forty-seven, after a long illness and a life devoted to scientific and humanitarian causes.

In the years since, I have come to understand some of the primal forces that might have motivated my mother, forces that lie at the core of my arguments in this book and that ordinarily work for the good of our societies. Natural selection has equipped us with the capacity and desire to join groups, and to do so in particular ways. For instance, we can surrender our own individuality and feel so aligned with a collective that we do things that would seem against our personal interests or that would otherwise shock us.

Nonetheless, our ability to be charitable to members of our social groups provides us with something very profound: we can see ourselves as *all* being part of the same group, which means that, in the extreme, we can see that we are all human beings. We can efface the tribalism of small groups and find a kindness for large groups. Knowing my mother's values and her commitment to the common humanity shared by all people, I choose to see her statement this way: she was pleading for forbearance. Clearly, not all Americans were bad people; some were just young boys, like her beloved children.

A few years later, when I was fifteen, I saw another volatile crowd, this time while on a trip to Crete with my socialist grandfather, who was also named Nicholas Christakis. We watched the leader of the Panhellenic Socialist Movement, Andreas Papandreou, whip a vast crowd into a frenzy during an election. By then, I had seen movies of huge political rallies, and I could not believe I was witnessing such a phenomenon myself. We stood way in the back of the crowd, quite safe, but still I could feel its power. My grandfather took me aside and explained that leaders could sometimes draw on people's sense of community and their xenophobia simultaneously, and he also explained the origins of the word *demagogue.* In ancient Greece, it meant the use of exceptional oratorical skills to advance the interests of the common people. I was tremendously roused by the experience, and, despite my grandfather's benign etymology, I can still remember the disturbing feeling that such a crowd might also be capable of injustice.

In his classic work *Extraordinary Popular Delusions and the Madness of*

*Crowds*, published in 1841, Scottish journalist Charles Mackay argued that people "go mad in herds, while they only recover their senses slowly, and one by one."[1] People in crowds often act in thoughtless ways—shouting profanities, destroying property, throwing bricks, threatening others. This can come about partly because of a process known to psychologists as *deindividuation*: people begin to lose their self-awareness and sense of individual agency as they identify more strongly with the group, which often leads to antisocial behaviors they would never consider if they were acting alone. They can form a mob, cease to think for themselves, lose their moral compass, and adopt a classic us-versus-them stance that brooks no shared understanding.

Despite my mostly negative personal experience with crowds, it is clear that they can be a force for good. Even nonviolent crowds can threaten dictators and authoritarian governments—as in Greece in 1974, in China in 1989 (at Tiananmen Square), in Tunisia in 2010 (the Arab Spring), and in Zimbabwe in 2016 (the anti-Mugabe demonstrations). Crowds are especially feared by those in power when they emerge organically, without explicit organization, as they frequently do. In recent years, governments have tried to control access to the internet precisely to keep people from more easily organizing themselves.

Consider the famous organized civil rights marches in the United States, from the March on Washington in 1963 (where Martin Luther King Jr. delivered his famous "I Have a Dream" speech) to the march at Pettus Bridge in 1965 (where the Alabama police brutally beat African-American protesters demanding voting rights). The coalescence of concerned and aggrieved individuals into larger groups reinforces their own beliefs, but it also demonstrates to outsiders a power not available to a similar number of isolated persons acting independently.

For good or ill, forming crowds comes so naturally to our species that it is even seen as a fundamental political right. It is codified in the First Amendment to the U.S. Constitution, which notes that "the right of the people peaceably to assemble, and to petition the Government for a redress of grievances" shall not be infringed by law. The right to assemble is similarly encoded in constitutions of countries all over the world, from Bangladesh to Canada to Hungary to India.[2] As with the capacity for empathy, the inclination to assemble into groups and deliberately choose friends and associates is part of our species' universal heritage.

### *Mutual Understanding*

As I write, the United States seems riven by polarities—right and left, urban and rural, religious and a-religious, insiders and outsiders, haves and have-nots. Analyses reveal that both political polarization and economic inequality are at century-long peaks.[3] American citizens are engaged in vocal debates about their differences, about who can and should speak for whom, about the meaning and extent of personal identity, about the inexorable pull of tribal loyalties, and about whether the ideological commitment to the melting pot in the United States—and to a common identity as Americans—is feasible or even desirable.

Lines appear sharply drawn. It may therefore seem an odd time for me to advance the view that there is more that unites us than divides us and that society is basically good. Still, to me, these are timeless truths.

One of the most dispiriting questions I have encountered in my own laboratory research is whether the affinity people have for their own groups—whether those groups are defined by some attribute (nationality, ethnicity, or religion) or by a social connection (friends or teammates)—must necessarily be coupled with wariness or rejection of others. Can you love your own group without hating everyone else?

I have seen the effects of overidentifying with one's group and witnessed mass delusions up close, and I have also studied them in my lab using experiments with thousands of people and by analyzing naturally occurring data describing the behavior of millions. The news is not all bad. Human nature contains much that is admirable, including the capacity for love, friendship, cooperation, and learning, all of which allows us to form good societies and fosters understanding among people everywhere.

I first began to think about this issue—of how humans are fundamentally similar—almost twenty-five years ago during my work as a hospice doctor. Death and grief unite us like nothing else. The universality of death and of our responses to it cannot help but impress human similarity upon any observer. I have held the hands of countless dying people from all sorts of backgrounds, and I do not think I have met a single person who didn't share the exact same aspirations at the end of life: to make amends for mistakes, to be close to loved ones, to tell one's story to someone who will listen, and to die free of pain.[4] The desire for social connection and interpersonal understanding is so deep that it is with each of us until the end.

My vision of us as human beings, which lies at the center of this book, holds that people are, and should be, united by our common humanity. And this commonality originates in our shared evolution. It is written in our genes. Precisely for this reason, I believe we can achieve a mutual understanding among ourselves.

In highlighting this, I want to be clear that I am not saying that there are no differences among social groups. It's obvious that some groups struggle with social, economic, or ecological burdens that other groups can only imagine. It's not immediately obvious what modern-day hunter-gatherers in Tanzania's Rift Valley might share with software engineers in California's Silicon Valley. But a focus on the differences among human groups (fascinating and actual though they might be) overlooks another fundamental reality. Our preoccupation with differences is akin to focusing on variations in weather between Boston and Seattle. Yes, one will find different temperatures, amounts of precipitation and sunshine, and wind conditions in these two cities, and these can matter (possibly a lot!). Nevertheless, the *same* atmospheric processes and underlying physical laws hold in both of them. Moreover, weather around the world is inextricably linked. We could even say that the fundamental point of studying the planet's diverse microclimates is not to enhance the understanding of local weather conditions but rather to have a fuller understanding of weather in general.

Therefore, I am less interested in what is different among us than in what is the same. Even though people may have varied life experiences, live in different places, and perhaps look superficially different, there are significant parts of others' experiences that we can all understand as human beings. To deny this would mean abandoning hope for empathy and surrendering to the worst kind of alienation.

This fundamental claim about our common humanity has deep philosophical roots as well as empirical foundations. In his essay "The Culture of Liberty," Peruvian novelist Mario Vargas Llosa notes that people who live in the same place, speak the same language, and practice the same religion obviously have much in common. But he points out that these collective traits do not fully define each individual. Seeing people only as members of groups is, he says, "inherently reductionist and dehumanizing, a collectivist and ideological abstraction of all that is original and creative in the human being, of all that has not been imposed by inheritance, geography, or social pressure." Real, personal identity, he argues, "springs from the capacity of human beings to resist these influences and counter them with free acts of their own invention."[5]

True enough. But the exercise of individual freedom and the focus on our individuality is just one way to efface tribalism. We can also broaden our perspective to the level of our *universal* heritage. As human beings, we have a shared inheritance, shaped by natural selection, regarding how to live with one another. This inheritance gives us a mechanism to abandon a dehumanizing perspective that privileges difference.

Think of how exposure to a foreign culture can be both a bracing and a reassuring experience. What starts as a heightened sensitivity to differences in attire, smells, appearances, customs, rules, norms, and laws yields to the recognition that we are similar to our fellow human beings in numerous fundamental ways. All people find meaning in the world, love their families, enjoy the company of friends, teach one another things of value, and work together in groups. In my view, recognizing this common humanity makes it possible for all of us to lead grander and more virtuous lives.

Ironically, many people come to this realization during war, which is the starkest manifestation of between-group animosity. There is a poignant demonstration of this in *Band of Brothers*, a 2001 television series based on the experiences of Easy Company in the 101st Airborne Division during World War II. One of the real-life soldiers, Darrell "Shifty" Powers, speaking late in his life in documentary footage that accompanied the show, made the following observation about a German soldier: "We might have had a lot in common. He might've liked to fish, you know, he might've liked to hunt. Of course, they were doing what they were supposed to do, and I was doing what I was supposed to do. But under different circumstances, we might have been good friends."[6] Not just friends but *good* friends. In a 2017 documentary about another war—a series called *The Vietnam War,* directed by Ken Burns and Lynn Novick—a Vietcong veteran named Le Cong Huan came to a similar realization. As a young soldier, he looked through the trees at the Americans after a bloody battle, and he had the sudden sense of our shared humanity: "I witnessed Americans dying. Even though I didn't know their language, I saw them crying and holding each other. When one was killed, the others stuck together. They carried away the body, and they wept. I witnessed such scenes and thought, 'Americans, like us Vietnamese, also have a profound sense of humanity.' They cared about each other. It made me think a lot."[7]

### *A Blueprint for a Good Society*

Where does this cross-cultural similarity come from? How can people be so different from—even go to war with—one another and yet also be so similar? The fundamental reason is that we each carry within us an evolutionary blueprint for making a good society.

Genes do amazing things inside our bodies, but even more amazing to me is what they do *outside* of them. Genes affect not only the structure and function of our bodies; not only the structure and function of our minds and, hence, our behaviors; but also the structure and function of our societies. This is what we recognize when we look at people around the world. This is the source of our common humanity.

Natural selection has shaped our lives as social animals, guiding the evolution of what I call a "social suite" of features priming our capacity for love, friendship, cooperation, learning, and even our ability to recognize the uniqueness of other individuals. Despite all the trappings and artifacts of modern invention—our tools, agriculture, cities, nations—we carry within us innate proclivities that reflect our natural social state, a state that is, as it turns out, primarily good, practically and even morally. Humans can no more make a society that is inconsistent with these positive urges than ants can suddenly make beehives.

I believe that we come to this sort of goodness just as naturally as we come to our bloodier inclinations. We cannot help it. We feel great when we help others. Our good deeds are not just the products of Enlightenment values. They have a deeper and prehistoric origin.

The ancient tendencies that form the social suite work together to bind communities, specify their boundaries, identify their members, and allow people to achieve individual and collective objectives while at the same time minimizing hatred and violence. For too long, in my opinion, the scientific community has been overly focused on the dark side of our biological heritage: our capacity for tribalism, violence, selfishness, and cruelty. The bright side has been denied the attention it deserves.

## CHAPTER 1

# The Society Within Us

After World War II, when my mother, an ethnic Greek raised in Istanbul, was a little girl, she spent her summers on the island of Buyukada, a short ferry ride from the coast. Many years later, in 1970, she took her children to visit. The Greeks had always called the island Prinkipos (the Princes' Island) and resented its Turkish name. The place had changed little since Leon Trotsky went into exile there in 1929. Then, as now, it did not allow motorized transport, and people got around on foot, on donkeys, or in horse-drawn carriages that slipped on the cobblestones. In 1970, it had been two decades since my mother had been there because she and her parents, like other minorities, had been driven out of Turkey in the 1950s during a period of substantial interethnic strife.

My younger brother, Dimitri, and I were only eight and six, and although we could speak Greek, we could not speak Turkish. Still, we ventured out and found a dozen boys with whom to play. In the pine-covered hills behind my grandfather's abandoned time capsule of a house, with its wood-fired water heater and blistered green shutters, we boys initially organized ourselves into a large group, working together to explore the terrain and communicating, via pantomime, the urgent necessity of accumulating large piles of pinecones. Eventually—inevitably—we decided to split into two teams and engage in combat, lobbing cones at each other and attempting to steal them via furtive raids. A simple market economy emerged in parallel with the brigandage: small green cones that were easier to throw were exchanged for enormous beautiful cones with brittle, exploding petals that we imagined were grenades. Since our ordnance was not destroyed upon being fired, each attack fueled the opponents' supply of weapons. The games—with their petty warfare, barter economy, group solidarity, and occasional cheating—lasted hours.

The Turkish boys were different from my brother and me in some ways, of course. They had shorter haircuts and wore vests. They threw their pinecones sideways from the hip rather than overhand across the shoulder, as we did. And they knew the terrain better. But these differences seemed minor and were easily ignored. The social play in which we engaged was wordlessly comprehended by all. Separated by a significant cultural and linguistic distance, we were able to jointly create a little social order with familiar features that we all enjoyed.

One purpose of play is for children to ape adult behaviors and practice grown-up roles. But play is not just about children being taught, explicitly or implicitly, to act like adults. In many forager societies, adults leave children to play by themselves, and they are often only vaguely aware of what their children are up to. Play arises spontaneously, without any guidance. And play like this—a purely voluntary, intrinsically motivated, and eminently enjoyable experience—very often involves the "experiments in social living" that my Turkish friends and I pursued on the island.[1]

Here is one anthropologist's description of a long-term playgroup consisting of thirteen children in Ua Pou in the Marquesas Islands. The children, who ranged in age from two to five years old, were observed every day for several months playing for prolonged periods without adult supervision in an area near the beach (a spot with "strong surf" and "sharp lava-rock walls" as well as "machetes, axes, and matches" nearby for good measure). They "organized activities, settled disputes, avoided danger, dealt with injuries, distributed goods, and negotiated contact with passing others—without adult intervention."[2] A more systematic set of landmark longitudinal studies of play in places around the world (Nyansongo, Kenya; Khalapur, India; Juxtlahuaca, Mexico; Tarong, Philippines; Taira, Japan; and "Orchard Town," a pseudonym for a town in New England), spearheaded by anthropologists Beatrice and John Whiting and their colleagues from the mid-1950s to the mid-1970s, concluded that, while there was much notable variation by gender, age, and culture in children's typical companions, activities, toys, and venues for play, children's *social* behavior and interaction styles while playing were always extremely similar.[3]

Societies themselves might even be seen as just scaled-up versions of such children's games. In *Homo Ludens,* the classic 1938 book about humans and play, social historian Johan Huizinga goes so far as to argue that "human civilization has added no essential feature to the general idea of play."[4] Children's behavior often involves innately making a kind of miniature and temporary society. From an early age, humans cannot help themselves.

### *Child's Play*

Looking back more than forty years later, I can see that the games my brother and I played with the Turkish boys involved a high degree of social organization with features I have since come to recognize by technical terms: *in-group favoritism, trade complementarity, social hierarchy, collective cooperation, network topology, social learning,* and *evolved morality.*

I have my own laboratory now, but I am still playing with and thinking about these sorts of things. My group has devised specialized software to recruit thousands of adults from around the world and then to track their behavior as they participate in miniature societies that we create online. I manipulate the social interactions in these societies—for instance, randomly assigning people to be rich or poor or surreptitiously dropping in programmable robotic agents who pretend to be real people to see what mischief they cause—in order to peer more deeply into the origins of human social living and to understand where cooperation, cohesion, hierarchy, and friendship come from. My group also explores the evolutionary biology of these phenomena, searching for the ancient origins of social life even as we concoct thoroughly modern examples.

One of the more dispiriting phenomena that we have observed is the in-group favoritism mentioned above—that is, people's preferences for their own groups. It's that warm feeling of belonging to a team that I experienced on Buyukada. In-group favoritism is seen even in preschool children, and many researchers have explored whether this preference is innate. In one experiment, five-year-old children were given T-shirts of different colors (red, blue, green, orange) and then shown pictures of other children wearing T-shirts of the same or different color as their own. The children understood that their shirt colors were randomly assigned. And there was indeed no specific difference among the children in the photographs other than their shirt colors. Still, the children preferred the kids wearing the same T-shirt color; they allocated more of a scarce resource (toy coins) to them; and they reported more positive thoughts about them.[5] They also felt that the kids in their shirt-color group would be more likely to be kind and share toys. And they were better able to remember and recall positive actions of their in-group, encoding favorable information describing those of their own type. All this arose simply because of randomly assigned T-shirt colors. Other studies of in-group bias at even younger ages, at three or five months, further support its innateness.[6]

But this is not the only socially relevant sensibility that we are born with. Humans also appear to have a rudimentary moral sense from birth. And, like the construction of the whole of Euclidean geometry from its few axioms, our inborn moral principles provide a foundation for social behavior that is only later shaped by experience and education.

For instance, psychologist Paul Bloom and his colleagues have documented sensitivity to fairness and reciprocity—which are crucial for cooperation—in babies as young as three months old, using a variety of ingenious experiments.[7] In one experiment, three-month-old babies were shown a blue square "helping" a red circle up a hill and a yellow triangle pushing the circle down. The babies reliably chose the blue square when given a choice (colors and shapes were varied to be sure that those features were not driving the preferences).[8] In other experiments, babies could tell the difference between puppets who helped or hindered actions attempted by other puppets. Babies preferred the good guys, and they disliked the jerks. Still other experiments involving puppets showed that thirteen-month-old babies have a "theory of mind," meaning that they have an understanding of the mental states (knowledge, beliefs, intentions) of others, which is obviously crucial for moral reasoning and helpful for social life.[9] In another set of experiments, toddlers spontaneously and without any prompting helped adults who were pretending to struggle with opening a cabinet.[10] In short, at a very young age, humans appear pre-wired (in the sense of having a strong, innate proclivity) to interact in positive ways, with insight into the intentions of others and with a tendency to care about being fair. It is hardly surprising, therefore, that, although details vary from place to place, every society values kindness and cooperation, defines acts of cruelty, and categorizes people as either virtuous or nasty.

Why are humans this way? Why, even from birth, do we manifest such consistent, socially relevant behaviors? Where do the social principles that guide children's play and shape adult lives come from? And how do humans in every society come to create a similar kind of social order with important and familiar features that are universally regarded as good?

### *Cultural Universals*

It's easy to lose sight of what human societies have in common because, when we look around the planet, we see such wondrous and compelling diversity in technology, art, beliefs, and ways of life. But focusing on the differences among societies obscures a deeper reality: their similarities are greater and more profound than their dissimilarities.

Imagine studying two hills while standing on a ten-thousand-foot-high plateau. Seen from your perch, one hill appears to be three hundred feet high, the other nine hundred feet. This difference may seem large (after all, one hill is three times the size of the other), and you might focus your attention on what local forces, such as erosion, account for the difference in size. But this narrow perspective misses the opportunity to study the other, more substantial geological forces that created what are actually two very similar mountains, one 10,300 feet high and the other 10,900 feet.

In other words, what you see depends on where you stand. And very often, when it comes to human societies, people have been standing on a ten-thousand-foot plateau, letting the differences among societies mask the more overwhelming similarities. Extending the metaphor, consider how specifically human activities, such as farming and mining, reshape the landscape. These human actions might modify the details of the appearance of the hills, but they do not fundamentally change the mountains themselves—their origins relate to deeper forces outside of human control. The same might be said of human culture: it reshapes certain aspects of human social experience, but it leaves many other features solid as a rock.

A broader perspective helps us appreciate this. Astronauts—who are not chosen for their sentimentality—very often come to appreciate how trivial human differences really are. Cosmonaut Aleksandr Aleksandrov put it this way: "We were flying over America, and suddenly I saw snow, the first snow we ever saw from orbit. I have never visited America, but I imagined that the arrival of autumn and winter is the same there as in other places, and the process of getting ready for them is the same. And then it struck me that we are all children of our Earth." When Donald E. Williams, a space-shuttle commander, saw the blue orb suspended in the darkness of space, he observed: "The experience most certainly changes your perspective. The things that we share in our world are far more valuable than those which divide us."[11]

Most experiences that induce such a sense of awe prompt us to feel as if we are transcending our usual frame of reference. Some scientists believe (though it is hard to prove) that awe is an evolved emotion intended to cause a cognitive shift that reduces egocentricity and makes people feel more connected to others. Responding to powerful natural phenomena—like thunderstorms or earthquakes or vast expanses of ice or desert—with a loss of selfishness and an increase in group bonding might have had survival value to ancient humans. A key feature of awe, psychologists Dacher Keltner and Jonathan Haidt have argued, is that it quiets self-interest and makes individuals feel part of the larger whole.[12] According to primatologist Jane Goodall, chimpanzees experience something similar—they can be amazed by things outside themselves and gaze dreamily at waterfalls and sunsets—which suggests a possible evolutionary origin for this feeling.[13]

Still, the perspective of the few souls who have ventured into space notwithstanding, there is a long history of ferocious arguments between those who think there are cultural universals that bind humanity together and those who think that the sheer variety in human experience means that no traits can be truly constant. *Culture* may be defined as the whole set of ideas (and artifacts) produced by a group, ideas that are usually transmitted socially and that are capable of affecting individual behavior. Cultural universals are traits shared by all peoples around the world. The traits' very universality suggests that they were likely shaped by evolution. For example, the fact that people are uniquely identified in every culture (almost always through the use of personal names) suggests that there is something fundamental about personal identity.[14]

Some critics believe that claims about cultural universals are both scientifically and morally suspect. The search for universals is seen as problematic because it would seem to impose standard categories (often Western ones) on all people, and thus it is seen as disrespecting, rather than just looking beyond, human diversity. Some fear that accepting the reality of any particular cultural universal might allow observers to have a position from which to judge alien cultural practices and label them as aberrant.

Some extreme critics see even a single exception to a claimed universal as negating its universality. But universal capacity is not the same as universal expression. And these critics typically overlook the fact that exceptional cases usually have required tremendous pressure to reshape the natural order. For example, there is (as far as we know) only one society in the world—the Baining people of New Guinea—that manages to suppress the innate tendency to play. But this does not mean that Baining children are not pre-wired for play. In fact, subverting the natural urge to play requires a great deal of cultural force, with the Baining adults devaluing play and actively discouraging children who attempt it.[15]

The debate about universals also evokes broader tensions in the sciences. The most famous tension, to which we shall return, revolves around the relative contributions of nature and nurture as explanations for human experience. Those advocating for the existence of universals are generally seen as belonging to the nature camp. Another tension arises between "lumpers" and "splitters." Lumpers seek to group similar things together; splitters identify fine distinctions in the natural world.[16] Still another tension is between those focusing on the average tendency of a phenomenon (such as the average price of a house in a market) and those interested in its variation (for instance, the range of house prices and the forces that contribute to inequality in prices from place to place). But these different agendas—searching for consistency or studying variation—should be seen as complementary, rather than opposing, ways of studying natural phenomena, including our species, scientifically.

In the first half of the twentieth century, social scientists such as Émile Durkheim, Franz Boas, Margaret Mead, and Ruth Benedict held that culture could not be explained by psychological or inherited biological traits. Culture was seen as something deliberately and thoughtfully produced by humans and not reducible to deeper causes.[17] In the 1970s, cultural anthropologist Clifford Geertz argued that, while underlying universals did exist, they were uninteresting in comparison to the variegated ways in which such traits found expression. The degree of abstraction required to identify universals was too great to be of use, he felt.[18] At best, human nature provided an undifferentiated and extremely malleable raw material that was of negligible importance.[19] Hence, in this line of thought, cultural variation was the central focus of scientific inquiries.

Other social scientists have held different views. In 1923, anthropologist Clark Wissler described a "universal pattern" of cultural features, and he proposed that these universals—related to speech, food, shelter, art, mythmaking, religion, personal interactions, and attitudes toward property, government, and war—were rooted in human biology. In 1944, the famous anthropologist Bronislaw Malinowski also discussed the dependence of culture on the "organic needs of man," matching a set of basic needs (such as safety, reproduction, and health) with respective cultural responses (such as protection, kinship, and hygiene).[20]

In his 1945 essay "The Common Denominator of Cultures," anthropologist George Murdock offered an alphabetically ordered "partial list" of universals that was, in actuality, exhaustive and alarmingly detailed (and, in my opinion, tedious and arbitrary). It included everything from personal adornment to sports activities, dream interpretation, sexual practices, soul concepts, and even weather control. Murdock conceived of these universals as being specifics of *classification*, rather than content. In other words, while the precise details of human behavior in any of these domains might differ from place to place, they constituted a common foundation that was rooted in "the fundamental biological and psychological nature of man and in the universal conditions of human existence."[21]

In 1991, anthropologist Donald Brown challenged what he described as the broad taboo against the search for universals in the field of anthropology. He outlined the three broad mechanisms by which cultural features could have become universal: (1) they might have started in one place and diffused widely (like the wheel); (2) they might reflect commonly discovered solutions to challenges that are imposed by the environment and that all humans face (such as the need to find shelter, cook food, and ensure the paternity of offspring); and (3) they might reflect innate features common to all humans (such as the appeal of music, the desire to have friends, or the commitment to fairness). Some universals, although not all of them, must be a product of our evolved human nature.[22]

In a detailed description of a hypothetical "Universal People," Brown enumerated dozens of surface-level linguistic, social, behavioral, and cognitive universals, similar to Murdock's long list, that have been noted by ethnographers:

> In the cultural realm, human universals include myths, legends, daily routines, rules, concepts of luck and precedent, body adornment, and the use and production of tools; in the realm of language, universals include grammar, phonemes, polysemy, metonymy, antonyms, and an inverse ratio between the frequency of use and the length of words; in the social realm, universals include a division of labor, social groups, age grading, the family, kinship systems, ethnocentrism, play, exchange, cooperation, and reciprocity; in the behavioral realm, universals include aggression, gestures, gossip, and

> facial expressions; in the realm of the mind, universals include emotions, dichotomous thinking, wariness around or fear of snakes, empathy, and psychological defense mechanisms.[23]

These fundamental categories of universals are clearly important, and they come into relief when we step off the ten-thousand-foot plateau and move to lower ground.

Variations in seemingly disparate cultural traits may be connected. For example, societies with alphabets have more complex religions than those without. Such a pattern of correlated traits suggests that there is indeed a deeper organizing force that shapes the complexity of human societies, and this has been documented by a study of 414 societies spanning ten thousand years from thirty regions around the world.[24] Many key features of human societies are functionally related; they are not independent; they coevolve in predictable ways; and a single underlying metric can be used to capture them. Methodologically, this is similar to the way that the single underlying metric of expense might explain why disparate features of a car (its acceleration, safety, instrumentation, and amenities) go together.

The evidence for innateness in multiple aspects of human experience has been building across many domains. Psychologist Paul Ekman has proposed a universal connection between core emotions and many facial expressions—particularly for happiness, anger, disgust, sadness, and fear—and suggested an evolutionary basis for them.[25] Such expressions are innate, even if their exact manifestations upon the human face can sometimes be culturally shaped.[26] The study of the universal features of language, championed by linguist Noam Chomsky, psychologist Steven Pinker, and others, provides another fertile area for discerning universals.[27] And ethnomusicologists have verified another category of cultural universal: musical forms.[28] A sample of three hundred and four musical recordings from around the world yielded numerous "statistical universals" (meaning there were few exceptions to the patterns) across nine geographic areas; these spanned features related to pitch and rhythm as well as performance style and social context.

These musical universals may be so fundamental that even other species manifest them; for instance, cockatoos make music by drumming with rhythms similar to our own.[29] Moreover, music's function—whether in birds, elephants, whales, or wolves—may be deliberately social. This observation regarding the appearance of human universals in other species is itself a very powerful idea. If a phenomenon (like friendship or cooperation, for example) is present in our species and also in others, then that phenomenon is an especially good candidate for a universal across groups *within* our own species. If we share a trait with animals, then we can surely share it widely with one another.

Still, the problem with many inventories of universals is that they often seem more like exhaustive lists of features that cultures *can* include rather than a core set of items that cultures *must* include. The latter is my concern here. In addition, I focus on those universal features that are specifically *social* in nature, that have to do with how groups of people function. And, finally, I am interested in universals that have evolutionary rather than ecological origins. That is, I am focused on universals that are encoded in our genes rather than those that have arisen (independently, in multiple locations) simply as an immediate response to the environment the humans are in (for instance, the potentially universal presence of fishing nets in cultures exploiting rivers and seas for food). In this regard, an evolutionary perspective obliges us to focus on traits that evolution can actually act on. The practice of medicine is not something encoded in our genes, even if virtually every society has a tradition of healing, but a desire for health and survival (both in ourselves and in those we love) and also the motivation of one person to help another are indeed hardwired within us.

My own list of universals is thus more focused and more fundamental than these prior efforts. Centered on a crucial set of specifically social features, it is related to why humans make what we believe are good societies. It is derived, the evidence will show, from our species' evolutionary heritage. And it is, at least partially, encoded in our genes. I call this list of universals the *social suite.*

### *The Social Suite*

Human societies are so vibrant, complex, and encompassing that they take on lives of their own. They may seem to be built by others, by powerful people, or by historical forces beyond human comprehension. In the 1970s, when I was a child, some people were so impressed by the apparent sophistication of ancient civilizations in Egypt and the Americas that they fantasized that aliens must have fashioned them. But human societies do not come from somewhere else. They come from within us.

The capacity to band together to make societies is indeed a biological feature of our species, just like our ability to walk upright. This innate capacity—so rare in the animal kingdom—has also made possible what evolutionary biologist E. O. Wilson has called "the social conquest of earth."[30] It's not our brains or brawn that allows us to rule the planet. And, like other behaviors that have helped our species to survive and reproduce, the human ability to construct societies has become an instinct. It is not just something we *can* do—it is something we *must* do.

At the core of all societies, I will show, is the social suite:

(1) The capacity to have and recognize individual identity
(2) Love for partners and offspring
(3) Friendship
(4) Social networks
(5) Cooperation
(6) Preference for one's own group (that is, "in-group bias")
(7) Mild hierarchy (that is, relative egalitarianism)
(8) Social learning and teaching

These features arise from within individuals but they characterize groups. They work together to create a functional, enduring, and even morally good society.[31] Individual identity provides a foundation for love, friendship, and cooperation, allowing people to track who is who across time and place and to faithfully repay kindness offered by others. Love is a particularly distinctive human experience (built on a trait seen in only a few other mammals, namely, the practice of bonding with mates). Love also paves the way, evolutionarily speaking, for us to feel a special connection not only to our kin, but also, ultimately, to unrelated individuals. That is, humans have friends, and this, too, is a crucial part of the social suite. We form long-term, nonreproductive unions with other humans. This is exceedingly rare in the animal kingdom, but it is universal in us. As a consequence of having friends, we assemble ourselves into social networks, and here, too, the particular ways we do this are universal. The mathematical patterns of friendship are the same around the world. Humans everywhere also cooperate with one another. And this cooperation is supported not only by the fact that we reliably interact with friends rather than strangers within the face-to-face networks we fashion, but also by the fact that we form groups whose boundaries we enforce by coming to like those within the group more than those outside of it. People everywhere choose their friends and prefer their own groups. In turn, cooperation is a crucial predicate for social learning, one of our species' most powerful inventions. No human has to learn everything on his or her own; we can all rely on others to teach us, a hugely efficient practice present in all cultures. And friendship networks and social learning, finally, set the stage for a kind of mild hierarchy in humans in which we accord more prestige to some group members—typically, those who can teach us things or who have many connections—than to others.

These features relate to banding together, and they are highly useful for survival in an uncertain world, offering us ways to acquire and transmit knowledge more efficiently and allowing us to pool risk. These traits are evolutionarily rational, in other words, enhancing our Darwinian fitness and advancing our individual and collective interest. By endowing us with social sensibilities and behaviors, our genes help to shape the societies we make on both small and large scales.

This manufactured social environment in turn creates a feedback loop across evolutionary time. Throughout history, humans have lived surrounded by social groups, and the presence of our fellow humans—people we must interact with, cooperate with, or avoid—has been as powerful as any predator in shaping our genes. Evolutionarily speaking, our social environment has shaped us as much as we have shaped it. Moreover, although the physical, biological, and social environments have all been pivotal in our evolution, they differ in one substantial respect. Aside from the (hugely important) mastery of fire over a million years ago, it is only in the past few thousand years that humans have been able to significantly shape their physical and biological environments—by damming rivers, domesticating plants and animals, generating air pollution, using antibiotics, and so on. Prior to the invention of agriculture and cities, humans did not build their physical environments; they simply chose them. By contrast, humans have always made their social environments.

Living socially places special demands on us, and many cognitive capacities and behavioral repertoires evolved in order for us to cope. For example, we are innately equipped to cooperate, and living in cooperative groups favors certain genetic predispositions related to kindness and reciprocity. We are wired to have friends, not just reproductive partners, and when we form friendships, we modify the social world around us in a way that makes friendliness useful. Individuals lacking these prosocial capacities are not as successful in their efforts to survive and reproduce as others. Our genes guide us to create a social environment that feeds back and favors particular kinds of genes that are useful in the environments we have created.

Over evolutionary time, for this reason too, humans have genetically internalized universal social axioms.

Core features of human societies are guided by a *blueprint* that our species has helped to sketch over eons. Some evolutionary biologists bristle at the metaphor of a blueprint.[32] The reason for this is partly that they consider blueprints to be fixed and deterministic. Another issue, though, is that if you construct a building based on a set of plans, someone else can inspect that building and then work backward to create its blueprint, whereas with the genetic code that provides instructions for an organism, no one can inspect the organism and then easily work backward to the code. You can go in both directions with a blueprint but in only one direction with genetic code. As a result, these scientists prefer metaphors like programs or recipes, but one can indeed make some predictions about a recipe by inspecting a prepared food. And in any case, one cannot always re-create the exact original blueprint from inspecting a building. Blueprints are not necessarily fully realized or even complete. They are open to interpretation. Though blueprints are specific guidelines, they can be modified—revised by the architect, interpreted by the builder, or changed by the occupants.

More important, from the point of view of using this sometimes controversial metaphor, when I use it here, I do not mean that genes *are* the blueprint. I mean that genes act to *write* the blueprint. A blueprint for social life is the product of our evolution, written in the ink of our DNA.

Our evolutionary past compels us universally to make a basic, obligatory sort of society. This blueprint also means that societies have some shapes that they *cannot* assume and some constraints that they must observe, both of which we will explore. Humans can deviate from the blueprint—but only up to a point. When they deviate too much, as we will see, society collapses.

### *What Unites Us*

Cross-cultural variations in traits and behaviors have attracted tremendous interest for a long time, and these variations have often been used in a deplorable way to justify disdain for, or oppression of, "outsiders." These cultural variations are sometimes linked to observations about genetic variations in human physical traits (such as different types of hemoglobin, which can confer benefits such as tolerance of high altitudes or resistance to malaria).[33] This might make it seem reasonable to search for genetic causes for variation in cultural practices, and there is some limited evidence for this with respect to traits such as violence, novelty seeking, risk aversion, and migratory behavior.[34] But genes surely explain very little of the variation among cultural groups in the long lists of traits propounded by anthropologists. There are no genes for surgery or idolatry that explain why some societies cut people open or make images of gods. Such variation is due to culture.

Nevertheless, even if our genes do not explain cultural *variation,* they can explain cultural *universals.* Moreover, genes can explain why culture exists at all. Evolution provides the underlying foundation for human culture by equipping us with the ability to cooperate, make friends, and learn socially. We manifest cultural variation precisely because we evolved to have this capacity in the first place.

When some scientists describe the evolutionary basis of behaviors, whether at the individual or societal level, they often focus on the differences between humans that can divide and even fracture us. But when I speak of a blueprint, my interest here is altogether different. I am not saying that *differences* across societies are based on our genes. Rather, I am saying that the *similarities* across societies—instantiated in the social suite—are based on our genes.

I am interested in the deep social features all humans share, in where these features came from and what biological and sociological purpose they serve, and in how they continue to shape our societies no matter the cultural details. A relatively small set of universal features supports the self-assembly of humans into societies. If we took groups of people from anywhere on the planet and let them form societies on their own, without any formal guidance or authority, what would they do?

## CHAPTER 2

# Unintentional Communities

Most social scientists would leap at the chance to conduct a series of experiments resembling what BBC television producers organized for the reality program *Castaway 2000*. The show filmed thirty-six people (including couples and families) who were stranded together on an uninhabited island for one year, the goal being to "find out what happens when a cross-section of British try to create a new society."[1] The challenge the group faced was to build a cohesive, sustainable, and functioning community from scratch. Newspapers around the world breathlessly hailed the show as "a bold social experiment for the new millennium."[2]

Of course, the program's artificiality shaped the decisions and behaviors of the castaways. But still, many of the participants themselves noted that, in their eyes, the project was primarily a scientific experiment. "This is going to sound incredibly naive," participant Julia Corrigan later explained, "but we were blissfully unaware at first of the importance of the filming of the project, especially in the very early days. When we were at application stage, the emphasis seemed to be fully on the 'social experiment' side of things."[3]

Plucked from their lives and planted on the small, remote Scottish island of Taransay, these thirty-six people were required to cultivate their own food, raise their own animals, maintain their own shelters, and organize and run an effective community. Upon arrival, they were assigned pods to live in, allowed a crate of personal items, and given a few weeks' supply of food to last them until their crops grew. Everything else was in their hands. Ron Copsey, one of the participants, later summarized the initial stages of building their community:

> Most of the early days were spent getting to know each other and having endless meetings about how we were to live on Taransay. Many [arguments] ensued regarding the work rota and how to spend the community budget—even some punch-ups between some of the men.[4]

Twenty-nine of the participants stayed the course, developing an affinity for one another and for Taransay, but seven people (three individuals and one family of four) voluntarily left the island for various reasons.[5] Copsey, one of these seven, summed up his feelings about the experiment:

> We were given a wonderful opportunity to live differently, and all we did was replicate how we lived at home: people wanted rules, cliques, some kind of permanent, secure structure. It was disappointing that the Taransay community appeared to reflect society.[6]

Why did Taransay turn out to resemble regular society so closely? Why were the participants unable to create something new despite their desire to do so?

In an ideal world, researchers would conduct the Taransay experiment multiple times over sustained periods and with real scientific rigor. But it is hard to imagine how to actually do this, given the logistical and ethical impediments. One possibility is to conduct such experiments on a smaller scale. My lab has developed one way of doing this online over short periods and with simplified types of interactions. Another possibility would be to examine the fascinating history of *intentional* small-scale efforts to make society anew. Many times over the past few centuries, groups of people—motivated by utopian, philosophical, or religious visions or by practical exigencies—have voluntarily set themselves apart in order to form a different kind of community. These familiar utopian strivings have especially strong roots in the United States, which has a history rich in communal groups (think of the Puritans and the Shakers and, in more recent times, the communes of the 1960s). Still another way to study the development of societies would be to examine *unintentional* efforts to create social order from scratch, such as by shipwrecked sailors who found themselves facing the challenge of working together to make a functional community in order to survive.

We will explore these different angles in the chapters to come, but it's worth noting right now that the most striking feature of these advertent and inadvertent occurrences is their thoroughly predictable outcome. Most efforts to form societies with radically different rules either collapsed altogether or, like Taransay, came to resemble the society from which they originated. Despite the extraordinary and uniquely human capacity for innovation, illustrated by the great variety of cultures around the world and the unending social changes everywhere, human beings are drawn to some fundamental and universal principles—namely, the social suite. Attempts to abrogate these principles typically end in failure.

### Natural Experiments

Before we delve into people's spontaneous efforts to fashion societies, let's consider what large-scale experiments with social systems might look like, at least in scientists' dreams. Scientists might want to describe all possible types of societies, being as imaginative as science fiction authors, and then do experiments with real people placed into these societies in order to see which ones "worked" according to some definition they would specify—like whether the inhabitants were happy or avoided fratricide. Some variations on this type of idea are feasible. Scientists can systematically manipulate social arrangements and observe the effects on people and groups in the short term (as we will see in chapter 4), and they can deliberately manipulate the social organization of other social species by, for instance, removing a leader from a macaque monkey troop (an experiment we will discuss in chapter 7).

A different kind of experiment directed at exploring humans' inborn propensity to make societies might involve raising children without any cultural exposure at all in order to see what sort of society they create as adults. Such a conceit has been imagined for a very long time by people eager to understand the origins of language. It's called the "forbidden experiment," in fact, because it would be patently cruel and immoral.[7] According to Herodotus, the Egyptian pharaoh Psamtik I (who ruled from 664 to 610 BCE) gave two newborn babies to a shepherd to raise without language in order to discover whether they would speak on their own. He was not the only king with that idea. Frederick II (1194–1250), James IV of Scotland (1473–1513), and the Mogul emperor Akbar (1542–1605) allegedly attempted the same experiment.[8] The exercise has also been central to science fiction stories, like the 1960 classic "The First Men."[9]

Another hypothetical experiment might be to introduce mutations into genes related to social activity (for instance, genes that regulate how people choose their friends) and then see how populations of such mutated humans interact. Might people with different genes build different sorts of societies? Conducting genetic experiments with humans is obviously impossible, although, as we shall see in chapters 6 and 10, such experiments are possible with rodents.

But I am unaware of any scientific experiments in which whole, complex social systems involving humans are created for sustained periods along with control groups and different "treatments"—an experimentalist's term of art that means deliberate modifications of the conditions to which subjects are exposed.

Given the restrictions on experiments with human groups, it is very difficult to gather data on societies that humans build from scratch. However, at various times and places in history, there have been *natural experiments* that have approximated this. In these situations, communities of people have been thrust together accidentally or deliberately, albeit without explicit scientific manipulation. To what extent did groups like stranded sailors or self-isolating utopian sects wind up reproducing the crucial aspects of the societies from which they fissioned? And to what extent were they capable of realizing a new form of social organization in a sustained way? Did their success or failure have anything to do with how they lived, socially speaking? Before considering what such examples teach us, let's think about why experiments—whether of the deliberate or natural variety—are helpful at all.

Suppose some doctors believe they understand the physiology of a disease and want to see if a certain new drug is helpful in treating the disease. They administer the new drug to some patients and observe that people who take the drug are more likely to die. They might be tempted to conclude that the new drug is harmful. But perhaps they chose to give the drug only to relatively sick people, and, of course, sicker people are more likely to die, regardless of the medication they take. So, if only sicker patients are given a drug, how can scientists know whether it is helping or harming the patients? They would need a group of similarly sick patients who were *not* given the drug as a comparison. Furthermore, it's possible the doctors could have the opposite problem. Maybe they chose to give the new drug only to relatively young, healthy patients. This approach might make the drug seem *safer* than it is. The best way to ensure that health, age, and other factors are not confounding the scientists' assessment of the efficacy of the drug is to *randomly* assign a pool of patients so that some get the drug and some do not and then compare the outcomes in both groups. This type of experiment—in which exposure to the drug is controlled by the scientists, thus minimizing the impact of extraneous factors—is the gold standard of scientific research.

Science encompasses diverse practices, and the role of experiments remains paramount.[10] Still, experimentation should not be conflated with the scientific method in general. The *scientific method,* widely practiced by scientists since the seventeenth century, refers to a way of studying the natural world; it is characterized by systematic observation, careful measurement, and, *sometimes,* actual experimentation, all of which is coupled with the formulation, testing, and revision of hypotheses. There are many situations in which scientists cannot do experiments, and not just in fields like astronomy or paleontology. For example, we cannot experimentally assess whether the loss of a spouse increases a person's risk of death (known as the widowhood effect, or "dying of a broken heart") because we cannot kill or randomly remove people's spouses! Nor can we experimentally evaluate what tobacco exposure does to humans by randomly assigning people to be smokers, as we already know that it's deadly. In such circumstances, scientists resort to other, statistical approaches to find answers.

In addition, scientists can take advantage of what are known as *natural experiments,* where the treatment has been assigned to different groups of subjects by outside forces, seemingly by chance. Natural experiments can sometimes very closely approximate real experiments. For instance, in the 1980s, there was a debate about whether military service increased or decreased soldiers' wages after they left the service. Or perhaps any effect simply had to do with who signed up. Were men who enlisted more capable than men who did not enlist? Or did men sign up to serve because they had few skills or job prospects? Taking into account the qualities of those who had enlisted, did service in the military harm men financially or not? In an ideal experiment, we would randomly assign men to serve in the army and then examine their wages some years after discharge. In reality, this is not possible. But economist Joshua Angrist used the natural experiment of the 1970s-era Vietnam draft lottery instead, and he showed that serving in the army reduced subsequent earnings.[11]

Historians, biologists, archaeologists, and diverse social scientists have used natural experiments to study everything from the long-term impact of British colonial institutions in India to (famously) the evolution of beak morphology in Darwin's finches on the various Galápagos Islands.[12] Natural experiments vary a lot, however, in the extent to which the experimental treatments really are randomly assigned (as was the case in the military draft example). Randomization in most natural experiments is rarely so perfect.

However, the key idea is always that the treatment is assigned by some force other than the scientist and in a manner that does not predict the outcome. In one natural experiment, economist Daron Acemoglu and his colleagues concluded that the parts of Germany that were invaded by the French army following the French Revolution were quicker to abandon feudal governance.[13] These same regions in Germany then went on to experience greater prosperity and greater urbanization in subsequent centuries. This sort of natural experiment helps shed light on how social institutions and practices affect diverse economic outcomes in a way that simply is not otherwise possible. Researchers clearly could not randomly assign different regions of Europe to have different forms of government in order to study how this affects their economies over the following decades. A scientist could never do that. But the French army could.

Natural experiments allow scientists to circumvent practical impediments, mitigate ethical obstacles (like killing spouses), and study large-scale phenomena that are impossible to replicate (such as the effects of military invasions). Still, researchers cannot always be sure that interventions are truly allocated by chance. Maybe the French army specifically chose to invade certain parts of Germany that were somehow destined to be more prosperous in the future!

Natural experiments with social order can take many forms. Let's start by considering people stranded in remote places.

### *An Archipelago of Shipwrecks*

Survivor camps established after shipwrecks provide fascinating data about the societies that groups of people make when it's left up to them, about how and why social order might vary, and about what arrangements are the most conducive to peace and survival. An archipelago of shipwrecks, formed over centuries, more or less at random, has resulted in people participating, unintentionally, in multiple trials of this experiment.

Shipwreck survivors have had a special hold on the human imagination for thousands of years, beginning at least since Homer crafted the *Odyssey* and stretching through when Shakespeare penned *The Tempest,* Cervantes described Don Quixote's marooning, and more modern authors wrote *Robinson Crusoe, Swiss Family Robinson,* and *Lord of the Flies.* In fiction, the castaway narrative tends to feature an idyllic state of nature, following Jean-Jacques Rousseau, or a state of anarchy and violence, following Thomas Hobbes—two philosophers with rather conflicting ideas about human nature.

Hobbesian examples abound in real-world shipwreck situations. Consider the crew of the *Batavia,* who in 1629 systematically planned the mass murder of women and children to conserve resources.[14] Or consider the crew of the *Utile,* a French slave ship that wrecked in 1761 on Tromelin Island in the Indian Ocean. The sailors managed to get off the island, but they left behind sixty enslaved persons. They promised to send help but failed to do so for fifteen years. When a ship finally arrived at the island, only seven women and a baby were still alive.[15]

Some shipwrecks reflect a notably dysfunctional, if grimly familiar, breakdown of social order that includes not only murder but also cannibalism (which is not too uncommon). The extreme circumstances of the shipwreck may overwhelm people's innate tendencies to behave well. The wreck of the *Medusa* in 1816 (which left one hundred and forty-six people on a large, unstable raft, only fifteen of whom lived to be rescued thirteen days later) and the wreck of *Le Tigre* in 1766 (involving four people, three of whom survived for two months) both exhibited murder and cannibalism as survival means.[16] In the case of *Le Tigre*—as described in a book that was an international bestseller in the eighteenth century—special consideration was given to the lone female survivor, and the male survivors made provisions for her protection. The lone black survivor was murdered and cannibalized first, because of his lower perceived status. No such niceties applied in the case of the *Medusa*, where, as far as can be determined, male, female, black, and white survivors all killed and ate one another indiscriminately.

The relationship of cannibalism to the breakdown of social order depends, of course, on the reasons for the cannibalism—whether the individuals involved ate the bodies of those who had already died because otherwise they would have died of starvation themselves (as in a twentieth-century case involving a plane crash in the Andes) or whether people were deliberately murdered.[17] Contemporary readers regarded the two shipwrecks in different lights. *Le Tigre*—ironically, given its fealty to sexism and racism—was seen as a remarkable story of resourcefulness and endurance, while the *Medusa* was held up as the epitome of depravity and animalistic barbarity.

We know about these events because of a quirky literature, marketed to armchair thrill-seekers, of first-person accounts of these disasters. These seem to have peaked in the nineteenth century.[18] The genre had wonderful titles, including:

- *Remarkable Shipwrecks, Or, A Collection of Interesting Accounts of Naval Disasters with Many Particulars of the Extraordinary Adventures and Sufferings of the Crews of Vessels Wrecked at Sea, and of Their Treatment on Distant Shores: Together with an Account of the Deliverance of Survivors* (1813)
- *The Mariner's Chronicle Containing Narratives of the Most Remarkable Disasters at Sea, Such as Shipwrecks, Storms, Fires and Famines: Also Naval Engagements, Piratical Adventures, Incidents of Discovery, and Other Extraordinary and Interesting Occurrences* (1834)

And we can supplement these accounts with more formal evaluations of shipwrecks undertaken by historians and archaeologists in the twentieth century.

During the period of European exploration of the globe, from the sixteenth century through the advent of modern navigation and communications in the twentieth century, there were more than nine thousand shipwrecks. In the great majority of wrecks, all souls were lost to a watery grave. Occasionally, survivors endured at sea in small vessels; for example, the *Essex* went down in 1820, and its crew drifted in narrow whaleboats for weeks, eventually resorting to cannibalism. (Their story inspired Herman Melville to write *Moby-Dick*.) But for our present purposes, we need cases in which survivors made landfall and set up camp, and those are rare.

**Figure 2.1: The Shipwreck Archipelago**

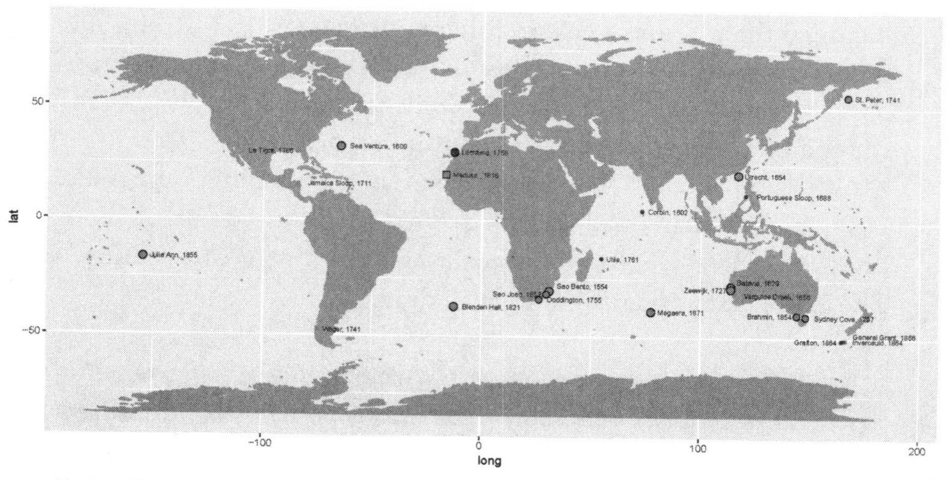


The locations and dates of twenty-four shipwrecks in the period 1500–1900 CE are shown. Circle symbols indicate the twenty wrecks that are part of the core sample considered here (involving at least nineteen castaways stranded for at least two months); square symbols indicate four further wrecks discussed in the text. Open symbols are shipwreck communities lasting less than one year, and filled-in symbols indicate communities lasting more than one year. Symbol size corresponds to the number of castaways, with small size indicating wrecks with fewer than nineteen people, medium size indicating nineteen to fifty people, and large indicating more than fifty people.

**TABLE 2.1 SMALL-SCALE SHIPWRECK SOCIETIES FROM 1500 TO 1900**

| NAME | YEAR | INITIAL SURVIVAL COLONY POPULATION | FINAL NUMBER OF SURVIVORS | DURATION |
|---|---|---|---|---|
| CORE SAMPLE | | | | |
| Sao Joao* | 1552 | 500 | 21 | 5 months |
| Sao Bento* | 1554 | 322 | 62 | 2.5 months |
| Corbin** | 1602 | 40 | 4 | 5 years |
| Sea Venture | 1609 | 150 | 140 | 10 months |
| Batavia | 1629 | 280 | 190 | 2 months |
| Utrecht | 1654 | 94 | 89 | 2 months |
| Vergulde Draek* | 1656 | 75 | 7 | 6 months |
| Portuguese Sloop | 1688 | 20 | 16 | 6 years |
| Zeewijk | 1727 | 208 | 88 | 9.5 months |
| Wager*** | 1741 | 101 | 10 | 8.5 months |
| St. Peter/Sviatoi Piotr | 1741 | 74 | 46 | 9 months |
| Doddington | 1755 | 23 | 22 | 7 months |
| Litchfield** | 1758 | 220 | 220 | 18 months |
| Utile | 1761 | 60 | 7 | 15 years |
| Sydney Cove | 1797 | 51 | 24 | 5 months |
| Blenden Hall | 1821 | 82 | 70 | 4 months |
| Brahmin | 1854 | 41 | 25 | 5 months |
| Julia Ann | 1855 | 51 | 51 | 2 months |
| Invercauld | 1864 | 19 | 3 | 1 year |
| Megaera | 1871 | 289 | 289 | 3 months |
| ADDITIONAL CASES | | | | |
| Le Tigre (Pierre Viaud) | 1766 | 4 | 3 | 2 months |
| Medusa | 1816 | 146 | 15 | 13 days |
| Grafton | 1864 | 5 | 5 | 19 months |
| General Grant | 1866 | 15 | 10 | 18 months |

Some numbers are approximate.
* Sources disagree.
** Hostile contact (violence, enslavement, etc.) with local people.
*** A subgroup was stranded for approximately five years.

One accounting of more than eleven hundred shipwrecks occurring near Tasmania revealed that only fifteen incidents (1.4 percent) resulted in survivors who established a campsite and stayed in one location for more than a week.[19] Fewer shipwrecks still would involve sufficient survivors or enough time to build some semblance of a small society. Many of the people who made it to land died due to scurvy, malnutrition, exhaustion, or injury soon afterwards. Mortality rates over the course of the disaster typically exceeded 50 percent. We also want situations in which the survivors were left alone and not attacked, enslaved, or incorporated by local groups. And, of course, we need at least one of the survivors to have lived to tell the tale.[20]

For an informative social experiment, we need a group of at least nineteen survivors who set up camp for at least two months.[21] Very few shipwrecks meet those criteria. I was able to identify twenty such cases between 1500 and 1900 (see table 2.1 and figure 2.1). And measurements of survivor counts and camp duration are often complicated by the extraordinary fact that survivors were sometimes shipwrecked twice—first after the original shipwreck and then after a party set sail to get help only to be shipwrecked somewhere else.

We must acknowledge that, even in these twenty examples that fit our criteria, the survivors are not strictly representative of humanity. The people who traveled on ships were not randomly drawn from the human population; they were often serving in the navy or the marines or were enslaved persons, convicts, or traders. Shipboard life involved exacting status divisions and command structures to which these people were accustomed. Survivor groups were therefore made up of people who not only frequently came from a single distinctive cultural background (Dutch, Portuguese, English, and so on), but who were also part of the various subcultures associated with long ocean voyages during the epoch of exploration. These shipwreck societies were, consequently, mostly male. Furthermore, the majority of our research subjects had narrowly escaped death and were psychologically traumatized, arriving at their islands nearly drowned and sometimes naked and wounded.

So, clearly, shipwreck survivors are not ideal experimental material. What scientists pursuing the forbidden experiment would really want is people who are strangers to one another, who lack any cultural background, who are comfortably taken to an isolated, plentiful environment, and who are then left to establish a new society that researchers could secretly watch unfold. Nevertheless, we can still learn from the few precious natural experiments that have occurred.

We have already discussed some shipwrecks that went badly, devolving into murder and cannibalism. But what factors were shared by shipwreck societies that were most successful? In our sample, the groups that typically fared best were those that had good leadership in the form of mild hierarchy (without any brutality), friendships among the survivors, and evidence of cooperation and altruism—all key elements of the social suite.

Survivor communities manifested cooperation in diverse ways: sharing food equitably; taking care of injured or sick colleagues; working together to dig wells, bury the dead, coordinate a defense, or maintain signal fires; or jointly planning to build a boat or secure rescue. In addition to historical documentation of such egalitarian behaviors, archaeological evidence includes the nonseparation of subgroups (for example, officers and enlisted men or passengers and servants) into different dwellings and the presence of collectively built wells or stone signal-fire platforms.[22] Other indirect evidence is found in the accounts of survivors, such as reports of the crew being persuaded, because of good leadership, to engage in dangerous salvage operations. And we have many hints of friendship and camaraderie in these circumstances. Violence and murder were not typical.

One shipwreck in which altruism involving resource sharing and risky volunteerism was particularly evident was the case of the *Julia Ann*. The ship wrecked in the Isles of Scilly, a reef in the Pacific, on September 7, 1855, stranding fifty-one people for two months. The misadventure was brought to a close when the captain and a crew of nine volunteered to row three days into the horizon to reach Bora Bora, 217 miles to the east, in order to get help.

Five lives were lost when the *Julia Ann* struck a reef, but all of the fifty-one survivors were eventually rescued. A newspaper later reported:

> Capt. Pond's chief desire throughout the whole sad affair seemed to be to save the lives of the passengers and crew, as the following noble act illustrates: While the crew were engaged in getting the passengers ashore [using a lifeline from the wreck offshore], Mr. Owens, the second mate, was going to carry a bag containing eight thousand dollars belonging to the Captain, ashore. The captain ordered him to leave the money and carry a girl ashore.... The child was saved, but the money lost.[23]

This visible act of altruism at the outset powerfully established an example for the group to cooperate and work together. Half the *Julia Ann* castaways were of the Mormon faith, and this may have helped the group cohere. The captain noted that they were "so easy to be governed" and "always ready to hear and obey my counsel."[24]

The availability of local resources and specific expertise among the castaways clearly helped as well. The *Julia Ann* survivors found turtle eggs, coconuts, and fresh water. They fashioned a forge and bellows and repaired a boat (the ability to make and use a bellows appears frequently in successful stories). And the men who volunteered to row the rescue boat risked their lives to save the group, as a passenger later recalled.

> We invoked God's blessing on the captain and the nine brave men who accompanied him, who boldly risked their lives in an open, crazy boat, to cross an open ocean, to endeavor to bring us succor and relief. As we watched the boat recede from the land... there was not one amongst us but was aware that on that boat... depended our very existence.[25]

Or consider the *Blenden Hall*, which wrecked on July 22, 1821, on the aptly named Inaccessible Island in the middle of the South Atlantic. This wreck was also marked by heroism and cooperation as in the case of the *Julia Ann*, though arguments and some acts of theft and violence marred the experience. Seventy of the eighty-two people on the ship made it to shore and survived for four months.

In the initial days, the survivors were able to salvage wood, canvas, and fabric to build shelters, as well as some liquor and a surgeon's kit that contained some phosphorus they could use to start a fire. The salvaging of liquor was a mixed blessing—it offered calories and temporary mood elevation, but it also prompted violence and threats.

After the incessant rain from the storm that wrecked them had subsided, the castaways began to feel a bit more optimistic. The *Blenden Hall* survivors had relatively easy, though not unlimited, access to food and water. But the availability of food and materials needed to be coupled with effective collective resource allocation for the group to succeed and survive. The survivors took care to share resources such as penguin meat, wild celery, and clothing so no one starved or froze.[26]

The captain, Alexander Greig, showed leadership and tact, keeping the peace at crucial moments and organizing a division of labor (work parties for salvage, exploration, and firewood collection). Unfortunately, the survivors often formed distinct groups at odds with one another (sometimes along the lines of class, rank, sex, and race), reflecting the sort of in-group bias that is also a part of the social suite. Tensions erupted, and in late September, the crew attacked the passengers. They were repelled by twelve men organized by the captain. Afterward, he attempted to organize punishment for the ringleaders, but the lashing was staved off by the entreaties of one of the women whom the crew had attacked.

The eighteen-year-old son of the captain, who himself showed great leadership during the ordeal, kept a diary in penguin blood written in the margins of salvaged newspapers. He perceptively described their predicament:

> I must acknowledge that to me it was always incomprehensible what could induce such a feeling of hostility to exist at this period...among the passengers generally. It is true that our troubles were calculated to ruffle our tempers and render us irritable; but at the same time one would have imagined that in our extreme exigency, with starvation almost inevitable, the common dictates of humanity would have been sufficient to suppress outbreaks and induce each to commiserate with his fellow-sufferer.[27]

Group divisions remained prominent even after the crew members' attack. Rather than working together and pooling the scarce available salvaged materials, three separate groups competed to build boats to leave for Tristan da Cunha, an island twenty miles away. One party of six left on October 19; they were not heard from again. But another party made it to Tristan da Cunha on November 8 and sounded the alarm. The others were then quickly rescued. Would the survivors of the *Blenden Hall* have fared better if they had not been plagued by aggression for their entire four-month stay? Probably. But in the end, their access to necessary resources likely reduced the extent and impact of the strife, and their capable leadership and evident cooperation were crucial too.

In the case of the *Sydney Cove,* wrecked on Preservation Island off Tasmania on February 9, 1797, fifty-one people initially made landfall. The documentary and archaeological evidence suggests the creation of substantial social order, including collective digging of a well and construction of a common dwelling; and surviving accounts indicate altruistic acts ranging from saving a shipmate from drowning to going to search for help.[28]

On February 28, seventeen men set out in a longboat for Port Jackson, on the mainland, but wrecked *again,* this time on the southeastern coast of Australia on March 1, at which point they began walking to Port Jackson, nearly four hundred miles away. Here is how the supercargo, William Clark (the person charged with overseeing the cargo and its sale), who led this secondary expedition, put it:

> Imagination cannot picture a situation more melancholy than that to which the unfortunate crew was reduced—wrecked a second time on the inhospitable shore of New South Wales; cut off from all hopes of rejoining their companions; without provisions, without arms, or any probable means either of subsistence or defense, they seemed doomed to all the horrors of a lingering death, with all their misfortunes unknown and unpitied. In this trying situation, they did not abandon themselves to despair.... Danger and difficulty lessen as they approach—the mind, as if its ultimate strength were reserved for arduous occasions, reconciles itself with calm resignation to sufferings from which, on a more distant view, it would recoil with horror.[29]

The success of their journey depended not only on this mental fortitude but also on the fact that the Aboriginal inhabitants of Australia showed noteworthy altruism to the strangers. On several occasions, Clark reported that they were befriended by locals who escorted them up the coast, gave them fish and other food, and even rowed them across rivers. In his diary entry of March 29, 1797, Clark wrote that there appeared to be "nothing human" about the Aborigines "but the form," and he frequently referred to them as "savages"—a reflection of our universal tendency toward in-group bias. But Clark soon changed his tune:

> We came to a pretty large river, which, being too deep to ford, we began to prepare a raft, which we could not have completed till next day had not three of our native *friends*, from whom we parted yesterday, rejoined us and assisted us over.

> We were much pleased with their attention, for the act was really kind, as they knew we had this river to cross, and appear to have followed us purposely to lend their assistance.[30]

And then again, a few days later, on April 2:

> Between 9 and 10 o'clock we were most agreeably surprised by meeting five of the natives, our old friends, who received us in a very amicable manner, and kindly treated us with some shellfish, which formed a very acceptable meal, as our small pittance of rice was nearly expended.[31]

A couple of other encounters with the indigenous people were more hostile, and one meeting resulted in three men sustaining wounds. Still, the overall interaction hardly supported Governor John Hunter's later summary of the trek in which he described the "savage barbarity of the natives."[32] If anything, it appears that the Aboriginal people saved the strangers' lives. Clark's party viewed them as both barbarians and friends. I would not be surprised if the indigenous Australians felt similarly about Clark and his men.

There were just three survivors of Clark's group, but they were able to sound the alarm about the others stranded in the first shipwreck, and they were then rescued (twenty-one of the thirty-four left behind had survived).

Cooperation was key to survival in the wreck of the *Doddington*. In the middle of the night of July 17, 1755, after the ship had rounded the Cape of Good Hope and sailed eastward for a day, it struck a rock in Algoa Bay in the Indian Ocean. As usual, the disaster was swift and brutal, as the journal of William Webb, the third mate, recorded:

> The first stroke awoke me, being then asleep in my Cabin. I made all the haste I could to get upon deck, where I found

> everything in the most terrifying condition imaginable; the ship breaking all to pieces and everyone crying out to God for mercy, as they were dashed to and fro by the violence of the sea.[33]

Within minutes, he was battered by waves, breaking "the lesser bone in [his] left arm" and sustaining a blow to his head that knocked him unconscious. When he awoke sometime later on a plank, he found a nail in his shoulder. He made it to nearby Bird Island, although he almost drowned. Only twenty-three people (all crew members) got to this small rocky island alive. The other two hundred and forty-seven people, both crew and passengers, perished.[34]

Bird Island, just forty-seven acres and rising just nine meters above sea level, lacks any fresh water. Now part of Addo Elephant National Park, the island is still home to a large breeding colony of seabirds on whose eggs the castaways gorged. The men, who could actually see the mainland in the distance, survived for seven months on the island, exploiting salvaged food and supplies from the wreck (including, as usual, casks of liquor) and eating fish, birds, seals, and eggs.[35]

One of the castaways, Richard Topping, was a carpenter. Assisted by the other men and, especially, by the inventive Hendrick Scantz (a seaman who had training as a blacksmith and could fashion tools), he was able to build a sloop. The men named it the *Happy Deliverance* and sailed away from the island on February 16, 1756. All but one of the men who made it ashore survived to leave the island (though nearly half died subsequently during their efforts to sail north along the African coast).[36]

Webb's detailed account of their seven months stranded on Bird Island, their subsequent journey up the coast, and their final deliverance in late April focuses on their efforts to get food and build a sloop and gives detailed descriptions of wind and sea conditions.[37] We can piece together glimpses of their collective lives. Some hierarchy was evident in the group. Certain castaways were singled out for special treatment, as noted in Webb's journal: "Our brandy all expended, except two gallons, which we keep for the Carpenter."[38] When supplies of water and food ran low, provisions were rationed fairly and amicably. The only mention of any friction was one unsolved and quickly forgotten episode of a theft from the ship's treasure chest.[39] There are many descriptions of cooperative efforts to care for the wounded, salvage the wreck, fish and forage for food, make rope and repair fabric for sails, and build the sloop in which to escape. The men also built a few catamarans and a small fishing boat. On several occasions while using these watercraft, the men had to rescue one another from drowning or from being stranded on another nearby islet. They were united in a common purpose.

The men were also kind to one another. On July 20, three days after their stranding, the body of the second mate's wife, Mrs. Collet, washed ashore. Mr. Collet apparently had "a most tender affection for his wife," so the men decided to conceal the discovery from him at first and reveal the news to him later. They distracted Collet and took him to the other side of the island, then buried his wife (alas, in the bird dung that covered the ground) and read a burial service for her out of a copy of the *Book of Common Prayer* they had salvaged. When they told Mr. Collet about it a few days later, he reportedly could "hardly believe" it until he was shown his wife's wedding ring.

### *Two Wrecks at the Same Time and Place*

The closest we come to an almost perfect natural experiment involves two ships, the *Invercauld* and the *Grafton,* wrecked on opposite sides of Auckland Island in 1864. The island, which lies two hundred and ninety miles south of New Zealand, is twenty-six miles long and sixteen miles wide. It was the site of so many shipwrecks in the nineteenth century that castaways would sometimes stumble on evidence of previously wrecked crews. For example, two years after the *Grafton* crew was rescued, ten survivors of the wreck of the *General Grant* spent eighteen months on the island and discovered the cabin made by the men of the *Grafton* (and took up residence in it). Eventually, the New Zealand government started placing supplies and signs on the island to assist people washed up on its shores.

Still, although they struggled for their lives on the same island at the same time, the crews of the *Invercauld* and the *Grafton* were not aware of each other. In the case of the *Invercauld*, nineteen of twenty-five crew members made it ashore, and only three survived, for just over a year, until their deliverance. In the case of the *Grafton*, all five people on board reached land, and all five made it off the island nearly two years later. What explained the different survival outcomes? Comparing these two cases allows us to explore the impact of the social suite and the role of friendship, cooperation, hierarchy, and social learning.

The *Invercauld* wrecked on May 11, 1864, in a rough cove on the northwestern part of the island and was reduced to "atoms" within twenty minutes.[40] Nineteen men swam away from the ship; they landed on a rocky beach at the foot of tall cliffs with no shoes, few clothes, and only what was in their pockets—some matches and a pencil. Just two pounds of sea biscuits and three pounds of salted pork were salvaged, enough to feed them for only a few days. They remained at the foot of the cliffs for four days and then climbed to the top, with great difficulty, leaving one weaker man behind to die.

Over the coming year, the men formed groups that splintered and reunited as they crossed the island, ultimately reaching Port Ross to the north, where they found traces of seal hunters' huts and even the remains of an earlier, failed European settlement. Along the way, members of the party who were too weak or injured to go on were left behind or, in one case, eaten. Three remaining survivors, including Captain George Dalgarno, were rescued by the Portuguese ship *Julian* on May 20, 1865. All of the lower-ranked seamen, except for a very resourceful sailor by the name of Robert Holding, had died, perhaps because the officers were better fed prior to the wreck or perhaps, as seems likely from available evidence, because they were more selfish afterward.[41]

Captain Dalgarno probably suffered from post-traumatic stress disorder after his rescue. A note in the *Otago Witness* on October 28, 1865, observed, "His health is still very delicate, owing to the extraordinary privation to which he had been exposed, and his medical advisor has forbidden him to speak to any one on the subject of the wreck, as any recurrence on his part in the sad event always brings on a nervous attack."[42]

The wreck of the *Grafton* occurred four months before the *Invercauld*'s, on January 3, 1864, in Carnley Harbour on the southern part of the same island. This crew of five men came from five different countries: Captain Thomas Musgrave (age thirty) was American; François Édouard Raynal (age thirty-three), the mate, was French; Alexander McLaren (age twenty-eight) was Norwegian; George Harris (age twenty) was English; and Henry Forgés (age twenty-eight), the cook, was Portuguese. This diversity is noteworthy, though it's hard to know what role it played in their success.[43] This group was much smaller than the group of survivors from the *Invercauld*, and they were able to salvage more supplies and food from their wreck (including guns, navigation equipment, tools, and, crucially, a dinghy). Both Musgrave and Raynal kept detailed journals, written in seal blood, which were later published.[44] Their stories are riveting. A correspondent from the *Times* described Musgrave's account as "almost as interesting as Daniel Defoe, besides being, as the children say, 'all true.'"[45]

Like William Clark of the *Sydney Cove*, Raynal was initially filled with despair, noting in his diary: "How and when should I escape from this island, hidden in the midst of the seas, and lying beyond

Figure 2.2: Crew of the *Grafton*, from the Frontispiece of François Raynal's 1874 Book

the limits of the inhabited world? Perhaps never!... I was almost suffocated; tears which I could not restrain filled my eyes, and I wept like a child."[46]

The men stuck together and worked collaboratively from the very beginning. Despite some disagreements, there was tremendous cohesion. At the time of the wreck, Raynal had been very ill, and, when the ship foundered, the men did not abandon him but contrived to bring him (and valuable supplies) ashore via a rope. This highly visible altruistic act at the start served to unite and motivate the survivors, and it signaled their intent to cooperate and invest in reciprocal relationships. It also provides a clear contrast to the *Invercauld* crew's abandonment of a man at the bottom of the cliffs, an act that set the stage for a different collective fate.

The leadership and communal spirit of the crew of the *Grafton* were also superior. The experienced Raynal was extraordinarily resourceful. He guided the crew to build a twenty-four-by-sixteen-foot cabin with a stone chimney next to a stream close to the shore. Eventually, he also built a forge and bellows (made of sealskins), which the men used to fabricate nails and tools from salvaged metal. Using a Roman recipe, he made concrete by baking seashells and combining the result with sand. He even taught himself how to tan leather and make shoes (also from sealskins) in their first year on the island.

During their encampment, Raynal crafted chess pieces, dominoes, and a pack of cards that they later wisely discarded because Musgrave was a bad loser and fights often erupted. The crew taught one another foreign languages and mathematics. One advantage of their impromptu school, Raynal noted, was that it equalized the men: "We were alternately the masters and pupils of one another. These new relations still further united us; by alternately raising and lowering us one above the other, they really kept us on a level, and created a perfect equality amongst us."[47] Of course, these activities reflected teaching and learning within the group, another part of the social suite. Still, the men did not lack all hierarchy, and they shared particular respect for Raynal.

Not long after the crew had become somewhat settled, in February, Raynal proposed that they should vote and choose one of themselves as "not a master or a superior, but a 'head' or 'chief of family,' " and that this person's duties would include "to maintain with gentleness, but also with firmness, order and harmony."[48] They further agreed that this person could be replaced by some other member of the crew on a future vote if this was deemed necessary. Raynal suggested Musgrave for the role; the men unanimously elected him, and he remained in this capacity throughout their two-year ordeal.

At one point, Musgrave fell ill, and Raynal realized that "the death of any one of us, in our present circumstances, would more injuriously affect the morale of the others, and perhaps be attended with fatal consequences for all of us. So my constant prayer is, that in our already severe afflictions, God would spare us this trial."[49] Ultimately, three of the men (Musgrave, Raynal, and McLaren) set sail in the repaired dinghy on July 19, 1865, and made it to Stewart Island in New Zealand after five days. Musgrave and others returned immediately to rescue the remaining two men. Afterward, the crew members of the rescue ship searched the island and found the body of James Mahoney, one of the crew of the *Invercauld*, though they did not realize it.[50]

Individual character clearly played a role in the *Grafton* survival experience. Captain Dalgarno of the *Invercauld* seemed mostly interested in his own survival, but Captain Musgrave showed real leadership throughout the ordeal. A few months after his own rescue, he returned yet again to the island on the off chance that there were other men who had been shipwrecked there, noting that, "having suffered myself, I would gladly have gone to the pole to have succored others under similar circumstances."[51]

On November 7, 1865, Raynal read a newspaper article titled "Narrative of the Wreck of the *Invercauld* on the Auckland Islands," by Captain Dalgarno, and this was the first anyone on the *Grafton* knew of another shipwreck on their island at the same time.[52] Musgrave offered his thoughts on Dalgarno's lack of leadership in a letter to a merchant friend in Invercargill to whom his own memoir was later dedicated. In the letter, he noted that Dalgarno's own account "proves that there has been no unity amongst them, neither has the Captain attempted (or he has not been able) to hold any authority or influence over them; to which cause I attribute a great number of their deaths."[53]

The differential survival of the two groups may be ascribed to differences in initial salvage (though, as noted, the *Invercauld* crew found abandoned huts and tools within a month of their arrival) and differences in leadership, but it was also due to differences in social arrangements. Among the *Invercauld* crew, there was an "every man for himself" attitude, whereas the men of the *Grafton* were cooperators. They shared food equitably, worked together toward common goals (like repairing the dinghy), voted democratically for a leader who could be replaced by a new vote, dedicated themselves to their mutual survival, and treated one another as equals. In all these regards, the *Grafton* crew had many features in common with the *Julia Ann* survivors, including the fact that their ordeals began with the saving of a life. Both groups also had technical expertise, selfless leadership, an ethos of cooperation, and a risky trip taken by a few in order to seek help for the others.

### *Pitcairn Island*

These shipwreck societies, while unintentional in their creation, all had a desired end: their participants wanted to rejoin the broader world. In contrast, the founders of some other unintentional communities had no such desire. One of the most notorious natural experiments of people thrown together in isolation involved the mutineers of the *Bounty* in 1789, who subsequently established a small society on Pitcairn Island that still endures today. This case is widely studied for insights into everything from space colonization to constitutional governance.[54]

Fletcher Christian, the master's mate, led a group of eighteen other mutineers and took over the *Bounty* by seizing the captain, William Bligh, a protégé of the British explorer Captain James Cook. Notwithstanding popular depictions of Bligh as a tyrant, he was, by many accounts, an enlightened and humane captain, and he had actually been a friend of Christian's. Bligh and eighteen loyal men were cast off the ship in a twenty-three-foot open launch. They navigated four thousand miles and landed on Timor forty-seven days later.[55] The mutineers headed to Tahiti, where the ship had made landfall not long before the mutiny. A leading theory of the cause of the mutiny is that the mutineers simply wanted to return to Tahiti and resume the pleasant and sexually adventurous lives they had led there instead of continuing with their perilous and uncomfortable lives at sea under the thumb of the British navy. After the mutineers got back to Tahiti, however, nine of them, including Christian, decided to settle at a place that British authorities would not discover. After kidnapping some Tahitian men and women, they set out to find a remote yet habitable island.[56]

At some point in the voyage, Christian decided to read some of the books aboard the *Bounty*. The description of one island, "isolated on the outer edge of island clusters...halfway between South America and Australia," caught his attention.[57] It had been sighted only a couple of times in recorded history and was said to have steep, unwelcoming cliffs and dense foliage; furthermore, it had only a single, dangerous space for docking ships. But when they reached the designated geographic coordinates, they found nothing. Assuming that the island was likely mismarked, Christian began searching the general vicinity, and not long after, Pitcairn Island rose above the horizon. The mutineers were thrilled by the outside world's inaccurate knowledge of the island's location, which added another layer of protection and concealment. And to further reduce the probability of being found, they decided to burn their ship so that any other ships sailing by would not spot it. Pitcairn Island would become their permanent home and the site of an entire community that they would build from scratch.

It was immediately clear that the island could comfortably accommodate the small community. Uninhabited, though with traces of a disappeared population, its four square miles had trees for lumber, fresh-flowing water, and fertile volcanic soil covering an eighty-eight-acre plateau. The bordering shores supported fishing for rock cod, red snapper, mackerel, and lobster. The climate was warm, largely pleasant, and, with eighty inches of rain annually, conducive to year-round farming.[58] The nine white mutineers split the island into nine equal shares; the six Tahitian men were entirely excluded from land ownership. The mutineers each claimed one Tahitian woman, while the six Tahitian men were allotted three women to share (I do not know how else to describe these grossly predatory actions).

Initially, in spite of these racial and gender inequalities, the group lived in relative peace. The mutineers managed to cooperate to secure their new home. They established rules to conceal the community's presence and make use of the land. They stripped and salvaged all they could from the *Bounty* before setting it aflame. Houses were to be built inland. The cutting of trees near the shore was forbidden. A lookout was maintained and fires were doused if a passing vessel was sighted.[59] With the remaining stores from the *Bounty* (including pigs, goats, and chickens) and the natural produce they found on arrival (coconuts, fish, seabirds, and eggs), the island's new population had food enough to last a few months. The men set about turning over the soil on their lots and planted crops from seeds they had (bananas, plantains, melons, yams, and sweet potatoes).[60]

Despite this auspicious beginning in an environment of plentiful resources and good weather, "the community (if it can be called that) appears to have very soon drifted into as close to a state of pure anarchy as ever existed in any human society," according to one historian.[61] When decisions affecting the entire population had to be made, they were reached via consensus or majority vote at irregularly convened meetings that included only the European men, each of whom had an equal say. Cooperation was limited for a number of reasons. Each mutineer had absolute dominion over his share of the island, and life was characterized by shifting private alliances. Social organization was fluid and unpredictable. The community had no legal code, no central political authority, and no monopoly on the use of force. Hence, given the absence of ties of friendship and meaningful ongoing cooperation, there was no reliable way to mediate disputes or enforce collective decisions.

An early indication of the tenuousness of this social contract appeared in the spring of 1791 in an episode that, to me, recalls Agamemnon taking the epically mistreated slave Briseis from Achilles during the Trojan War. When John Adams's wife, Paurai, died after falling off a cliff while collecting seabirds' eggs, Adams determined that he would take one of the women previously allotted to the Tahitian men. Another of the mutineers, the ship's armorer, John Williams, had lost his wife, Pashotu, to a throat disease shortly after arriving on the island and had also demanded another woman; he had been unsuccessful, but Adams was a more forceful figure. And he got his way. The two men thus drew lots for two of the Tahitian women.

Angered by this further insult to their status, the Polynesian men hatched a plot to kill the mutineers responsible. The plot was discovered by Fletcher Christian and resulted in the Europeans forcing the Tahitians to execute two of their own, Tararo and Oher. In an interview after leaving the island in 1819, Jenny, one of the Tahitian women, captured the pain and anger expressed by the Tahitian men: "Tararo [the man whose wife was taken by Williams] wept at parting from his wife, and was very angry. He studied revenge, but was discovered, and Oher and him were shot."[62] Pitcairn had witnessed its first murders.

Following the executions, life on the island stabilized. The small population grew; seven babies were born in those first three years. The islanders consolidated their livestock and crops and built a settlement of sturdy wooden buildings thatched with palm leaves. The status of the few remaining Tahitian men worsened, however, and they were basically reduced to slavery.[63]

On September 20, 1793, the simmering resentments of the Tahitian men finally erupted. After the women had gone to the mountains and the Europeans were scattered, the Polynesian men shot and bludgeoned six of the white men, killing five and injuring one (John Adams).[64] Jenny described the further bloodbath that took place over the next seven years. It began with a murder among the Tahitians and culminated in the death of every single adult man on Pitcairn save John Adams, who had started the troubles, in a kind of Hobbesian anarchy of "war of all against all":

> Inflamed with drinking the raw new spirit they distilled; and fired with jealousy, Manarii killed Teimua by firing three shots through his body. The Europeans and women killed Manarii in return. Niau, getting a view of McKoy, shot at him. Two of the women went under the pretense of seeing if he was killed, and made friends with him. They laid their plan, and at night Niau was killed by Young. Taheiti, the only remaining native man, was dreadfully afraid of being killed; but Young took a solemn oath that he would not kill him. The women, however, killed him in revenge for the death of their husbands. Old Matt [Quintal], in a drunken fit, declaring that he would kill F. Christian's children, and all the English that remained, was put to death in his turn. Old McKoy, mad with drink, plunged into the sea and drowned himself; and Ned Young died of a disease that broke out in his breast. Adam Smith [an alias adopted by John Adams] therefore is the only survivor of the Europeans.[65]

Though this was a conflict primarily between the men, the women also played a decisive role over the course of the slaughter, which, according to Jenny, stemmed from their anger over the bloody deaths of their white partners.[66]

In 1808, nearly twenty years after the mutiny, a ship on a seal-hunting expedition happened on Pitcairn Island. Captain Mayhew Folger and the crew of the *Topaz* at first thought it was uninhabited. They spent ten hours on the island with the thirty-five people who lived there—the surviving mutineers of the *Bounty*, their Polynesian captives, and their progeny. Folger was awed by the community's order and its members' ability to live without conflict in such a confined space. Referring to surviving mutineer John Adams, Folger wrote that he "lives very comfortably as Commander in Chief of Pitcairn's Island, all the children of the deceased mutineers speak tolerable English, some of them are grown to the size of men and women, and to do them justice, I think them a very humane and hospitable people, and whatever may have been the errors or crimes of [Adams] the mutineer in times back, he is at present in my opinion a worthy man."[67] Alas, a far less sanguine portrait emerged two hundred years later, revealing an apparently centuries-long culture of sexual predation of girls by older men. The island was still extraordinarily isolated, with only rare supply deliveries by ship and fewer than fifty inhabitants, but in 2004, many descendants of the mutineers were convicted of rape and child abuse following an explosive trial that uncovered the practice of "breaking in" virtually all of the island's ten- to twelve-year-old girls.[68]

The original settlers of Pitcairn were unable to form a functional society. If there is a blueprint for a basic, functional society that has been shaped by evolution and that is part of our genetic heritage, why do societies ever fail? Broadly, the blueprint specifies the shape of a society that humans form, but only if they are able to form one at all. Many barriers can stand in the way. First, humans have a parallel propensity for animosity and violence, and this can, of course, contribute to collapse. The social suite serves as a check on these tendencies (and it's ordinarily an extremely successful one). Limitations in environmental circumstances also play a role in social disaster, as do especially disruptive individuals and dysfunctional cultural elements (such as the deeply embedded sexual violence on Pitcairn).[69] Not all attempts, however organic, to create social order succeed. There can be stillborn societies.

So why did Pitcairn in particular fail? The usual causes of social collapse in larger states—such as bureaucratic mismanagement or corruption, emigration, war, environmental degradation, and population pressure—do not apply here. Nor were there resource constraints. Some have argued that the extreme isolation fostered total anarchy, but other isolated groups have coped well in similar circumstances, as we saw with the shipwrecks. In my view, the root cause of the anarchy in early Pitcairn was the initial inability of the colonists to sustain any cooperative impulses, an inability fueled by explicit racism, by intoxication with locally distilled alcohol (as happened in the early stages of the *Blenden Hall* wreck), and by the competition among the men for the smaller number of women. There was also notably ineffective leadership. Christian was a good mutineer but not a good governor of a colony, and the plans for fully democratic governance became unwieldy within a few years.

Sociologist Max Weber argued that one definition of a state is an entity in a given area that claims a monopoly on the *legitimate* use of force (as judged by the people themselves).[70] As states fail, they no longer protect individuals on equal terms, and factionalism often results in unrestrained violence. The features of the social suite stand in opposition to such violent disorder. Having upended one social order, the Pitcairn colonists were unable to invent another.

## *Stranded in Antarctica*

Leadership—as part of what I call mild hierarchy—is clearly important in the success and survival of these isolated social groups, especially when the leader works to foster solidarity and, perhaps ironically, to reduce hierarchy and ensure egalitarianism and cooperation within the group. We can see the importance of leadership through the Pitcairn colony's lack of it and in the contrast between the *Grafton* and *Invercauld* wrecks. Let's consider one final example of an isolated group that succeeded, in part due to this sort of leadership.

In 1914, seasoned polar explorer Ernest Shackleton is said to have placed an advertisement in a London newspaper: "MEN WANTED for hazardous journey, small wages, bitter cold, long months of complete darkness, constant danger, safe return doubtful, honor and recognition in case of success."[71] Shackleton, who had ventured to Antarctica twice before, was assembling a crew to accompany him on his Imperial Trans-Antarctic Expedition, the goal of which was to sail across the entire continent via small and shifting gaps of open water. However, on January 18, just forty-five days after the *Endurance* departed from South Georgia Island, the water surrounding the ship froze. The twenty-eight men aboard were trapped in a wasteland of ice. The promise of bitter cold, complete darkness, and constant danger had been realized.

For nine months, the ship, firmly wedged in an ice floe that was slowly drifting away from Antarctica, served as a home for the stranded men. Realizing that their survival, not the original expedition, was the new goal, the men began to make the necessary preparations to endure the harsh winter months, personalizing small living spaces within the ship, organizing the ship's food supply, and occasionally venturing out onto the ice to exercise or hunt penguins and seals.

On September 2, the *Endurance* started to buckle under the crushing pressure of the encapsulating iceberg. On October 27, the men reluctantly abandoned the ship and pitched their tents directly on the ice. The ice floe was drifting toward Elephant Island, which had not previously been visited by humans due to its inaccessibility and its inhospitable weather and landscape. When the island came into view, on April 9, the crew set out in three small boats (one whaleboat and two cutters), traveled through some of the world's coldest and most turbulent waters, and successfully landed there seven days later.

Given the crew's dire circumstances, Shackleton made a decision: he and five other men would take one of the small boats, sail the eight hundred miles back to South Georgia Island, and then hike across snow-clad mountains to reach the small whaling station there. Astonishingly, four months after leaving Elephant Island, Shackleton returned in a small steamer to rescue the twenty-two men who had stayed behind. Thus, all twenty-eight crew members stayed together as a community for a total of five hundred and thirteen days, and twenty-two men spent an additional one hundred and twenty-eight days together before Shackleton returned to rescue them. No one died.[72]

How did these men, confined and isolated for almost two years, organize themselves into a functional community and interact on a day-to-day basis? How did their social arrangements contribute to their success?

The work required to sustain this community was constant and overwhelming—hunting penguins and seals, building cabins, pitching tents, preparing meals, moving supplies, taking care of the dogs, and standing watch in awful conditions. But by and large, it was shared equally and amicably among crew members, just as it was with the *Grafton* survivors. The men chosen for this expedition—biologists, carpenters, physicists, surgeons, navigators—came from specialized backgrounds and different levels of society, but they cooperated and worked together effectively. Frank Worsely, the commander of *Endurance*, noted the interpersonal dynamics of the group in his diary:

> We are now six months out from England, and during the whole of this time we have all pulled well together and with an almost utter absence of friction. A more agreeable set of gentlemen and good fellows one could not wish for shipmates. Any and every duty is undertaken cheerfully and will-

> ingly and no complaint or whining is ever heard no matter what hardship or inconvenience may be encountered. The principal credit of this is due to the tact and leadership of the Head of the Expedition [Shackleton] and the cheery happiness and bonhomie of Wild [Second-in-Command]. They both command respect, confidence, and affection.[73]

Many have echoed Worsely and attributed this success in building a cohesive and cooperative group to Shackleton, who asserted that whether they lived or died, they would do it together. Shackleton required that all men, regardless of profession or status, yield to his authority and contribute to all forms of labor. Meals and meetings were strictly scheduled and mandatory, labor was allocated in a clear and fair manner, and food rations were split equally among the men (though, tellingly, Shackleton often gave his designated allotment to his crew). The men also taught and learned from one another, like the *Grafton* crew had, evincing a key feature of the social suite.

Strikingly, the men spent a lot of time on organized entertainment, passing the time with soccer matches, theatrical productions, and concerts. On one occasion, the men etched out a track in the snow, placed bets, and raced on dogsleds in a competition they dubbed the "Dog Derby."[74] Frank Hurley, the expedition photographer who produced the now-iconic photographs of their journey, noted in his diary: "Great fancy dress gathering and betting today on the Antarctic Derby Stakes. All available chocolate and cigarettes, the local currency, have been brought into requisition.... All hands are given the day off to see the race."[75] On the winter solstice, another special occasion, Hurley reported a string of thirty different "humorous" performances that included cross-dressing and singing. In his journal from the ordeal, Major Thomas Orde-Lees (who later became a pioneer in parachuting) noted: "We had a grand concert of 24 turns including a few new topical songs and so ended *one of the happiest days of my life.*"[76]

In sum, in this group stranded in Antarctica, there was not a totally egalitarian distribution of power. But there was friendship, cooperative effort, and an equitable distribution of material resources. The key to its cohesion and survival was not only the capable leadership of Shackleton and the abilities of the men, but also their ability to express so many features of the social suite.

### *The Settling of Polynesia*

The islands of the Pacific Ocean have provided other natural experiments of longer duration than the shipwrecks and even Pitcairn; whole societies have been established and endured for centuries, growing to substantial size. This was the great and well-studied case of the Polynesian expansion, where, over the course of more than a thousand years, settlers—by design or by accident—landed on islands throughout the Pacific, spreading eastward from their ancestral home. The Polynesian settlement of the Pacific islands illustrates a variety of historical and anthropological principles, and it is an especially powerful case study of the role of environmental constraints on social order. Being able to compare outcomes across widely separated islands with variable features really does feel like an experiment.[77]

When the Polynesians settled the Marquesas Islands around 700 CE, they initially lived in isolated coastal hamlets and hunted and gathered their food. But over a period of centuries, they settled the interior, invented agricultural practices, developed a tradition of large-scale feasts, dramatically expanded their population, and fashioned stone monuments and complex sociopolitical arrangements. By the time of European contact, the society on the islands was characterized by "incessant raiding, production of ever grander feasts, and emphasis on human sacrifice."[78]

In the larger and more distant Hawaiian Islands, perhaps discovered as early as 124 CE but settled between 600 and 1000 CE, political rule evolved from chiefship to kingship with ascribed divine descent by 1600 CE. The respective populations of the two kingdoms in Hawaii numbered perhaps sixty thousand to one hundred thousand. By the time Captain Cook arrived in 1778, full-scale feudalism had arisen, with commoners farming land in exchange for payment of labor and tribute to the king—a radical evolution from the ancestral Polynesian political system. Religious systems also evolved and became connected to a system of taxation, resembling places like ancient Egypt.

Using a sample of thirty Polynesian islands as a natural experiment, anthropologist Marshall Sahlins famously argued that different ecological circumstances were the primary factors shaping the political arrangements and cultural practices that emerged from an ancestral Polynesian system.[79] In general, bigger islands with more plentiful environments that could support bigger populations resulted in political arrangements with greater hierarchy and more formal institutions. On islands with little rainfall, where irrigation systems were invented, warfare emerged to capture these fixed assets, and societies were increasingly organized to support such warfare; they also developed human sacrifice to appease war gods. We even have archaeological evidence of cannibalism and specialized ovens for cooking humans.[80] All over the world, irrigation systems seem to be associated with the emergence of social stratification and a divergence between elites and the populace (in fact, contemporary societies that historically depended on irrigation rather than rainfall are still less democratic today).[81]

But ideally, if we want to identify a universal society and study bedrock, innate social features rather than the impact of environmental constraints, we should observe the emergence of a natural social organization in areas *without* severely limited natural resources. Even this would not guarantee success, of course, as we saw in the case of Pitcairn. But our imagined experiment, unlike Sahlins's Polynesian expansion, would involve taking a population, dividing it into groups of founders, and dispersing the groups onto many islands that had similar, plentiful resources, and then addressing questions such as: What sort of society would the people make? How great or small would the variations be among this set of societies? What features would be observed consistently?

In very resource-poor environments, in fact, plausibly *unnatural* social systems might emerge, such as the cannibalism and human sacrifice in Mangaia, another island settled by Polynesians but unable to support a community larger than five thousand persons. What did this group of people thrust into such a hostile environment do? They started to eat one another.[82] An analogy might be the way the human body adapts to severe food shortages—it is modified and nudged away from its usual, natural trajectory and manifests stunted growth. Stunted humans do not illustrate standard human physiology any more than extreme social arrangements in response to resource scarcity illustrate innate social order.

### *Living Socially*

What can we make of all these stranded groups? Two overarching observations are clear. First, it's significant that some groups fared so much better than others. Success was especially likely among groups that were particularly able to manifest the social suite. Second, we find commonalities in social behavior, expressed as the social suite.

However, equally notable is what we do *not* see: given the chance, isolated small-scale communities do not invent wholly new sorts of effective social order. No doubt, this relates partly to the fact that the marooned men and women were products of their own culture, which shaped their expectations about what society should look like. As we saw in chapter 1, psychologists researching social perception often try to work with infants as young as three months old, precisely so as to minimize such effects of cultural background. It's the same reasoning that prompted consideration of the forbidden experiment with feral children.

Examples of other shipwrecks and strandings from non-European settings are scarce. The examples from Asia that I am aware of involved mostly coastal seafaring and too-quick returns to civilization.[83] And I could not find any cases originating in Africa or the Americas, in part because of more limited seafaring technology and in part because of the absence of records. Still, the Polynesian expansion, in addition to illustrating the effect of environment, also illustrates the universal emergence of the social suite, despite the diverse political arrangements that arose over the centuries and that have otherwise preoccupied social scientists and historians. Once again, when it comes to our way of living socially, we are more similar than different.

The existence of failed societies, like Pitcairn, and societies featuring cannibalism, like Mangaia, do not subvert the centrality of the social suite. The social suite offers a successful, evolutionarily time-tested strategy for group living. Sometimes, groups cannot coalesce to express the social suite. Nevertheless, they do not have any viable alternatives to it.

Our observation about the sensitivity of social arrangements to the environment (for example, how plentiful food is) raises a subtler issue. We have seen that environmental constraints can lead to misshapen societies. But the role of the environment in shaping social interactions, both over the course of a person's lifetime and over the course of the evolution of our species, highlights a more profound point. If environmental variation causes cultural variation, then it is possible that any unvarying, universal features of human societies would be attributable to specific, consistent features of the environment itself. Maybe the reason that people form basically similar core social arrangements everywhere is that there is something consistent about the environment to which our species is responding. What might that be?

In a fundamental sense, there is indeed one aspect of the environment humans face that does *not* vary. That constant element is *the presence of other humans*. As we will see in chapter 11, humans have evolved to be social in a particular way precisely because they have been social in the past. The social systems our ancestors created became a force of natural selection. And once our species started down the path of living socially, humans set into motion a feedback loop that continues to shape how we live with one another today.

When you put a group of people together, if they are able to form a society at all, they make one that is, at its core, quite predictable. They cannot create any old sort of society they want. Humans are free to make only one kind of society, and it comes from a specific plan. Evolution has provided a blueprint.

## CHAPTER 3

# Intentional Communities

Near the end of March 1845, Henry David Thoreau borrowed an ax from a friend and shipwrecked himself, so to speak, on the shore of Walden Pond, in Concord, Massachusetts. He wanted to conduct an experiment in solitary living. He started by felling trees to make a small cabin furnished with three chairs—"one for solitude, two for friendship, three for society"—though his extra chairs were rarely occupied.[1] He grew his own food, considered eating woodchucks raw, read widely in multiple languages, and harbored runaway slaves. He also wrote *Walden,* a book on the merits of self-reliance, nature, and transcendental philosophy that was so influential, it remains widely read today.

The yearning to return to a mythical, bountiful state of nature with the goal of creating a new kind of social order where humans thrive has motivated dreamers and cranks, both individuals and groups, for thousands of years. Thoreau was focused on the benefits of solitude—which is paradoxical, in my view, given how natural our social state is. When all alone, he felt he grew "like corn in the night," with nature itself as his companion.[2] "Every little pine needle expanded and swelled with sympathy and befriended me," he wrote.[3] He did not have much use for the company of other humans, about whom he said:

> We meet at very short intervals, not having had time to acquire any new value for each other.... We have had to

> agree on a certain set of rules, called etiquette or politeness, to make this frequent meeting tolerable, and that we need not come to open war.[4]

Thoreau did not much like formal institutions either:

> One afternoon, near the end of the first summer, when I went to the village to get a shoe from the cobbler's, I was seized and put into jail, because...I did not pay a tax to, or recognize the authority of, the state which buys and sells men, women, and children, like cattle at the door of its senate-house.... Wherever a man goes, men will pursue and paw him with their dirty institutions, and, if they can, constrain him to belong to their desperate odd-fellow society.[5]

He was let out of jail the next day when an unknown friend apparently paid his poll tax for him.[6] He had not paid it in years because he objected to the use of the funds to wage war and expand slavery, as he later explained in his famous essay "Civil Disobedience"—which would go on to inspire the work of Mahatma Gandhi and Martin Luther King Jr.[7]

More than a century later, I visited a reconstruction of Thoreau's cabin at Walden, and it was as austere as a jail cell. I had moved to Concord with my wife and three children in 2001. In a coincidence that my family calls a "low brush with fame," we had acquired the former home of Sam Staples, the town constable whose duty it had been to jail Thoreau but who was nonetheless Thoreau's friend.

### *Attempting New Societies*

Thoreau's observations regarding social interactions of both a personal and institutional nature have also been explored by other thinkers. According to a theoretical classification introduced by philosopher Ferdinand Tönnies in 1887 and later advanced by sociologist Max Weber, people's social connections are of two general types: *Gemeinschaft* and *Gesellschaft*.[8] *Gemeinschaft* refers to personal interactions and their accompanying roles, values, and beliefs, which roughly correspond to the notion of a face-to-face community. But social ties can also involve more indirect interactions, impersonal roles, and the formal norms and laws about such connections. These interactions with a broader and impersonal society are known as *Gesellschaft*.

The distinction highlights a key problem with modern life. Many people have wondered how a sense of community can be preserved or regained in a large, impersonal society. Thoreau and other hermits occasionally found social interactions so unsatisfying and oppressive that they abandoned them altogether (at least for a while). But another response to the changing scale and quality of social order has been to form entirely new, smaller communities. Since at least Roman times and on every continent, communal movements have arisen with the objective of breaking away from the *Gesellschaft* of modern living to move back toward a society more firmly based on *Gemeinschaft*.[9] People who join communes often aim to forsake impersonal interactions and establish greater authenticity in their personal relationships.

Utopian communal experiments are typically more idyllic than the accidental communities of people thrown together by shipwrecks or other unforeseen or traumatic events. Still, they are not always more successful. From countless nineteenth-century American utopias to twentieth-century Israeli kibbutzim and other examples we will now review, most such experiments have failed, usually within a year or two. It can hardly be insignificant that the great majority of these experiments have been utter flops.[10] Even so, these natural experiments help us see which features of social organization recur and are crucial for success. While intentional communities have sometimes succeeded in forming social arrangements that temporarily deviate from the social suite, most have not. Few, if any, have achieved anything radically alien.

### *American Utopian Experiments*

In 1516, Thomas More coined the word *utopia* from Greek root words that mean "no place" but that in English also sound like the roots for "good place"—a telling ambiguity, given the failure of so many attempts at utopian societies.[11] America has been especially fertile ground for communal utopian efforts, and they have left a mark on society. Many people are aware of them through the products made by such communities—Shaker furniture, Amana appliances, and Oneida silverware, for example. Others may have visited tourist sites such as Fruitlands and Brook Farm in Massachusetts and marveled at self-sufficient, bygone lifestyles. Still others may remember 1960s-style communes or even apocalyptic cults like the Branch Davidians.

Over the years, people's reactions to these efforts have ranged from bemused to admiring to condemnatory. As historian Donald Pitzer has noted, "Communal experimenters have often been portrayed simply as colorful 'freaks,' psychological misfits outside the 'mainstream' who inevitably 'failed' because they allegedly were out of step with American life and values."[12] There is a large inventory of books, dating back at least to the nineteenth century, that, like the shipwreck anthologies we saw earlier, describe contemporary communitarian movements, catalogs with titles such as *The Communistic Societies of the United States, from Personal Visit and Observation*, published in 1875.[13]

These efforts got an early start after the European settlement of our continent. In 1694, forty celibate male scholars formed a community near Germantown, Pennsylvania, calling themselves the "Society of the Woman in the Wilderness." Since 1780, there have been at least four sharp peaks in the establishment of communes, including 1790 to 1805, 1824 to 1848, 1890 to 1915, and 1965 to 1975.

And in the late 2010s, we may be seeing another upsurge in the formation of intentional communities.[14]

It's not surprising that thousands of utopian communities have dotted the American landscape over its history.[15] The United States has always offered social and geographic mobility, commitment to free association, absence of intrusive, despotic institutions, openness to novel ideas, and endless opportunities for self-improvement and renewal. While the impulse to form these communities may run counter to Americans' Thoreau-type sense of rugged individualism, it taps a pioneering spirit of discovery and exploration and builds on an equally important tradition that Alexis de Tocqueville famously described as America being a nation of "joiners."[16]

Historically, communitarian movements flourish during periods of major social or cultural disruption. People growing into adulthood at a time when norms and expectations are being called into question are especially likely converts. Today's information revolution and growing robotic automation may be providing an impetus to communalism just as the Industrial Revolution and the Great Depression did for earlier generations. Banding together to share property is often a survival strategy for groups of poor individuals in times of change (though the communal ownership of property is a common feature of utopian communities even when their members are not poor).

Communitarianism had a particular efflorescence in New England in the 1840s, a time when the citizens of a fairly new nation were still steeped in arguments about social reform. The communitarians of the era fervently believed they could create circumstances fostering cooperation among people for the benefit of all, and they pushed back against hierarchies based on age, sex, or ethnicity. Rejecting their contemporaries' social strictures—most of them, anyway—they were convinced that people could voluntarily and happily suppress self-interest in the name of collective interest and that it was possible to free themselves from a corrupt past and start history anew.

The urge to rebuild a nation that had yet to celebrate even one centenary became almost a fetish among some philosophers, writers, and clergy even as they worried that their idealism might be seen as all talk and no action. "We are all a little wild here with numberless projects of social reform," wrote Thoreau's friend, transcendental philosopher Ralph Waldo Emerson, in 1840, and "not a reading man but has a draft of a new community in his waistcoat pocket."[17] Many experiments were attempted in Emerson's neighborhood, most notably the famous effort at Brook Farm.

### *Brook Farm*

The utopian community of Brook Farm, in West Roxbury, Massachusetts, was typical of its time. Brook Farm was the brainchild of George Ripley (1802–1880). As a young man, he hoped to study in Europe, but his plans were thwarted by his lack of means, and he was forced to go to "the cheapest stall where education can be bought," which was the Harvard Divinity School.[18] He graduated in 1826 and served as a Unitarian minister in Boston for more than a decade, accumulating a library of books on European philosophy that were so valuable, he was able to use them as collateral for a four-hundred-dollar loan when he founded Brook Farm. In 1836, before starting the commune, Ripley had founded the Transcendental Club, of which both Thoreau and Emerson were members. He had been motivated to abandon his work as a minister and put transcendental philosophy, of which he had become increasingly enamored, into practice. Transcendentalism emphasizes the inherent goodness of nature and humans, is suspicious of formal social institutions that can constrain people's independence, and rejects empiricism for subjective experience. In April of 1841, Ripley and his wife, Sophia, and a dozen others bought and settled on 192 acres and started an experiment in social living that lasted nearly six years before petering out and finally collapsing.[19]

The beginning was auspicious. Within a year or two, the community had increased in size to around ninety inhabitants, about half of whom were students or boarders and not formal members. One of the key organizing principles was the value of both intellectual and manual labor. Ripley observed that "we should have industry without drudgery, and true equality without its vulgarity."[20] That's not to say that life at Brook Farm was easy, however. Residents were expected to work sixty hours per week in the summer and forty-eight in the winter, though they were free to choose the work they liked.

Brook Farm had many of the qualities we have come to associate with intentional communities: (relative) gender parity, modest hierarchy, and a charismatic leader. Like Shackleton, Ripley led in part by sharing all the chores, including the traditionally female chores, such as the laundry. The rotation of jobs as part of work groups also served to equalize the members.[21] Several of the founders enjoyed the concept of authors and thinkers toiling alongside workers and farmhands, all laboring together in somewhat forced democratic equality.

The membership and leadership of overlapping work groups were fluid. Residents served in the Washing Group, the Farming Group, the Knife-Cleaning Group, the Onion Group (for children tasked with harvesting onions), and so on. Frederick Pratt was only a boy when his parents joined Ripley's experiment; sixty years later, he would observe: "I have seen Mr. Ripley and Mr. Hawthorne shoveling over manure, without a murmur, though I believe Mr. Hawthorne did not relish the job."[22] Among the farm's early investors and residents was the writer Nathaniel Hawthorne, who apparently quickly regretted the one thousand dollars he contributed. He would later happily leave Brook Farm behind.

Still, although everyone performed manual labor, Brook Farm had the feel of a middle-class undertaking in its conception and realization. It was founded as a joint-stock company rather than a communal society, and it had a written constitution at its founding (these articles of agreement had sixteen sections, including a preamble stating that the founders sought "to substitute a system of brotherly cooperation for one of selfish competition").[23]

The Farmers (as they were called) had a good time despite the heavy labor, and many residents and visitors noted their playfulness. One observer accompanying Bronson Alcott (father of Louisa May) described finding "80 or 90 persons playing away their youth and daytime in a miserably joyous and frivolous manner."[24] One resident opined, "Enjoyment was almost from the first a serious pursuit of the community."[25] Hawthorne's novel *The Blithedale Romance* was based in part on his experiences at Brook Farm and included a trenchant depiction of an outdoor masquerade party during his time there.

All these sybaritic commitments to play—from tobogganing to dancing to joking to acting—may have had a deeper function: binding the members of the community together in an effort to create a new kind of communal consciousness. Psychologist Jonathan Haidt, an expert on breaking down barriers between ideological foes, places a great premium on the unifying function of dancing and athletic events for the harmonious workings of groups.[26] Activities requiring acting were especially important to the Brook Farm community and seem to have had a very deep effect on the Farmers, just as they did on members of the Shackleton expedition.

The commitment that each Farmer made to multiple roles at Brook Farm and to dramatic play was a distinctive feature of this community, but there was still room for Farmers to express their individuality. As historian Richard Francis observed, "From the very beginning, the Farmers were conscious that the individual's identity, or rather his sense of it, was unduly restricted by the social role that was forced on him in 'civilization.'"[27] Every Farmer worked hard to establish both a core community identity and, somewhat paradoxically, an authentic individual personhood untethered to arbitrary functions imposed by society.

Intentional utopian communities have always struggled with the problem of individuality. One Farmer would recall, many years later, "It seems to me one would scarcely find forty persons with more strongly marked individuality; not loudly proclaimed and only after much study to be understood, but contributing a peculiar influence to the place."[28] For a time, at least, the Brook Farm residents were able to strike the right balance in this regard, respecting individuality but fostering community. In chapter 9, we will consider the paradox that individuality is actually essential for community and discover how this plays a crucial role in the social suite and in human evolution. Communal efforts like Brook Farm's, which respected the individual identity that is a part of our ancestral inheritance, generally fared better than efforts that did not.

Schooling at Brook Farm was strikingly progressive and organized to elicit children's good qualities and insights rather than to beat an education into them. Nora Schelter Blair would recall, nearly half a century later, the "harmonious blending" of the diverse student body and "the entire freedom of speech between pupil and teacher." Children addressed their teacher Sophia Ripley by her first name, and as Blair recollected, "She seemed to permeate her pupils with a joyous confidence in their individual ability to tread the path so clearly and pleasantly indicated by her."[29] Most profoundly, the children were seen as having an absolute *right* to education—"not doled out to him as though he was a pupil of orphan asylums and almshouses—not as the cold benefice and bounty of the world—but as his right—a right conferred upon him by the very fact that he is born into this world a human being."[30]

Unlike other utopian experiments in the nineteenth century, Brook Farm did not require participants to abandon their nuclear families or sever all ties with the outside world, so the happenings at Brook Farm were breathlessly reported in the popular press of the time. One Farmer would later note that "there were thousands who looked upon us as little less than heathens, who had returned to a state of semi-barbarism."[31] Many of the leading lights of the transcendentalist movement, including Ralph Waldo Emerson, Bronson Alcott, Henry David Thoreau, and Theodore Parker, passed through the commune. Early feminist and writer Margaret Fuller was also a frequent visitor and resident. Frederick Pratt recalled "the jolly time we children had together, and the use made of our boys' wheelbarrows or wagons in carrying the girls about; 15 or 18 years afterwards, my brother John married Annie Alcott. Louisa became a writer, and John Brooks and the *Little Men* became famous."[32]

So what, exactly, went wrong with such an appealing tableau? In early 1844, Brook Farm underwent a dramatic conversion to the then increasingly popular and more radical doctrines of French utopian thinker Charles Fourier, attempting to transform itself from a transcendental picnic into a regimented *phalanx*, which was Fourier's word for an ideal community. Fourier's theories were strange, rigid, and convoluted, and many members of Brook Farm objected to the shift.[33] Factions emerged within the community, and tensions rose. Ripley nonetheless embraced Fourierism, and the group built a 175-foot structure, the Phalanstery, in accordance with the doctrine. But it burned to ashes in two hours on the night of March 3, 1846, in a spectacular fire, and this proved the death knell of the community. They lacked the ability to recover.

The Farmers had hoped to set an example for the rest of society. As one of them, Amelia Russell, put it, "I even thought that the whole nation would be charmed by our simple, unobtrusive life, and that in time... [our] laws and government should extend and finally annihilate the existing executive of the country."[34] The members of Brook Farm were of course unable to annihilate any existing order except their own. The end came inevitably after the fire, as Russell would later note: "All lingered as long as they could, unwilling to give up the life, which was a sacred idea to them, or rather the poor remains of it, for each now lived on personal resources, and it became, as it were, the shadow of what it had been."[35]

### *The Shakers*

Some communal efforts did endure for quite a long time. One of the most famous was the United Society of Believers in Christ's Second Appearing, also known as the Shakers. The Shakers would become the most organized, economically successful, and longest-lived intentional community to emerge from the early American utopian experiments.

The Shakers were originally founded as a religious sect in England in the late seventeenth century. The movement had its most charismatic leader in "Mother" Ann Lee, an impoverished young woman born in Manchester, England, who joined the Shakers when she was just twenty-three and assumed its leadership nine years later, in 1768. Her own experiences changed the course of the faith. After marrying Abraham Stanley, a blacksmith, at age twenty-six, she had four children, all of whom died in infancy. She reacted to the physical and mental suffering of this heavy loss by solidifying her ideas about sex as a path to sin and all human suffering. Absolute celibacy became a key part of the faith. Shaker theology also saw God as both male and female in nature and interpreted the life of Jesus as only one phase of development in Christianity on earth. Many Shaker practices explicitly emulated what they believed to be practices of the early Christian church, including not only celibacy but also common property, pacifism, and confession. The Shakers saw women and men as equals, and often accepted African-Americans as well.

Lee was briefly imprisoned in England for blasphemy, and in 1774, she, her husband, and seven followers fled to New York City. She worked in the city as a domestic before the small group moved upstate to a tract of land later called Watervliet. Slowly, they began to recruit converts. From 1781 to 1783, Lee undertook an astonishing mission in Massachusetts and Connecticut to preach and recruit followers; during this effort, she was accused of treason, lewdness, blasphemy, and witchcraft, among other lesser offenses. Her fellow

Shaker missionaries were thrown from bridges or beaten with clubs by people that Shaker historians later called, in an allusion to the Inquisition, the "Disciples of Torquemada."[36] This extreme reaction stemmed partly from a fear that the Shakers might break up marriages, take property, or depend on public assistance. Some resented the Shakers' pacifist views or feared that they were British agents.[37]

After Ann Lee's death, in 1784, the Shakers were led by two of her primary followers, and by 1794, ten communities of thirty to ninety people were settled in five different states. Two male and two female Elders served as leaders in each community, and these "families" were the basic unit of Shaker social and economic life. Their numbers were never large (1840 was the high-water mark; at that point, they had 3,608 members), and by 1900, they had dwindled down to 855 believers.[38]

The Shakers emphasized order, harmony, and utility, and their communities were described as quiet and serene. Religious practice involved as many as a dozen meetings per week with distinctive dances and marches. Many members lived to be octogenarians or older, and one study found that, in 1900, 6 percent of Shakers were older than eighty, whereas only 0.5 percent of the U.S. population reached that age (though it might be that healthier people joined the Shakers in the first place).[39]

While individuals did not hold their own property, the Shakers relished the individuality of their members, just like the Brook Farmers did. Adherents could cultivate traits and abilities, form intimate and personal friendships, and make their own choices. The Shakers did not practice mindless conformity, and they valued autonomy.

Members of the same Shaker families lived close together, labored side by side in fields and shops, ate their meals in common, and even slept in communal beds in "retiring rooms" that each typically housed four residents. Relationships among the members were often very close, and they left behind many letters attesting to deep personal attachments.[40] Of course, romantic involvement was strictly forbidden—and nearly impossible. Men and women were not even allowed to pass each other on the stairs, leading to the construction of distinctive double stairways in many Shaker buildings. Even handshakes between the sexes were prohibited. Married couples were often assigned to different families to live and work.

The sort of altruism and solidarity seen among the Shakers is not often associated with substantial economic success, but the Shakers excelled precisely because of their communal ethos and practice. One analysis of the economic performance of Shaker communes in the latter part of the nineteenth century showed that, despite the communal ownership of property and the lack of compensation for workers' efforts, the communes performed as well as or better than similar enterprises operated in a noncommunal manner, both in agricultural and manufacturing activities.[41] This productivity arose from a number of features of Shaker life.

Like all other human groups, the Shakers had a division of labor. Women focused on domestic jobs and men worked in the fields and with machinery. Members were free to choose their own tasks and specialties, and variation in productivity was well tolerated, in keeping with the acceptance of individual differences. For instance, in one study of worker productivity, one woman produced just ninety bonnets a year while another made seven hundred and thirty.[42] It was not difficult for Shaker communities to incentivize work; they relied on residents' innate cooperativity supplemented by a shared ideology. Given the interdependence and physical closeness of members, norms of hard work were maintained by peer pressure and public shame as well as by shared goals and religious beliefs.

Children in Shaker communities went to school (boys in winter, girls in summer), until age fourteen, but the Shakers were suspicious of conventional schooling, and education focused on practical trades and skills. Of course, no Shaker children were born in the communities, but there were other sources of young people. Some were indentured to the group by parents or guardians for economic, personal, or disciplinary reasons, others were orphans or homeless, and some came with their parents when they joined the movement. Unsurprisingly, given the enforced celibacy, most young people left the community as they reached adulthood; one study found that only 5.7 percent of children stayed in the group in the period from 1880 to 1900; by contrast, during that same period, 28.7 percent of adults stayed.[43]

Along with attrition due to the enforced celibacy, many Shaker communities dispersed after calamitous fires and floods, again illustrating the vulnerability of small communities to such disasters. The Shaker movement also lost steam because the larger society outside became more appealing. Opportunities for economic advancement and self-determination grew during the nineteenth century in ways that made the Shaker lifestyle less compelling. As religious historian C. Allyn Russell argued, "The 'world' appeared to be saying to the Shakers, 'Whatever you can do, I can do better.' "[44] The broader society was also beginning to act more humanely toward the mentally ill and to treat women and men more equally, as the Shakers had long done. In 1968, there were just nineteen Shakers left, all women, in two communities: Canterbury, New Hampshire, and Sabbathday Lake, Maine. Interviews with these survivors in the 1960s revealed their remarkable equanimity in the face of the decline of their sect. They believed it required as much bravery to approach the end of an age as to start it.[45]

If not for the practice of celibacy, the Shaker movement might well have flourished, given its embrace of so many features of the social suite (cooperation, friendship, individual identity, mild hierarchy). Celibacy clearly reflects a rejection of the real world, not only because it is such a departure from all other ways of organizing social life, but also because it intrinsically means that this way of life cannot reproduce itself. Paradoxically, the Shakers thus required connection to the outside world, because that was the movement's only source of new adherents.[46]

### *Kibbutzim*

The impetus to found communal movements continued into the twentieth century. Israeli kibbutzim (the Hebrew word for "groups") are voluntary, democratic communities ranging in size from eighty to two thousand residents in which people live and work cooperatively. These provide additional examples of natural experiments in social life. The first kibbutzim were founded in Palestine in 1910, and, by 2009, there were 267 kibbutzim scattered throughout modern Israel. These groups account for only 2.1 percent of the country's Jewish population but 40 percent of the national economic agricultural output and 7 percent of the industrial output.[47] They have been long-lived because of both their economic success and their members' willingness to pragmatically modify ideology in response to practical and social exigencies.

During the first half of the twentieth century, members of the kibbutz movement, strongly motivated by an ideology based on Zionist, socialist, and humanist values, made a deliberate attempt to create something altogether new.[48] The founders believed that a change in external circumstances could profoundly reshape human behavior and human nature, and, as with the American nineteenth-century efforts, they also had a desire to remake society. But most kibbutzim eventually veered away from this extraordinary objective of creating a new kind of person and society—something they have not accomplished in any case.

The principles animating kibbutzim are similar to those of other communes: cooperation, communal self-sufficiency, shared labor and property, and egalitarianism. In early kibbutzim, all types of work were given equal value, and direct democracy (with a rotation of officeholders) was practiced, although these egalitarian features did not survive intact over the past century. The most striking objective of early kibbutzim was a radically novel structure for family life that centered on collective child-rearing. Parents resided in small apartments while their children ate, slept, and bathed in small houses (*bet yeladim*) with approximately six to twenty peers of a similar age. Children would spend only an hour or two each afternoon with their biological parents.[49] This arrangement also did not survive.

One reason for this feature of the kibbutz movement was to alter the patriarchal mode of family organization that was dominant in Eastern European Jewish culture. The aim of collectivizing child care was to free women from the burdens of domestic life and set them on an equal socioeconomic footing with men while simultaneously bringing men into a more nurturing role. The image of women in the early kibbutz emphasized equality with men, hard physical labor, modesty, and a downplaying of romantic relations.[50]

The idea of collective child-rearing was not unique to kibbutzim. It has been periodically attempted as a desired social disruption since antiquity. Plato believed that raising children communally would result in children treating all men as their fathers and thus more respectfully.[51] Communist societies have also been associated with collective child-rearing; the family is seen as a threat to state ideology because it fosters a sense of belonging to a family unit, and totalitarian ideology requires that family allegiance be subordinated to allegiance to the party or state. Liberal political theory has also struggled with the issue of the family being an obstacle to an egalitarian society (for example, because child care and family life generally impose greater constraints on women).[52]

But attempts to fundamentally restructure or minimize the bond between parent and child have very rarely, if ever, endured.[53] While mild forms of collective child-rearing are found in cultures all around the world (and in some other mammalian species, as we will see in chapter 7), they typically involve forms of alloparental care, whereby relatives share child-care duties. Dormitory sleeping arrangements for infants (of the kind initially attempted by the kibbutzim) are extremely rare. A 1971 survey of 183 societies around the world found that none maintained such a system.[54]

As in many utopian communities, the organization of child-rearing was motivated largely by adult imperatives. If men and women were to be treated truly equally, collective parenting might be seen as an obvious structural necessity, regardless of its implications for individual children and their development. Historian Steven Mintz noted in *Huck's Raft*, his sweeping work on American childhood, that almost every innovation in child welfare in the United States, including orphanages and subsidized child care, has been driven primarily by adult concerns. Of secondary importance were philosophical and pragmatic convictions about what was best for children.[55] As radical as communes may be in some key respects, they generally play by adult rules in regard to children, whose needs and concerns have never been, as far as I can tell, the primary motivation for any utopian community (even though some of them had amazing schools and treated children kindly). Setting up utopias seems to be like sex in at least one way: it is oriented to adult satisfaction.

The various facets of kibbutz life began to break down after the 1950s. Gender distinctions and pair-bonding between spouses (and between parents and children) were not easy to efface. The family gradually reemerged as a strong and central unit, and reports from both men and women suggest that physical appearance and mutual attraction were acknowledged once again as important features of relationships. Marriage took on renewed importance too. A gender-based division of labor—with men in production and women in services—grew stronger, and many kibbutzim returned children to their nuclear-family homes.[56]

Children in kibbutzim initially had lower rates of maternal attachment than peers raised in other settings, and this was one of the key concerns women had with the communal child-care arrangement.[57] Differences in maternal and paternal behavior with infants in kibbutzim also ultimately replicated those seen in other cultures; mothers were more likely to engage in caretaking behaviors for, display affection toward, laugh with, talk to, and hold their infants than fathers were.[58] By the 1970s, the shift from radical anti-familism back to a powerful familism was largely complete.[59] Women played a major role in the transformation of the structure of gender relations and the role of the family, often framing their arguments in terms of what they saw as the "natural needs" of womanhood and motherhood.

Collective child-rearing did provide some benefits for the children. Studies of kibbutz-raised and city children found that the former were more skilled social players, and they were more likely than the city children to engage in positive social play, spend more time in coordinated play, and be uncompetitive in group encounters.[60] One of the more intriguing consequences of collective child-rearing, however, was the virtual absence of marriage between peers. The longer individuals resided together on a kibbutz as children, the greater their aversion to sexual contact with one another. These findings support the so-called Westermarck effect, a psychological hypothesis proposed in 1891 by the Finnish anthropologist Edvard Westermarck. He suggested that childhood co-residence served as a kinship cue (people decide who their siblings are based on who they grew up with). This sense of kinship had two effects: first, it generated an incest taboo among unrelated individuals, and second, it increased the altruism between unrelated individuals.[61]

Hence, by the twenty-first century, collective sleeping arrangements for children and related practices had more or less died out, and most caregiving functions were transferred back to the family, mainly to women. As psychologist Ora Aviezer and colleagues noted,

> Collective education can be regarded as a failure. The family as the basic social unit has not been abolished in kibbutzim. On the contrary, familistic trends have become stronger than ever, and kibbutz parents have reclaimed their rights to care for their own children. Collective education has not produced a new type of human being, and any differences found between adults raised on and off the kibbutz have been minimal.[62]

In addition to discarding collective child-rearing, kibbutzim also eventually walked back some of their other special features. A transfer of chores to the private domain began in the 1970s, and communal dining rooms and laundries were shuttered.[63] And, since the 1990s, most kibbutzim have been shifting away from an equal-sharing economic model too; by 2004, only 15 percent maintained full equal sharing.[64] Like their nineteenth-century American counterparts, these utopian efforts reverted to the norms of their society of origin.

Kibbutzim failed to remake society wholesale. They were not even able to change gender roles. This latter failure is partly ascribable to how entrenched gender roles are, perhaps more so than any of the other features that kibbutz living was attempting to reverse.[65] In my view, however, it was the attempt to break the adult-child attachment bond that was most unrealistic from the start. Love of close family is one of the most important features of the social suite, as we will explore in more detail later. And even members of these idyllic, cooperative communities did not treat all people alike. In one experiment, kibbutz residents acted cooperatively when paired with other kibbutz members but not when paired with city residents, attesting to the psychological strength of in-group bias.[66] Though the pioneers of the kibbutz movement were successful in rejection of the European urban culture from which they hailed, conforming to the social suite appeared inevitable.

Our species' evolved psychology and sociology is the basis for these observations. Anthropologists Lionel Tiger and Joseph Shepher invoked the idea of a *biogrammar* to explain the reversion to certain traditional forms of social organization in kibbutzim.[67] They blended the notion of a universal grammar, outlined by linguist Noam Chomsky, with what they termed a *biogram*, or the basic, genetically encoded, and evolutionarily shaped form of an animal's social life. This is very similar to our notion of a blueprint.

The kibbutz experiments, like other modern communes, afford

us the luxury of relying on contemporary visits rather than just historical accounts. We can observe these efforts in real time. But, more important, the case of the intentional communities of the kibbutzim again highlights the crucial nature of the social suite and the difficulties that arise when people deviate too far from these key organizing principles. The kibbutzim have been able to survive, I believe, precisely because they instantiated crucial features of social life that are both universal and necessary.

### Walden Two *Communities*

Exactly a century after Thoreau moved to Walden Pond in 1848, Harvard psychologist B. F. Skinner published a utopian novel, *Walden Two*. The novel described a fictional rural community of about one thousand people embodying Skinner's theory of behaviorism, the idea that human behavior is primarily, if not solely, a product of the environment. Skinner argued that humans are *conditioned* to behave in particular ways, deemphasizing the importance of people's thoughts and feelings or their genes.[68]

Most people have heard of Russian physiologist Ivan Pavlov's famous experiments in which dogs were conditioned to salivate in anticipation of food at the sound of a stimulus (a metronome). A dog's or a rat's or a pigeon's behavior could be controlled by environmental changes, and human behavior, Skinner thought, could also be molded by modifying people's social environments. He believed that free will was much more limited than most people supposed and that almost any sort of social arrangement was possible. "Set [the utopia] up right and it will run by itself," he argued.[69]

Skinner was moved to write *Walden Two* by the return of soldiers after World War II. "What a shame," he observed, "that they would abandon their crusading spirit and come back only to fall into the old lockstep American life—getting a job, marrying, renting an apartment, making a down payment on a car, having a child or two."

Instead, he said, "they should explore new ways of living, as people had done in the communities of the nineteenth century." Skinner was aware of the many previous failures, like Brook Farm, but he thought that "young people today might have better luck."[70] In the postwar economic boom and ensuing cultural conservatism, *Walden Two* was a publishing flop, but sales rose in the 1960s, and by the 1970s, the book was selling two hundred and fifty thousand copies per year.[71]

Skinner had named his fictional utopia Walden Two to evoke the self-reliance and simple living championed by Thoreau. But the similarities did not extend much further than the name. The original Walden had just one inhabitant, and Skinner aspired to provide a guide for communal living. *Walden Two* is narrated by a Professor Burris, a university psychology instructor who takes a group to visit an intentional utopian group. There, they meet T. E. Frazier, who describes the workings of the community. The community is constantly testing strategies for living based on "behavioral engineering," which Frazier argues is crucial to avoid the failures of past efforts that were overly rigid. The governance of the community is based on a planner-manager system in which two professional and unelected boards, one of planners and another of managers, exclusively make decisions for the group. People in *Walden Two* work only about four hours a day, choose their work based on a point system, and raise their children communally; they have abandoned the nuclear family and have relaxed attitudes about sex (including thinking it quite natural for girls of fifteen or sixteen to have children). Professor Burris himself ultimately joins the community.

Skinner's objective in this work of fiction was not to advocate for these practices (though many of the practices in *Walden Two* were similar to extant communes'), but rather to advance a triumphalist belief that behavioral science could be collectively applied by ordinary people themselves to improve their lives. As Frazier observes: "The main thing is, we encourage our people to view every habit and custom with an eye to possible improvement. A constantly experimental attitude toward everything—that's all we need."[72] Critics of the book were concerned that Skinner's vision was more dystopian than utopian and that implementation of behaviorist principles might "change the nature of Western civilization more disastrously than the nuclear physicists and biochemists combined."[73] In this regard, Skinner's book joined an endless argument about whether science, including the social sciences, is more bane than boon in human affairs.

Although Skinner did not intend the novel to be an actual guide for building a real community, the imaginary society of *Walden Two* wound up inspiring dozens of real ones.[74] Two of the most successful and long-lived were Twin Oaks in Virginia and Los Horcones in Mexico.

At their first meeting, the eight founding members of Twin Oaks, established in 1967 and still in existence today, flipped through the pages of *Walden Two*, searching for ideas to guide them.[75] During the community's first five years, its members implemented many *Walden Two*–like systems, structures, and policies in their compound, including a labor-point system and the use of positive feedback. But, in a now-typical pattern, nothing seemed to work as anticipated. While the collapse of their original plan was less violent and dramatic than the events at Pitcairn, it was no less definitive.

A familiar point of contention at Twin Oaks was the communal child-care system, which was in a state of constant flux from 1967 to 1994. Co-founder Kat Kinkade concluded that it was unworkable, in large part because "parents want to be with the kids, they are attracted to the little things," and that it was simply too "radical [a] departure from the family concept."[76] Following this setback, the community tried and rejected different organized child-care programs before eventually adopting the still-prevailing model, which allows parents to spend more time directly with their own children and make personal decisions on whether to homeschool them or enroll them in public school.

People wanted to manage other aspects of their lives too. The planner-manager system of government described in the novel quickly disintegrated because members of the community entered with the expectation that they would be part of the decision-making process. "They would say," Kinkade recalled, " 'I don't know of any decisions I would have wanted to be different. The decisions were fine. I just wanted to be part of it.' "[77] Writing in 1983, Ingrid Komar, a member of the commune during its founding years, chronicled the ongoing "discontent around the area of government among the membership."[78] The community eventually accepted a democratic form of government that Kinkade characterized as "very eclectic and very scattered...very much in conflict all the time."[79] The balance of egalitarian consensus and benevolent authority that was key to the success of Shackleton's Antarctic crew was elusive at Twin Oaks.

Instability plagued Twin Oaks in the early years. Roughly a quarter of the population present at the start of each year moved away by the year's end, replaced by a like number of new arrivals.[80] With such high turnover, the perpetual and rapid reshuffling of social ties at Twin Oaks compromised the social cohesion and cooperation needed for the community.[81] Komar captured both the practical and visceral consequences:

> The disruption of friendships this almost always entails is emotionally wrenching for those left behind, and the unique contributions those leaving have made during their stays are often hard to duplicate. Departures tend also to be demoralizing and to call communitarian belief systems into question.[82]

Sustained (albeit not immutable) friendship ties are crucial for success.

Twin Oaks not only failed to enact its *Walden Two*–inspired plan but also took a decidedly rocky and decades-long path toward establishing *any* clear community structure. After years of messy trial and error, the arrangement that Twin Oaks (which nowadays has about a hundred members) eventually adopted bore little resemblance to *Walden Two.* In this sense, it kept the faith with Skinner's key agenda of community-directed experimentation. And the social order that resulted resembles those of other communal groups with high turnover. The residents live in dormitories and practice a collective labor scheme (each working forty-two hours per week); they participate in income-generating efforts (such as Twin Oaks Tofu, Twin Oaks Hammocks, or Twin Oaks Seed Farm) or do domestic chores (such as cooking, gardening, maintenance, and child care). Except for personal items, all property is held in common.

The other successful *Walden Two* commune was Los Horcones (which means "the pillars" in Spanish), founded in 1973 in Mexico. It has roughly thirty members. Skinner himself said that Los Horcones came the closest to the "engineered utopia" he described in *Walden Two.*[83] Its members said, "We are not a community based on Skinner's novel... but on the science on which the novel is based."[84] In published articles in the *Journal of Applied Behavior Analysis* and in regularly scheduled group discussions—in fact, in everything they did—the members of the Los Horcones commune placed behaviorism at the core of their existence.[85] Kat Kinkade, from Twin Oaks, observed during a visit to Los Horcones that the well-behaved children of the community were "just the sort of people you want your own kids to grow up to be. And they think they did it through behaviorism... [so] good for behaviorism."[86]

Just as depicted in *Walden Two,* the members of Los Horcones practiced communal child care, implemented a form of "ethical training," and operated a sort of labor-credit system. As was done at Brook Farm, the community founded a school open to outsiders from the nearby city of Hermosillo, thus deriving revenue. As at Twin Oaks, the members of the community shared all property (even clothes) and embraced principles of cooperation, equality, and pacifism. Though the community initially started with the planner-manager system, Los Horcones eventually embraced "personocracy," an original system that aspired to "promote participation from all members" and "increase the amount of [positive] reinforcement available for all."[87] Los Horcones even had a naturally charismatic leader who was much like the character Frazier in *Walden Two.*

Why did Los Horcones fare better than the many other *Walden Two* communities, most of which failed altogether? Key reasons were the tight connections at its founding along with the presence of good leaders. Unlike the superficial and fleeting connections that characterized the Twin Oaks community, the social ties among the members of Los Horcones started out (and have remained) deep and stable. Four founding members were already married before launching Los Horcones, and several founders had worked closely together on education projects for many years prior. The community also grew through a combination of procreation and the gradual absorption of close relatives and friends, not through lax admission standards.

### *Urban Communes in the 1960s*

Communal movements have ebbed and flowed over the centuries, but from 1965 to 1975 in the United States, their popularity reached a new high, with well over two thousand communes founded per year. Several historically specific factors likely accounted for this, ranging from the reactions to the Vietnam War, the extraordinary youth culture and alienation of the 1960s, the new freedoms enjoyed by women (arising from the women's movement and the invention of the birth control pill), and even the passage of federal food-stamp laws that ensured that people who joined communes would not starve. Still, no more than 0.1 percent of the U.S. population chose to participate in a commune at any one time.

In 1974, sociologist Benjamin Zablocki started a project that would follow sixty representative urban communes in the United States for twenty years (he also studied sixty rural communes).[88]

These urban communes (in New York City, Boston, Minneapolis–St. Paul, Atlanta, Houston, and Los Angeles) ranged in size from five to sixty-seven adults (with an average of 13.4) and adhered to various ideological goals.[89] Like their predecessors through the centuries, the members of these communes attempted to articulate and enact what they felt to be new beliefs and moral convictions. The residents were mostly unaware of prior American communitarian efforts and therefore were unknowingly repeating long-standing practices.[90]

Most of the members were white and educated, though many religions, occupations, marital statuses, ages, and backgrounds were represented. The median age was twenty-five, 54 percent were male, 72 percent were single, and 50 percent had college degrees. Most joined because they felt a sense of stagnation, a lack of purpose, or an alienation from the larger society. Typical descriptions of their lives were "something was lacking. What was important in life was not being confronted" and "life was comfortable but flat. Extremely bored, it was obvious that something was missing."[91] Surprisingly, when their levels of alienation and estrangement were quantified and compared to those of the broader U.S. population, the people who joined communes were found to feel *less* economic and political alienation. But they seemed to have a greater awareness of, and sensitivity to, personal meaninglessness. So most joiners were searching for meaning and had a desire to "reduce the world to a manageable size."[92] They were not, as some have claimed, seeking greater opportunities to practice socially deviant behavior.

While the reasons for joining these communes varied, the primary motivation for almost all members was to develop a consensual community based on shared values. Here is how some joiners described their motivations:

> I chose to live with a group of people I see as being support for the personal changes I have been and expect to continue going through. I wanted to be with people who share my

> political and personal perspective and who will allow me to participate creatively in their lives as well.

> We wanted to live together because we were the only hippies or freaks at the University and we all kind of wanted to live together and be a support group.[93]

Even when the subjects were interviewed ten years later, less than 10 percent of them dismissed their experiences as youthful exuberance or folly. Most felt it had been central to the formation of their adult identities.

Like many of the other communes we have considered, these societies could not survive without a flow of money arising from functional interaction with the outside world, whether in the form of gifts, rents, or wages earned by residents. The challenges communes faced were generally internal, not external. As one study concluded, "Many more communes went under because the dishes never got washed than were ever forced out of town by hostile neighbors or zoning boards."[94]

In fact, the dishes usually did get washed, but the workload in these urban communes remained heavily gendered, with women doing more cooking, cleaning, and babysitting (each spending 1.5 hours per week caring for children other than their own), and men doing more house maintenance and the "work" of "spreading the ideological message" (4.9 hours per week).[95] Some variation of a gendered division of labor is seen in all of the intentional societies we have been considering. This reflects the classic idea of comparative advantage—men and women specialize in different tasks (whatever they might be) and then complement each other's efforts by developing efficiencies related to expertise.

Leadership, along with mild hierarchy, often played a decisive role in the successful operation and endurance of the communes. In one commune, the charismatic leader motivated a team of ten hippie men without prior training in construction work to build cabins at the astonishing rate of one a day, and this despite poor equipment and terrible weather conditions. The grueling pace was not resented by the members but rather considered to be a valuable spiritual discipline.[96]

The turnover of people in the communes in this study was very high, higher even than at Twin Oaks. The study began in 1974, and in 1976, only about a third of the residents were still present. The main reason people left, they said, was that they did not feel "loved" by other members of the commune. Of the sixty communes at the start, forty-eight (80 percent) survived for one year, and thirty-eight (63 percent) survived for two years. Longer-lived communities tended to have more stringent entrance requirements or probationary periods for new members. In this sample of communes, 77 percent disintegrated for internal reasons (such as ideological schisms, leadership disputes, or sexual tensions) and 23 percent for external reasons (such as legal threats or disasters, including fires).[97]

Communes in this period were far from being sexually orgiastic, drug-hazed, lawbreaking enterprises, as is sometimes wrongly assumed. Membership in these groups came with normative pressures to be *less* extreme. Self-reports of drug use declined from 86 percent to 42 percent, public nudity from 40 percent to 32 percent, being in a sexual relationship with more than one person at a time from 24 percent to 14 percent, and participation in riots from 22 percent to 3 percent. Even participation in anti-war demonstrations declined from 57 percent to 9 percent.[98] Across the sample, 42 percent of residents did not have sex with other members of the commune (though some had sex with outsiders), and of those who did, 71 percent were monogamous. The incidence of group marriage or polyamory involving multiple people of both sexes was less than 1 percent.

Two broad factors typically determine group cohesion in these communes: ideology and structure. By *structure,* I mean not only the hierarchy in a group but also the pattern of social relations in the group (such as whether friendships are reciprocated and the extent to which people have friends in common). Both factors are important. Sociologist Stephen Vaisey analyzed the urban communes to see which groups were better able to generate a sense of *Gemeinschaft*, the sense of a collective self or a feeling of natural *belonging*. To what extent did this "we-feeling" in group members arise from structural factors, such as friendship ties, and to what extent did it arise from the sharing of ideas among the members?

Because the data regarding these communes were so detailed, we can map and analyze the actual social connections among the individuals in what is known as a *social network*, more examples of which we will see below. Commune residents were asked various questions about their relationships with others, including who they spent their free time with, worked with, had sex with, felt a loving feeling toward, and even disliked. Vaisey noted that certain measures of social structure were not enough to produce a sense of belonging on their own in the urban communes. But he found that a shared moral understanding—a unified set of beliefs and a common sense of purpose—was crucial.[99]

In these communes from the 1970s, we once again see manifestations of the social suite—manifestations such as the maintenance of friendship ties, the existence of mild hierarchy, and the respect for a sense of personal identity—and see its role in collective success. I'd like to consider one final example of an intentional community of a rather different sort, one that is physically isolated from the rest of the world and one that, having been established in still more recent times, has afforded scientists the opportunity to map the detailed social structure of the inhabitants in the form of social networks.

### *Scientists on the Pole*

A winter base in Antarctica is so isolated that it has long been used as a model for space travel.[100] Here is a 1902 diary entry of one explorer: "I could easily imagine we were standing not on Earth but on the Moon's surface. Everything was so still and dead and cold and unearthly."[101] Alfred Lansing, a journalist who wrote about the Shackleton expedition, claimed there was "no desolation more complete than the polar night. It is a return to the Ice Age—no warmth, no life, no movement."[102] Humans arrived at the geographic South Pole for the first time in 1911, and permanent stations, occupied by new groups of people every year, were set up there by 1956. Almost immediately, scientists and military officers responsible for these settlements began to consider the psychosocial features associated with the stations' smooth functioning.[103] This continent, originally devoid of humans, has, ironically, provided a valuable laboratory to study society.

Different countries maintain their own stations, many miles apart. The American facility at the South Pole, established by the U.S. Navy in 1956, is named the Amundsen-Scott South Pole Station, after the famous explorers who were the first to reach that pole. The facility included an iconic geodesic dome (fifty meters wide and eighteen and a half meters tall) that contained a number of modular buildings, but this was replaced with a major upgrade in 2003. The station is currently operated by the National Science Foundation, primarily for the purpose of research, especially in astrophysics and meteorology. As many as a hundred people inhabit the station during the summer. But during the winter, perhaps thirty people—the typical size of human groups we have been considering, from shipwrecks to communes—take up residence.[104] For about eight and a half months, they are cut off from the rest of the world.

During the southern winter, which stretches from March to September, the pole receives no sunlight at all. Despite its association with a vast expanse of snow, it is actually a desert, and it receives almost no precipitation. Still, it has high winds that can create snowstorms, with drifts that require a bulldozer to clear from the entrance to buildings. And if all this were not enough, the station is located at an altitude of 9,306 feet, sufficient to induce altitude sickness in new arrivals. The conditions are so harsh that even emergency rescue is not feasible, since flights to the pole in the winter are virtually impossible, and the next nearest American base, McMurdo Station, is eight hundred and thirty miles away.

In the past few decades, it has become increasingly possible to study groups using tools that were unavailable in prior eras (tools that, unfortunately, cannot easily be applied retrospectively to historical cases). For example, the detailed mathematical mapping of social ties in what is known as social-network analysis is very helpful in understanding the processes of group formation and function. I'd like to introduce the use of these tools by examining the case of a group of scientists self-stranded on the South Pole, the final natural experiment involving an intentional community that we will consider.

A few caveats are necessary. The people who choose to go to Antarctica, like those who joined communes or sailed on ships, are not a representative sample of the human population. They choose to be there and are vetted and supported by funders, military officers, psychologists, and others who clear them for participation. They are also selected for talents, abilities, and interests matched to their environment, an advantage not enjoyed by all isolated communities. And they have a prearranged time to be "rescued," which provides a bulwark against *Lord of the Flies*–style anarchy.

The people who winter over in the South Pole fall into two categories: the support personnel, known as the "trades" (a group that includes plumbers, electricians, mechanics, cooks, and scientific technicians who are responsible for daily operations of the station), and the research scientists, known as the "beakers." These divisions have existed from the beginning of these Antarctic camps.[105] After two months of joint training, winter-over crews go to Antarctica in October and remain there for thirteen months, until the following November. Other people come and go during the summer months, but during the winter, these crews are on their own, with no one arriving at or leaving the station.

People wintering at the pole can suffer sleep disturbances, hypoxia, altitude sickness, and various endocrine and immunological abnormalities. They sometimes also suffer from cognitive impairments and mild hypnotic states known colloquially as "long eye" or the "Antarctic stare." But crews report that the social and psychological stresses are actually harder than the physical ones, and depression can be a serious problem.[106] One scientist bemoaned the lack of "windows, privacy, living green things and animals, the sun, thick moist air to breathe, freedom to travel, or freedom to leave a rumor-infested isolated human outpost."[107]

As in the case of the shipwrecks, customary hierarchical structures can break down in such a small, isolated group. A cook might have more status than a senior officer, and a radio operator more authority than an advanced scientist. This can be difficult for highly trained scientists and career military officers to accept, but flexible authority and shared tasks among all members of the group are crucial for everyone's well-being.

Almost from the start of Antarctic social research in the 1960s, scientists took measurements of all possible social ties in the groups. For instance, early surveys of winter-over crews asked questions similar to the ones I use in my own lab half a century later: "Who have you found to be your closest friend(s) during the past few months?" "Which person, or persons, impressed you most by their ability to provide leadership for others when it was needed?" And "If you were given the task of selecting men to winter-over at a small station, which five men from this station would you choose first?"[108]

These basic questions yield data to map the interactions of people in a group in the form of a social network, which can then be visualized and analyzed mathematically. Sociologist Jeffrey C. Johnson and his colleagues collected data from three separate crews in the 1990s; the first year (year A) had twenty-eight people (nineteen men and nine women); the second year (B) had twenty-seven people (twenty men and seven women); and the third year (C) had twenty-two people (eighteen men and four women).[109] The station physician surveyed the residents on the fifteenth day of each month, and they rated the extent of their interactions with all the other residents and identified people they considered leaders in their group.

A study that evaluates essentially all possible ties among members of a group is known as a *sociocentric study*, and it yields images such as those of figure 3.1. Interestingly, the social architecture of these Antarctic groups resembled many others my lab has mapped among American college students, foragers in Tanzania, residents of Massachusetts towns, villagers in Honduras and India, and workers in firms around the world, as we will see in chapter 8.

**Figure 3.1: Social Networks of Antarctic Crews**

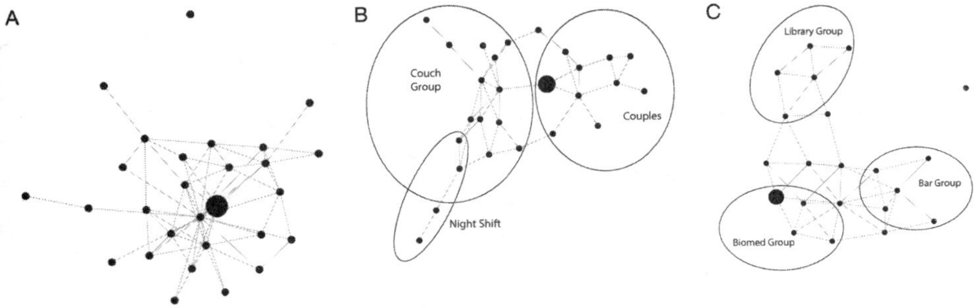

The three images represent the social networks of three annual winter-over crews of scientists and support personnel in the U.S. Antarctic station, labeled A, B, and C. People (nodes) and their primary friendships (ties) are shown. The manager in each case is the largest node. In year A, the manager is in the middle of the whole network, whereas in years B and C, he belongs to one subgroup. In year A, no subgroups were observed. In years B and C, three (labeled) subgroups were identified. The qualitative assignments to subgroups here do not overlap exactly with group assignments based on more formal mathematical algorithms.

### *Antarctic Networks*

In order to explain what such network diagrams (also called *maps*) mean, I need to digress a little and cover some of the basics of network science. Each person in a defined population (for example, an Antarctic station, a shipwreck colony, a firm, a school, a village, a whole country) is indicated by a circle, or *node*, in the diagram, and every connection between any two people (two friends, two co-workers, two relatives, or two spouses) is indicated by a line, or an *edge*. Connections among people are found by asking people questions with what is termed a *name generator*. For instance, people may be asked "Who do you spend free time with?" or "Who do you discuss important matters with?" Researchers might also ask people who they would borrow money from or lend it to, or they might ask them to name their close friends, siblings, co-workers, or sexual partners.[110] Sometimes, researchers can give people money or something else of value and ask each of them to actually give it to someone else anonymously, as a gift. Assuming that people generally do not give gifts to total strangers or people they loathe, the named gift recipients also identify noteworthy social connections.

A more direct way to measure human ties is, of course, to observe them. We could sit in a school cafeteria and note who sits with whom. Epidemiologist Marcel Salathe, political scientist David Lazer, sociologist Mark Pachucki, and others have fitted students with little radio trackers or other devices that capture such data, noting who is near whom and for how long (this method is also used with apes and elephants to map their interactions, as we will see in chapter 7).[111] Researchers could also set up video cameras at a particular location to track interactions and meetings between people. They could use e-mail or phone traffic or online social-network data to identify social ties. They could use data from workplaces about teams or groups of individuals sharing equipment—for example, who rides in a three-person snowmobile each day during work at the South Pole.[112]

After researchers have defined and collected information about the ties among the members of the target group, they can draw and mathematically analyze the network. The shape of a network, also known as its *architecture* or *topology*, is a fundamental property of any socially interacting community.[113] While the shape can be visualized in different ways, the actual *pattern* of connections that determines this shape remains the same regardless of how the network is drawn.

Here is why. Picture one hundred buttons strewn on the floor and imagine that you have four hundred strings you can use to connect the buttons. Next, imagine picking two buttons at random and, without disturbing them from their location on the floor, connecting them with a string. Now repeat this procedure, connecting randomly chosen pairs of buttons one after another, until all the strings are used up. In the end, some buttons will have many strings attached to them, having been, by chance, chosen frequently. Others will have just one string attached. And still others, also by chance, will have no strings at all. Some buttons might be connected in a group to one another but lack any connections to other, separate groups. These independent groups—even those that consist of a single unconnected button—are called the *components* of the network.

Imagine that the strings are fastened tightly to their buttons. If you were to pick any one button and lift it up from the floor, all the other buttons to which it was, directly or indirectly, attached (in its component) would follow it up into the air. The remaining components would be left behind on the floor. Now, here is the key detail: If you dropped this ensemble of buttons and string onto another spot on the floor, it would have a different appearance than it had when you picked it up. But each button would still have the same relational position to the other buttons that it had before; its location in the network would not have changed. In other words, the *topology* of the buttons—an intrinsic property of the network—would be exactly the same no matter how many times you picked up and dropped the mass of connected buttons.

A network can be drawn (or laid out on the floor) using algorithms that create a pleasing image of its structure, for example, by minimizing overlapping strings among the buttons (similar to the way you would try to untangle a ball of yarn by gently spreading it out on a table). Visualization software tries to reveal the underlying topology by putting the most connected buttons in the center and the least connected ones on the edges. But again, while the algorithms can draw the same network in various ways, it's important to understand that, regardless of how the network is drawn, it is still fundamentally the same object.

With this background in how to ascertain and visually represent networks, let's return to our Antarctic scientists. What does mapping their networks reveal? The three crews differed somewhat in their integration and whether there were discernible network subgroups within the whole group. Subgroups, which are also known as *cliques* or *network communities,* are not the same as the independent clusters (components) mentioned above; subgroups are sets of nodes that are highly connected among the members but not totally isolated from the rest of the population. The three crews also differed in the structural location and role of their leaders.

As we inspect the images in figure 3.1, we can see that year A has a single group with a core and periphery structure and no subgroups. Year B has visible subgroups within a somewhat coherent overall group. That year, ten individuals (a mix of both beakers and trades) frequently spent time together in the galley and came to be known as the couch group. There was also a subgroup of three people working the night shift who formed another tight-knit clique. Finally, there were four couples, several of whom had already spent prior winters in Antarctica and had preexisting connections. Year C shows still starker cliques, with three subgroups that self-isolated even more than the ones in year B. The clique names in year C come from the locations where these groups tended to congregate to watch videos together.

The leaders (or managers) identified by the winter-over personnel also occupy structurally different positions. In year A, the leader is near the center of the network for all the subjects; in year B, he is in one of the cliques but is still central to the overall group; and in year C, he is near the center of one of the cliques. Year A is the pattern most typically seen in well-functioning small groups. I suspect that, if we could have mapped the network in Shackleton's crew, this is the pattern we would have observed. By contrast, year C may be how the men of Pitcairn looked before their bloodbath.

Groups can have both *instrumental* leaders, those focused on practical objectives or tasks, and *expressive* leaders, people who work to build solidarity in the group. We already saw that Fletcher Christian was an effective instrumental leader (organizing a successful mutiny on the *Bounty*), but he was unable to resolve conflict among the mutineers after they landed on Pitcairn. Effective leaders have to help minimize group conflict, deal with troublesome individuals before they compromise group harmony, keep work on schedule, make rational decisions in emergencies, deal fairly with conflict, and facilitate communication. Sometimes it's not possible for the same person to serve both instrumental and expressive functions, which is why many societies have leaders who make war (generals) and leaders who talk peace (diplomats).

In addition to being more fragmented, the group in year C had identifiable troublemakers. We will discuss the role of animosity and antagonism in social structure in chapter 8, but for now, it's enough to note that the group evinced a vicious cycle of disconnection, gossip, avoidance, and consequent suspicions, all of which contributed to conflict and division.[114] In fact, groups in isolated environments often note the problem of "constant gossip" that adversely affects interpersonal relations.[115] In the Johnson study from the 1990s, several individuals also competed for the leadership role. Finally, a key issue in year C seems to have been excess alcohol consumption, another perennial threat to social harmony.

Despite the noteworthy variation in subgroup structure, there are still some consistent principles that guide the formation of friendship and cooperative interactions. Although a set of people and their connections could be arranged in various ways, real life imposes restrictions. Neither Antarctic scientists nor any other human group naturally forms networks such as those in figure 3.2. These networks are highly regular (each in different ways). For instance, in these three examples, every person in each network (except those on the outer edge in figures 3.2a and 3.2b) has the *same* number of social connections (three, eight, and eleven, respectively). In contrast, even though the three networks in the three winter-over years in figure 3.1 vary, they still have some key attributes in common (like having a structure with a center and a periphery, and variation in how many social ties each person has).

**Figure 3.2: Unnatural Social Networks**

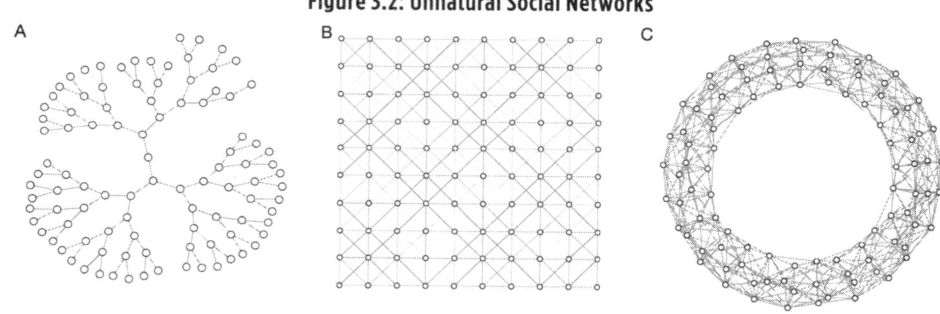


Social networks with the sorts of structures shown here, while in principle possible and while sometimes engineered by design, do not occur naturally. Nodes are individuals and lines indicate social connections. (A) One hundred people arranged as a telephone tree, with each person connected to two others (except at the edges). (B) One hundred people arranged in a regular lattice, with each person connected to eight others (except at the edges). (C) One hundred people arranged in a ring network with so-called neighborhood structure, with each person connected to eleven others.

As we noted above, year B had one clique, the couples subgroup, that arose in part because of the prior connections that existed among members of that group. The relevance of past interactions for the substructure of the networks highlights three points. First, networks are not static and (as we will see in chapter 8) friendships can form and break; connections usually take time to establish, and the strength of the ties is often related to their duration.

Second, the starting conditions matter. When groups of strangers are thrown together to confront a joint challenge or shared experience—as happens with Marine recruits, college students, religious pilgrims, or cruise passengers—there is a reduction in social awkwardness and an openness to making new connections, at least for a while. Everyone is in the same boat, facing the same novel experience. In the first week of college, it's normal for a stranger to sit next to you, introduce himself with "Hello, I'm Nicholas," and start a conversation; but it can feel creepy after a few months have gone by.

Third, the unfolding of a network over time is relevant to how the group comes together, or fails to. Yet another study of different winter-over groups in Antarctica provided a qualitative description of how ties are formed. Unsurprisingly, in groups this size, connections usually start as separate pairs and are initially based on shared tasks or interests (meteorology or music). As we saw with the Brook Farmers and the Shackleton expedition, collective activities such as games and songs are especially important when groups lack longstanding rituals or shared religious ideologies.[116]

### *Communal Success*

In describing people's responses to the scale of modern life and their disenchantment with the world, sociologist Max Weber observed in 1918 that "the ultimate and most sublime values have retreated from public life either into the transcendental realm of mystic life or into the brotherliness of direct and personal human relations."[117] The only cures for anomie and doubt are faith and a deep contact with reality. People who join communes seek *Gemeinschaft*, the sense of group identity and solidarity born of personal interactions. They seek authenticity through a reduction of scale.

Intentional efforts to form new communities have provided a series of natural experiments, shedding light on our species' social state and vindicating the importance of the social suite. These efforts did not result in successful new social forms. Many did not even survive more than a few years. But they illustrated timeless features of our social life, including friendship, cooperation, mild hierarchy, and a preference for an in-group.

New communities that respected the exigencies of the social suite fared better than ones that did not. Brook Farm and the Shakers, for instance, took seriously the idea that people were not a uniform, interchangeable mass that could be molded into any sort of society. Instead, people were thought to have individual personalities worthy of respect. Balancing group identity and individuality is key for successful social systems. With greater allowance made for the diversity of unique individuals (and even their property rights), the challenge then became how to reconcile a society that allows individuals to be themselves with one built to suppress competitive selfishness. Here, efforts to foster and exploit cooperative instincts and cultivate a sense of friendship and belonging to the groups were crucial. Moreover, effective leadership mattered as much in these intentional efforts as it did in the unintentional ones.

Different utopian communities adopted paradoxical approaches to sex. Some emphasized free sexual interactions among members of the group; others, like the Shakers, required complete abstinence. But both of these strategies shared the common aim of subverting the institution of marriage and eroding any deep personal connections between pairs of individuals; the objective of these strategies was to foster a sense of connection to the group *as a whole.* This is the reason that many communities also attempted to break down the nuclear family through communal child care and separate living quarters for parents and children, as happened in the kibbutzim. But as we have seen, such efforts almost always fail because they subvert our species' pre-wired instinct for love.

While deviating from the blueprint would seem to spell doom, strict adherence does not necessarily guarantee success. External forces also matter. And threats like natural disasters, fires, economic or environmental constraints—and even the availability of intoxicants—can very effectively destroy even well-established communities.

In short, though the specific circumstances vary, two broad sorts of forces serve to promote the success of, or hasten the collapse of, communalist dreams to make society anew: intrinsic biological pressures and extrinsic environmental pressures. Pushed by the blueprint within us, and even if pulled by the forces around us, it is not easy, or feasible, to abandon the social suite.

To more fully explore our natural social state, I would now like to turn to a set of unnatural social states created in the laboratory. These experiments with artificial societies composed of real people offer us tremendous control over, and insight into, what sorts of social worlds people can, and must, create.

# CHAPTER 4

# Artificial Communities

In 2005, the Amazon corporation developed a software system that allowed it to recruit and pay thousands of workers to perform small tasks aimed at improving its website (like identifying duplicate listings or checking product descriptions) for a few cents per task. Sitting at their own computers in their own homes at times of their own choosing, people could log on and work for as long as they wanted. The system distributed the tasks, collected the workers' input, and disbursed the payments.

Having solved its own problem, Amazon started allowing other employers, for a fee, to use the platform to hire workers, turning its invention into a profit center. The service was named Amazon Mechanical Turk, after a chess-playing machine first shown at the Austrian imperial court in the eighteenth century. That machine featured a mechanical man made of wood and wearing a turban, ostensibly driven by clockwork, who could play chess. The automaton was a hoax. It was actually controlled by a chess master of very short stature hidden inside, a human player good enough to beat Napoleon Bonaparte and Benjamin Franklin when they played against him, to their astonishment.[1] The Amazon Mechanical Turk system is similar in that it emulates a machine even though, in actuality, humans are under the hood. The system is especially good for work that is easy for humans (like transcribing handwritten documents), but not for computers; therefore, the jobs posted by employers on the platform are known as HITs, for "human-intelligence tasks."

More than five hundred thousand people from around the world have signed up to be Turk workers (sometimes called Turkers). About twenty thousand people are active at any given time, earning about six dollars per hour performing dozens of tiny chores that typically take a few minutes each. For instance, using this platform, one firm hired fifty thousand people to label the content of fourteen million images in order to create a database to train computers in image recognition.[2] After the humans did their work, the real machines could take over.

Amazon Mechanical Turk and other crowdsourcing platforms introduced in the past decade have revolutionized science as well as business. Scientists use these platforms to code data (like galaxies in astronomical images, proteins in biochemical images, or ancient ruins beneath jungles in satellite photographs), conduct marketing and scientific surveys, and get subjects for social science experiments. My lab was an early adopter of this platform, beginning our experiments in about 2008.

Having such a large and varied pool of subjects has transformed the social sciences in many ways, and despite a slow start of just a few scientific publications a year, now well over a thousand papers using Turk workers as research subjects are published annually.[3] No longer are scientists limited to using undergraduates in wealthy nations for their experiments, with sample sizes restricted to one hundred subjects. Now, scientists can do experiments with thousands of subjects who are more representative of a broader swath of humanity—people of all ages and many nationalities and backgrounds. And many studies have documented that the way Turk workers behave in various situations (for example, when they are faced with social dilemmas requiring cooperation or when they are assessing risk) is the same way that research subjects brought into the laboratory do.[4] These are real people acting in a normal, human fashion.

**Figure 4.1: The Mechanical Turk**

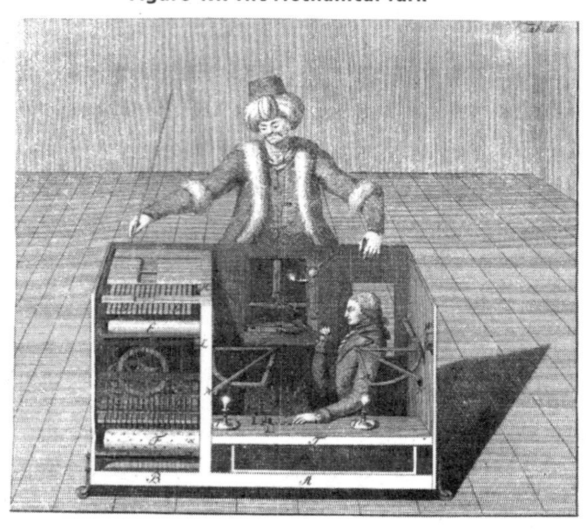

This is a 1789 copper engraving of a diagram of the Mechanical Turk, the chess-playing automaton. The automaton was not actually a machine. It had a chess master of very short stature concealed within the cabinet.

## *Building Small Societies*

It is hard to make confident inferences based on samples like three groups of Antarctic scientists, twenty shipwreck parties, or even sixty communes. Even if there were hundreds of people in each of these groups, that would not solve the problem. Since the features we are interested in are properties of *groups*, and not strictly of the individuals within them, we need a large number of groups to compare to one another in order to be more confident about what really matters here.

Furthermore, in natural experiments, whether with unintentional or intentional communities, we cannot control for all the factors that might be affecting what interests us. What resources were available to the wrecked crews or Polynesian settlers? Who chose to join which communes? Just having a large number of groups is not enough. Ideally, we would like to form the groups ourselves with more representative subjects. For all these reasons, as part of conducting real experiments, we need to be able to manipulate the composition, organization, and interactions of social groups, like TV producers on a large scale, and thus move beyond the natural experiments we have so far been considering.

To do this, my lab developed software to create temporary, artificial, miniature societies using Turk workers as subjects. We call this software Breadboard, after the wooden boards budding engineers once used to assemble and test electrical circuits (an exercise many of us can remember from childhood). But the components we manipulate are variables such as the structure of interactions between the subjects (that is, the topology of the social network into which we place the subjects) and the nature of the interactions (for example, whether the subjects are allowed to cooperate with one another, or how much information the subjects are given about their surroundings). We can also change individual attributes, such as how "rich" or "poor" the subjects are in an experiment (because we give them real money to play with in our games). Breadboard can be used to do experiments with other subjects too, such as employees at companies, students in classrooms, or thousands of ordinary citizens on panels that are maintained by commercial survey firms.

This may not sound revolutionary, but it is. With the exception of the field of psychology, the great majority of traditional work in the social sciences, including sociology, economics, anthropology, and political science, has involved observational studies, not the controlled experiments that are so common in the natural sciences. In 1969, the sociologist Morris Zelditch asked, rhetorically and skeptically, "Can you really study an army in the laboratory?"[5] Half a century later, the answer is yes. Over the past few years, experiments with thousands, sometimes even millions, of people have been conducted online. More than twenty-five thousand people have participated in our Breadboard experiments alone.

In one key experiment, involving seven hundred and eighty-five people arrayed in forty groups, we brought Turkers together and dropped them (at random) into a social network with a particular (random) structure, shown in color plate 1. Each participant was assigned to have between one and six social connections, emulating what is known about the number of ties people have in real life. And everyone in the group had a different set of neighbors than the others in their group.

Our goal was to develop a situation that re-created the challenges of producing public goods—items that people work cooperatively to fashion and that are of mutual benefit, like lighthouses or wells. Everyone has to pitch in and make a sacrifice in order to create something that helps the whole group, including each individual, and that repays each person more than he or she contributed.

In our experiment, participants were given chits that would be converted into real cash at the end of the game. The game was played in multiple rounds. In each round, participants were told they could either keep their money or make a donation to their neighbors. If they made a donation to their neighbors, we would double the money their neighbors received. They would pay a small price, but their neighbors would get a bigger benefit. Since many rounds of this game were played, a potential norm of reciprocity was set up: if you were generous to your neighbor in this round, your neighbor might be generous to you in the next. If so, you could reciprocate again, and you both would benefit repeatedly over time.

Of course, the best outcome from a selfish point of view would be for you not to contribute to your neighbors and for them to give money to you. Why not let everyone else build that lighthouse if you can benefit without helping out? But if everyone did that, the group would collapse acrimoniously. Everyone would stop cooperating, and no one would support the creation and maintenance of public goods.

And that is what our experiments confirmed. When people were assigned to their initial social connections, they usually began by being generous and cooperating with those other people. Yet sometimes, their newly assigned "friends" would not contribute to them (technically, they are called *defectors*). People did not like being exploited by the defectors. Since participants were not allowed to change their initial connections, if their neighbors defected, then all they could do to avoid being abused was to defect themselves (that is, stop being generous). In this branch of the experiment, we found that defection took over the societies we created. In the rigid (and leaderless) social worlds where subjects did not have any control over whom they interacted with (and thus were trapped with a group of friends we assigned to them), people stopped cooperating.[6]

However, in a different branch of the experiment involving other groups of subjects, we allowed people to exercise some control over whom they interacted with. During each round of the game, in addition to choosing whether to cooperate or defect, they could also choose with whom to make or break ties. Understandably, subjects chose to form ties with nice, cooperative people and to break ties with mean, defecting people. Allowing a certain amount of *fluidity* in social ties and some control over friendship choice made all the difference. Cooperation persisted in these societies, and people were nice to one another. We also found that the cooperative people wound up forming cliques, flocking together to avoid intransigent, exploitative neighbors. In short, even the possibility of being able to change social connections can shape communities for the better.

People often think that personality traits such as kindness are fixed. But our research with groups suggests something quite different: the tendency to be altruistic or exploitative may depend heavily on how the social world is organized. So if we took the *same* population of people and assigned them to one social world, we could make them really generous to one another, and if we put them in another sort of world, we could make them really mean or indifferent to one another. Crucially, this indicates that the tendency to cooperate is a property not only of individuals but also of groups. Cooperation depends on the rules governing the formation of friendship ties. Good people can do bad things (and vice versa) simply as a result of the structure of the network in which they are embedded, regardless of the convictions they hold or that the group espouses. It is not just a matter of being connected to "bad" people; the number and pattern of social connections is also crucial.[7] Aspects of the social suite, such as cooperation and social networks, work together.

An analogy is helpful. If you take a group of carbon atoms and connect them one way, you get graphite, which is soft and dark and perfect for making pencils. But if you take the same carbon atoms and connect them another way, you get diamond, which is hard and clear and great for making jewelry. There are two key ideas here. First, these properties of softness and darkness and hardness and clearness are not properties of the carbon atoms; they are properties of the *collection* of carbon atoms. Second, the properties depend on how the carbon atoms are connected. It's the same with social groups. This phenomenon, of wholes having properties not present in the separate parts, is known as *emergence,* and the properties are known as *emergent properties.* Connect people in one way, and they are good to one another. Connect them in another way, and they are not.

In another experiment, this one involving 1,529 people arranged in ninety groups, we assessed exactly how much social fluidity optimized cooperation by varying the rate at which people could rewire their ties to their new friends. Interestingly, we found a parabolic relationship between social fluidity and cooperativeness (see figure 4.2).[8] Neither too much rigidity nor too much fluidity is optimal for

**Figure 4.2: Manipulating the Rate of Interaction Affects Cooperation Levels in Groups**

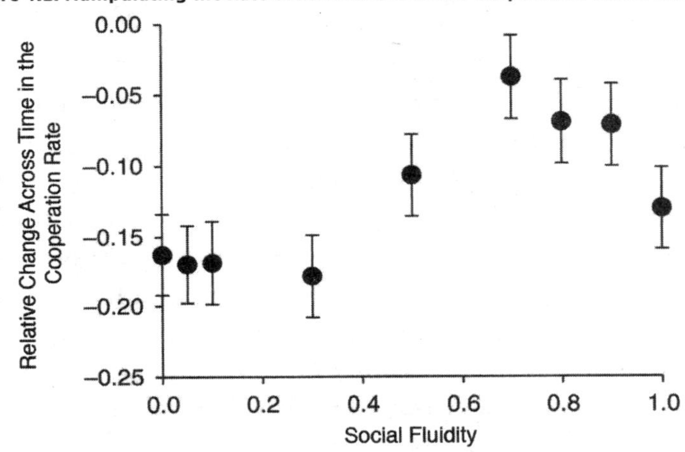


This graph demonstrates the relationship between social fluidity, or the rate at which ties may be re-wired, from 0 (never) to 1.0 (always) (on the x-axis) and the amount of cooperation shown by subjects in experimentally created social network groups (on the y-axis). When there is little or no social fluidity (that is, when people are stuck with their neighbors), cooperation is low. Similarly, when there is very high social fluidity (that is, when people's neighbors change very often), cooperation is low. Optimal cooperation in groups is achieved when there is intermediate fluidity.

the cooperation of groups. As described above, at one extreme, if people's connections are too rigid, they can lose the incentive to cooperate. If you have exploitative neighbors from whom you cannot escape, the only way you can respond is by stopping your own altruism. And if you have cooperative neighbors who cannot escape from you no matter how badly you behave, you also might be tempted to become noncooperative. At the other extreme, if people's connections change too frequently, you (and they) might also lose the incentive to cooperate. You have no reason to invest in your neighbors and treat them kindly if they might well be gone the next moment. We already saw this idea play out in the Twin Oaks community, which struggled with a high turnover rate, compared to Los Horcones, which was much more stable.

In another experiment, we created forty-eight miniature societies (with a total of 1,163 people) in order to explore the extent to which the benefits of cooperation had to exceed the costs before people willingly cooperated. Specifically, we wanted to know how the ratio of benefits to costs related to the number of social connections. It turns out that the more connections people have, the more they must be incentivized to cooperate.[9] We were even able to quantify this, showing that the cost/benefit ratio of cooperation must, on average, exceed the number of friends one is interacting with. For cooperation to arise in social groups, the benefits must be twice the costs if a person has two partners, four times the costs if a person has four partners, and so on. This makes sense; cooperation is harder in larger groups, so the payoff has to be worth it.

We also have explored how the distribution of wealth within groups affects them. We took 1,462 subjects and randomly assigned them to eighty groups and again gave the subjects small amounts of real money to play with.[10] Each group occupied a different environment that we engineered; some groups had total equality of wealth at the outset, others were placed in mildly unequal situations, and still others experienced high inequality. In the unequal settings, subjects were randomly assigned to be rich or poor (by being given relatively more or less money to play with).

The experiment looked at how much wealth the groups were able to collectively generate and how cooperative and friendly their members were with one another. While inequality itself might, in principle, be corrosive to group performance, we found that, in practice, this was not the case. The real threat seemed to be the visibility of the wealth; when people could actually see how much money others had, it subverted group cohesion, making people less cooperative, less friendly, and ultimately less able to work together to enhance their collective well-being. For instance, in miniature societies in which people's money was on display, subjects cooperated roughly half as much. This experiment suggests that one reason that so many communal utopias require simple, uniform styles of dress and communal property-holding is to simultaneously equalize and conceal status differences, thus fostering cooperation and friendship ties.

These experiments with real people performed by my lab (and other labs around the world) supplement the natural experiments with unintentional and intentional communities we reviewed in prior chapters. In a laboratory setting, we can be much more certain about causation (for example, that the pattern of friendship ties is a cause of the cooperation rate, rather than vice versa), and we can explore specific aspects of the social suite while keeping other aspects constant. No one research method is perfect; people might respond differently in real life than in laboratory settings. Still, in these artificial situations, people behave in very human ways, creating types of social order that comport with the rules of the social suite.

### *Massive Online Games*

Another rich source of online data on de novo societies and features of the social suite are the massive multiplayer games involving hundreds of thousands of people. In these games, players typically adopt avatars with vivid, customizable, three-dimensional appearances. To enhance the capacity for individual identity, the choices regarding avatar appearance are huge; in the virtual world of *Second Life*, for example, players can manipulate one hundred and fifty parameters and change everything from eye color to foot size to gender.[11] Games often run for many months, and players can acquire possessions, powers, money, even pets. Some games, such as *World of Warcraft, City of Heroes, EverQuest,* and *Second Life,* are built on the possibility, even the necessity, of social interactions, and players band together in groups to accomplish tasks, trade goods, form friendships, compete in adversarial relationships, or engage in warfare.

In 2016, for example, every month, at least five million people around the globe participated in *World of Warcraft*.[12] This game has so many players that, if it were a country, it would have more citizens than Norway or New Zealand. Players typically band together in groups known as "guilds" to defeat monsters and other enemies or to acquire resources. One study of three hundred thousand players found that guild size ranged from three people to two hundred and fifty-seven, but the average group was about seventeen people (and 90 percent of the guilds had thirty-five people or fewer).[13] This was roughly the size of the shipwreck parties or urban communes. Over three thousand guilds were observed. Although the guilds were short-lived (roughly 25 percent dissipated within a month), bigger groups lasted longer (just as with the urban communes). Modest levels of hierarchy also helped keep the groups together. Finally, social-network structure played a role; guilds that had higher densities of ties and were better connected lasted longer. These are all important features of the social suite.

It is perhaps not surprising that in one survey of nearly a thousand gamers from forty-five countries, players reported that they actually made "good friends" within the game (with an average of seven friends). About half the gamers believed that these online friends were comparable to their real-life friends, and nearly half discussed sensitive issues, like family problems, work problems, and sexuality, with them.[14]

In our online interactions, we still act in very human ways. We do not leave cooperation, friendship, or in-group bias behind when we cross over to the digital world. For example, people appear to follow racial stereotypes. One study examined the willingness of avatars in one virtual world to help out individuals of other races who made simple requests. Disturbingly, requests made by dark-skinned avatars were much less likely to be honored.[15] Avatars also obey gender norms consistent with the real world. For instance, pairs of male avatars (regardless of the genders of the real people controlling them) keep a larger interpersonal distance in the virtual world than female avatars.[16] These online interactions can be so realistic that the games have been proposed to treat some psychological problems, like social anxiety disorder.[17] Therapists based at Drexel University in Philadelphia created private rooms in the virtual world of *Second Life* and interacted with patients there; they eventually took these patients into the broader virtual world to practice social interactions (such as initiating conversations with strangers at a virtual bar or delivering a presentation in a virtual conference room).

Another study of three hundred thousand people in the online game *Pardus* found strong similarities between people's social interactions online and their interactions in the real world. Despite the fact that, in principle, people could interact in novel ways in the game, they chose not to.[18] The social networks in *Pardus* hewed to the line on several standard properties. The average number of friends in the game was 9.8; this number was higher than the number of real-life friendships because in this game, connections were more like acquaintances. The transitivity was 0.25, meaning that there was a 25 percent chance that a person's friends were also friends with one another, the same percentage found in real life. Even the phenomenon of "the enemy of my friend is my enemy" was observed. We will return to these properties of social networks in chapter 8, but for now, it is worth noting that despite the freedom of the online world, players generally re-create fundamental social behaviors.

### *Constructed Groups*

What can we learn from unintentional, intentional, and artificial communities? All of the cases shed light on the fundamental aspects of the social suite, expressing the traits that are required for functional societies. Yet each of the various examples has its limitations. For instance, the Antarctic scientists are healthy and arrive unharmed at their outpost by intention, but the shipwreck survivors do not. The isolation of both the scientists and wrecked sailors is pronounced, and social interactions can unfold with little perturbation from the outside world. But that is not the case with the urban communes or with communes like Brook Farm; they were far less isolated and were quite near the "real world." Unlike the communes of the nineteenth and twentieth centuries, which were generally intended to be permanent, groups of people in the Antarctic stations and in internet game worlds are time-limited. By contrast, unintentional groups like the Shackleton expedition and the shipwrecked groups functioned (to varying degrees) without any known end date. Finally, we can use modern scientific methods to study social interactions in Antarctica, the urban communes, and online experiments, but we cannot do so easily in some of the other situations we have examined.

To be clear, our experiments have limitations too. They can be unnatural or simplistic. But that is the whole point of experiments. Scientists study the pancreas of an inbred rat eating highly controlled food in a laboratory in order to glean insights about diabetes in humans. Experiments simplify reality, but they allow researchers to choose and manipulate variables of interest, focus on narrow and specific features of the natural world, and control the parameters of inquiry so that scientists can make robust inferences and demonstrate that one thing really is the cause of another.

We can learn a lot about the societies people make for themselves when left alone and about the functioning of social groups by studying all these examples. These studies complement one another and allow us to explore different social properties. But how might we unify and make sense of all these cases? Is there a general shape of all societies, something roughly analogous to, say, the way that all triangles resemble one another? Is there a way to synthesize our observations of these social groups?

### *A World of All Possible Shells*

I am going to use the study of seashell shapes to tackle this topic. This approach will allow us to imagine the myriad forms human societies could *conceivably* take and help us understand why so few of these forms have actually arisen, why the social rules in all of them are so consistent, and why the social suite appears universal.

In 1966, paleontologist David Raup published a paper on an extremely obscure topic, the geometric analysis of shell coiling. It attempted to develop a general theory of seashell shape.[19] Though the subject matter is arcane and specialized, the challenges it addressed apply to many scientific questions, including ours.

Raup explored an interrelated set of problems. First, he studied the diversity of shells found in the natural world. Second, he investigated whether there was a general way to formulate and describe the shape of shells so that all theoretically possible shapes could be summarized mathematically, even ones that did not occur in nature. Finally, he wondered what might account for the fact that only a small minority of these possible shapes had ever come into existence. Were there certain constraints that limited the possible range of seashell shapes?

Raup's work was part of a broader effort that he characterized as theoretical morphology (*morphology* is the study of the form and structure of living things).[20] One might think that there are countless possible shell shapes, but it turns out that only certain kinds of seashells have ever arisen—which raises the question of whether there are restrictions on the seashell shapes one can observe. We can mathematically array the range of both possible and actual seashell shapes, and that spectrum of shapes is known as a *morphospace.*

The mathematical modeling of the shape of shells is actually much older than Raup's work, stretching back at least as far as 1838 to the analyses of a British clergyman. Raup built on this tradition, and, by defining just three parameters, he was able to characterize the general extent of a "world of all possible shells." These three parameters were size, coiling, and elongation. *Size* refers to the rate of increase in the diameter of the shell's cross section (the cavity the animal occupies) as one traverses the interior of the shell (as it twists).

Imagine entering a tunnel at its mouth. If the tunnel is shaped like a cylinder, there will be no change in size as you move through it; if the tunnel is shaped like a cone, it will narrow as you proceed through. *Coiling* refers to the rate at which the shell coil spins away from its axis. Think of a tightly wound versus a loose roll of stamps. *Elongation* refers to the rate at which the coil itself moves (translates) along, or up, its axis—in other words, whether it is a tall or a squat shell. Imagine a compact Slinky toy on a table versus a stretched-out one.[21] Small changes in these parameters could convert a scallop to a nautilus, unifying all shells in a single equation.[22]

The next crucial point is that these three parameters (size, coiling, and elongation) define a three-dimensional space—the morphospace—into which all possible shells can be placed (as shown in figure 4.3, taken from Raup's paper). In explaining the notion of a morphospace, evolutionary biologist Richard Dawkins used the metaphor of a museum of all possible animals in which cabinets displaying forms of creatures are arrayed.[23] Every animal is placed next to those animals it most resembles. Galleries extend in every direction—left and right, front and back, up and down—and every direction captures a dimension along which animals might vary slightly. For example, as you walk down an aisle of herbivores, the neck length of the animals might slowly increase until their necks are as long as a giraffe's. As you walk across an aisle, the neck length might stay the same, but some other feature, like coat pattern, might vary. As you move up and down the levels, the size of the animals might vary. To be clear, the morphospace need not be limited to three dimensions.

Defining the morphospace leads to a second, and more profound, finding: Raup's mapping revealed how few of the possible shells have ever come into being. We can see that the portion of the cube occupied by shells that have existed in the real world is very small relative to the volume describing all possible shells. These unoccupied regions are deserts, the empty quarters of biology, devoid of

**Figure 4.3: Raup's Shell-Shape Morphospace**

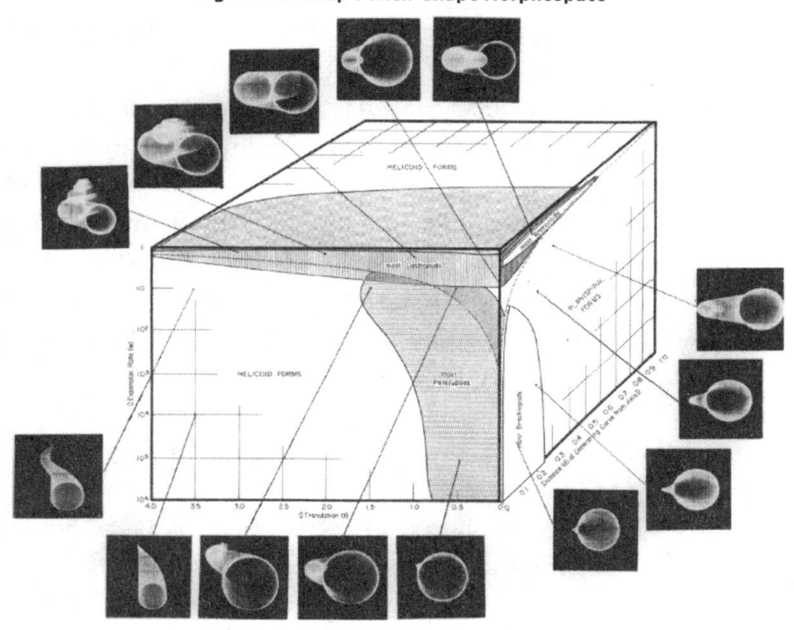


The three axes of the cube define three parameters that might vary among shells, giving rise to the morphospace of conceivable shell types. *Size* is the rate at which the shell's cross section (the cavity that the animal occupies) expands as one moves through the shell. *Coiling* refers to the rate at which the shell coil itself spins away from its axis. And *elongation* refers to the rate at which the coil itself moves (translates) along, or up, its axis — like a tall or squat shell. Example shells are shown in black. Only part of the morphospace (highlighted in gray) is occupied by any shells at all. The remainder of the cube has never had an example of a shell.

life. Determining the reason that only a small, well-defined part of the cube is occupied—for shelled and all sorts of other animals—has become one of the major challenges of evolutionary biology.

The same puzzle presents itself in other areas. Scientists have tried to use what is known about skeletons in the animal world to create a matrix of all possible skeletons.[24] Physicist Stephen Wolfram generated a "world of all possible leaves."[25] Others have considered ways of quantifying the world of all possible flowers, showing that perhaps only a third of imaginable flower types have ever evolved and that the remaining two-thirds are either structurally impossible or manifestly maladaptive.[26]

To understand this idea, consider an hourglass (as shown in figure 4.4). Imagine that a single parameter defines how pinched the middle is. If the area between the two parts of the hourglass (the top and bottom) is only slightly pinched, sand flows down, and the hourglass serves its purpose of telling time. The extent of pinching would modify the speed at which the hourglass empties. But if we pinch the glass so that there is no opening between the two bulbs, the sand cannot flow. Such an hourglass could exist, but it would be useless and so *would not* be built. Finally, if we pinch even more, we will sever the connection between the two bulbs. Such an hourglass would be unsound; there is nothing to hold the top bulb up, so this form is structurally impossible. Such an hourglass *could not* be built. Too much pinching would result in a form that cannot exist: it's just not an hourglass anymore.

Back to seashells: Why was so much of Raup's morphospace empty? There are at least three different explanations. First, some

**Figure 4.4: Real, Useless, and Impossible Hourglasses**

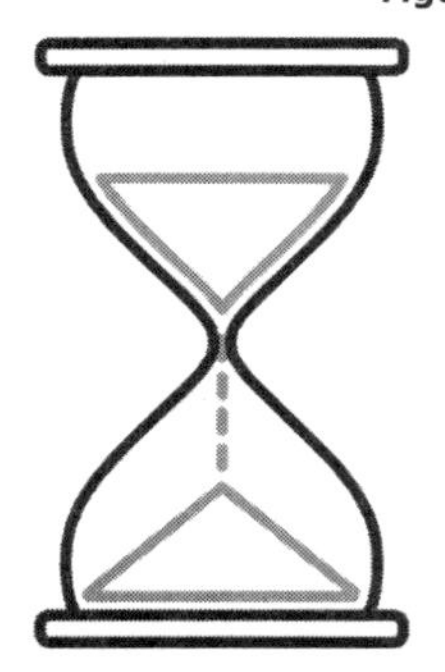

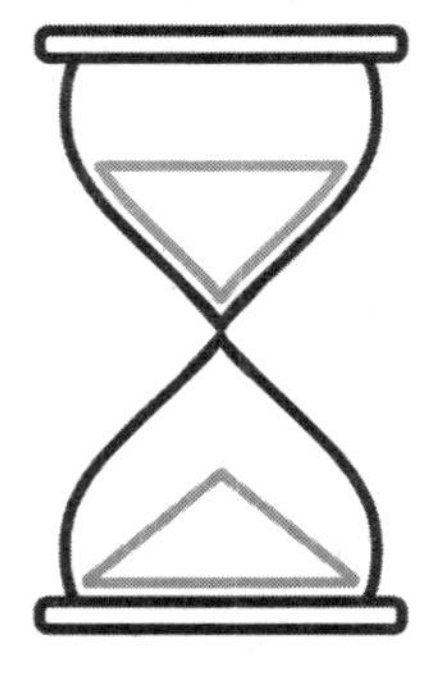

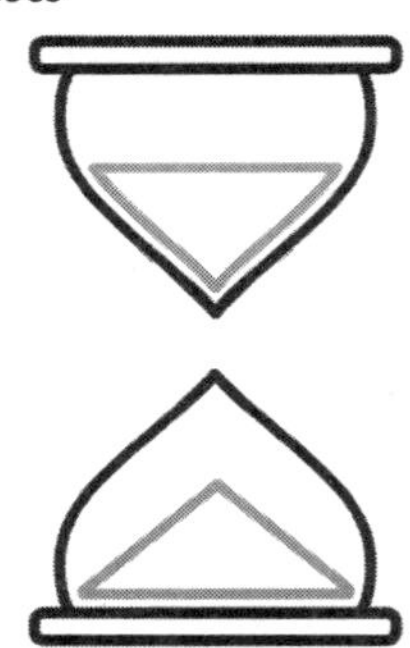

In these three possible hourglasses, when the pinching of the neck of the hourglass gets too tight, as defined by a parameter describing the diameter of the neck, the hourglass no longer functions (in the middle image). Eventually, when the pinching gets even more extreme, the hourglass assumes an impossible shape (the far right image).

forms of shells might have failed to evolve because there were no feasible mutations that permitted them to come into existence; that is, they were genetically impossible (or very unlikely). To return to our hourglass metaphor, it's as if there were no clear glass available, only opaque glass, which would defeat the whole point of an hourglass (to tell time). Second, some shell forms might not have evolved because they were not adapted to any environment that has ever been present on the planet. That is, they were too weak to be viable, let alone to endure. This is similar to the hourglass with a totally pinched center—it would be useless as a timekeeper, so there would be no reason to make one. Third, some shell forms might not have evolved because they violated the laws of physics and chemistry, which would be comparable to the two globes of the hourglass that are completely separated and thus cannot remain upright, let alone pass sand between them.

This last explanation means that physical constraints may have prevented the emergence of certain shells. For instance, a bivalve shell must have a particular shape and satisfy particular mathematical properties (having "nonoverlapping whorls," for one) if it is to have a functional hinge, and so shells not satisfying this constraint simply cannot exist.[27] Consider an analogy to a world of all possible snowflakes. Snowflakes are physical objects, not biological ones, but, for the sake of argument, let's set that aside. Suppose we imagine that there is variation in how many sides a snowflake can have—three, four, five, six, seven, and so on. Our mathematical model could specify this. Yet fundamental physical constraints having to do with the properties of water molecules dictate that snowflakes can have *only* six sides. So the failure to find snowflakes with four or five sides in other regions of the morphospace does not mean that those types of snowflakes do not work and so did not arise; rather, it means that they are simply not physically possible.[28]

The first two explanations for why so much of the morphospace is empty neatly demarcate two streams of biological thought about natural selection and variation. The first explanation was that certain forms of shells do not exist because there has never been enough underlying genetic variation to allow them to come into being (known as the genetic-availability argument).[29] This is the opaque hourglass idea. The second explanation is that there was no need for shells to explore these regions of the morphospace—that is, there was no pressure by the environment favoring shells of those types (this is the selection or adaptationist argument). These types of shells would have been, given all possible environments, just unsuccessful shells. This is the tightly pinched hourglass idea.

Let's apply this concept beyond seashells and see just how puny the inhabited morphospace for animals really is. Consider that no animal has ever evolved a means of locomotion involving wheels or a means of flight involving heating air in some sort of balloon.[30] These ideas strike us as ridiculous. But are alternative body plans—animals with wheels for feet or birds with pouches of air instead of wings—impossible? Given the enormous diversity of phenomena that natural selection can produce (including bizarre constructions like biological electric batteries, acid ejectors, light-bending lenses, air bladders for flotation underwater, and animals the size of enormous buildings), one must wonder whether the lack of wheels reflects some sort of intrinsic limit. Or perhaps the second explanation holds—maybe no environment has ever conferred an advantage on wheeled locomotion over other available alternatives. Navigating a natural, irregular terrain where there are no roads is much easier to do with feet than with wheels, in part because feet are less prone to slipping and offer greater ease of navigating obstacles. This is the adaptationist argument, that there has never been any functional utility to wheels. Indeed, there are still many highland villages in Greece and Turkey where it's easier to get around on donkeys than on wheeled carts, and the U.S. military is developing walking robots for this same reason.[31]

The genetic-availability argument is supported by a minority of biologists who, as Richard Dawkins puts it, "feel that large portions of the museum are forever barred to natural selection; that natural selection might batter eagerly on the doors of a particular corridor but never be admitted, because the necessary mutations simply cannot arise."[32] In contrast, the latter argument, the aforementioned adaptationist view, holds that organisms take certain forms because only certain forms are physically possible and, more important, *adaptively valuable.* The majority of biologists hold this view. In our case, only certain types of social organization—in humans and other mammals—might help us cope with the physical, biological, and social environments. Only certain types of social organization make sense, and that is the social suite.

### *A World of All Possible Societies*

The idea of a morphospace and the possible reasons that only part of it is occupied are very helpful in making sense of our observations regarding the reproducible nature of key aspects of social order. Imagine that—just as Raup did with his shells—we could arrange the social systems that various animal species might construct on a large, multidimensional grid that described all the social arrangements that could theoretically exist—a world of all possible societies. Or imagine that we did the same thing with the range of human societies. We might then unify and examine the societies of various species, or (more narrowly) the societies within our own species—such as the shipwrecks, communes, scientific settlements, online experiments, and many other communities, such as colonial outposts, monasteries, prisons, boarding schools, nuclear submarines, trapped miners, space-habitation experiments, and so on. We could include modern forager societies that resembled the kind of societies our species formed prior to the agricultural revolution ten thousand years ago too. Such an exercise would shed light on just how similar human social organization really is across all these situations.

To do this, we would have to define the key axes, just like Raup's three parameters. One important axis might be the hypothetical size of the society, perhaps defined as the size at which people actually know the others in the group well, even if they are not close friends; this could range from, say, zero (meaning that no one knows anyone in our imagined society) to two thousand (each person knows two thousand other people intimately). In reality, most people have about four or five close social contacts and know roughly one hundred and fifty people well—*well* being defined as familiar enough that they can pick up a conversation where they left off after an absence. This latter number is known as Dunbar's number.[33] Another axis we might focus on is the cooperativeness of the society or some measure of its proclivity for intragroup violence, perhaps quantified as the chance that two people would cooperate with each other when playing a public-goods game (using a percentage ranging from 0 to 100, 100 percent being the most cooperative). In real human societies, the chance is typically about 65 percent, meaning that roughly two-thirds of people are inclined to cooperate with a stranger when it comes to sharing a possible reward. But the extent of cooperative behaviors can vary somewhat across societies (as we will see in chapter 9).[34] A third axis might be related to the structure of the social ties—for example, the number of connections people have or the likelihood that their friends are themselves friends with one another (this is known as transitivity in the network, and it ranges from 0 percent to 100 percent). An alternative parameter for the third axis could be a measure of hierarchy or equality in the distribution of some key resource. Once we chose and defined our various axes, we could put all our examples—and, indeed, all known societies—into such a grid with three (or more) dimensions. And we would then see that, fundamentally, very few social arrangements have ever appeared in real-world societies.

In 1999, anthropologist Lee Cronk conceived of what he called *ethnographic hyperspace,* which encompassed all possible combinations of the many variables that he felt described human societies—a version of our social morphospace, as it turns out. He included not just attributes that might be core features of any society, such as the social suite, but also many attributes of a more cultural nature, such as types of body adornment and varieties of tools. He used the famous Human Relations Area Files, which is a comprehensive documentation of features of over eight hundred cultures, coded with uniform descriptors regarding dozens of social, political, economic, religious, reproductive, and other practices of these societies. Based on his calculations, he estimated that there would be an unimaginably high number of such possible combinations (perhaps $1.2 \times 10^{53}$), only a tiny fraction of which had ever been observed by anthropologists.[35]

Anthropologists Jack Sawyer and Robert A. Levine performed a similar analysis even earlier, in 1966, the year after Raup considered his own problem with shells. They concluded that a diverse sample of five hundred human cultures could be summarized with nine variables, among them sociopolitical stratification (ranging from rigidly caste-bound to totally egalitarian), subsistence practices (agricultural, pastoral, hunter-gatherer), the permissibility of polygamy, and so on. They, too, concluded that, although there were over one hundred million possible combinations of these variables, in their estimation, only a very small fraction of those had been explored by human beings over the course of history.[36]

As we saw in chapter 1, one cultural anthropologist, Donald Brown, attempted to describe a hypothetical tribe that he called the Universal People whose members had features shared by all human societies, a kind of fundamental social order similar to our idea of a pre-wired society based on the social suite. This society, he argued, would have gossip, music, taboos, belief in magic, rites of passage, attention to male aggressive impulses, fancy speeches for special occasions, as well as other features.[37] Human cultures, he concluded, were more similar than different. Indeed, such an exercise is required because, without an objective metric for evaluation, we might delude ourselves into believing that our differences are greater than they actually are. If we don't define the morphospace of human cultures and specify the most relevant axes, how can we know how different or similar human groups are? But when we do array our societies in a morphospace, we see how narrow is the range of ways of being social. This reinforces our similarity. And it brings our common humanity into relief.

### *Imaginary Societies*

Science fiction authors have done a better job than anthropologists of imagining wholly different—one is tempted to say "alien"—social arrangements, exploring forms that can be imagined but are not actually seen. These imagined social systems can occupy regions of the morphospace either near or far from those occupied by real societies. And societies too far outside the normal range may seem as implausible and strange as cubic seashells or wheeled animals.[38]

Science fiction writers often deliberately place groups of humans under extreme conditions and then envision how they would respond. What happens to a human society that has been confined to a space vessel for hundreds of generations? How might social organization change if women could reproduce without needing men? What would a world without love or friendship look like? What happens if social inequality or hierarchy become hereditary or very extreme?

But even in the farthest reaches of human imagination, alternative social orders are surprisingly pedestrian. There are the standard dystopian tropes, such as state control of reproduction, suppression of thought or emotion, or ant-like societies with rigid castes. Indeed, when science fiction authors attempt to describe a truly unrealistic and dystopian society, they often choose metaphors from social insects, making people more like ants, wasps, or termites. This is, of course, a radical departure for humans—if not a nightmare.

In contrast, the utopian versions imagine people living unfettered in just, safe, and healthy worlds. These stories often include one element that is missing in the real world, however, and that authors seem to crave for their utopias: a special kind of rapport among individuals.[39] Fictional utopian societies are generally characterized by the social suite, but a kind of empathy or trust is added—sometimes illustrated by an extreme vision of a telepathic connection among people.

Hence, in a validation of the *Anna Karenina* principle (that all happy families are the same but every unhappy family is unhappy in its own way), the utopian worlds are surprisingly similar, but every dystopian society seems uniquely dystopian. This is in keeping with the classic understanding of entropy—there are more ways for something to be broken than for it to work; more disordered states of nature than ordered ones; more dysfunctional forms of social organization than functional ones. Still, the science fiction universe remains surprisingly recognizable.

*The Time Machine*, the novella by H. G. Wells published in 1895, is a cornerstone of science fiction and popular culture (in fact, it introduced the term *time machine*).[40] This foundational work explores not only time travel but also novel social arrangements. In the story, despite a deceptive tranquility, one group of people so dominates another that it uses them as a source of food. The idea of such a shockingly stratified society is also mined in Aldous Huxley's dystopian novel *Brave New World*, published in 1932; George Orwell's classic and equally dystopian *1984*, published in 1949; and Robert Heinlein's *Orphans of the Sky*, which first appeared in 1941 and was published in book form in 1964.[41]

Charlotte Perkins Gilman's 1915 story *Herland* envisions a society without men.[42] It's not so much her feminist premise that is important; it's that Gilman constructs a society with the complete absence of conflict and competition and with extremely high levels of cooperation and social equality, thus manipulating the cooperation and hierarchy elements of the social suite. The inhabitants of Herland "had had no wars. They had had no kings, and no priests, and no aristocracies. They were sisters, and as they grew, they grew together—not by competition, but by united action."[43] But once again, the utopia created by Gilman is similar to real life in other respects. For instance, the people in her story differentiate into individuals with unique characteristics and identities.[44]

The theme of individual identity is also explored in *The Giver*, the 1993 novel by Lois Lowry well known to many American middle-schoolers and their parents. The story follows the life of an eleven-year-old named Jonas in a futuristic society called the Community (evoking the communal societies we saw in chapter 3).[45] The Community is characterized, above all, by "sameness," a kind of abrogation of individuality, achieved in part by the elimination of individual memory. Society has been carefully engineered to advance sameness. Everyone is genetically engineered to look the same and is taught to act the same. The individual identity element of the social suite is taken to its (negative) extreme. Family units are strictly regulated so that they support the mechanical needs of life but do not foster emotional closeness. Social relationships exist, but they are bland, since all uncomfortable memories have been eradicated.

At the other extreme is Rudyard Kipling's classic 1912 short story "As Easy as A.B.C.," which is set in 2065 in a world dominated by the Ariel Board of Control (A.B.C.).[46] In his imagining of an alternative type of society, Kipling chose to reject the very basic structure upon which human societies are founded. In his story, forming groups becomes the crime of "crowd-making," and individuals value independence and privacy above all else. Allusions to earlier ages—to the times of crowds and democratic rule ("crowd rule")—are dark and contain reminders of past plagues facilitated by association. Because their material needs are met by the A.B.C., citizens can conduct their lives without ever needing to congregate.

This is a very small selection of famous science fiction stories taken from a vast and entertaining literature, but it illustrates several points. First, violations of the social suite—moving to either extreme of any of the elements of the suite—are mostly seen as dystopian, or at least disturbing. I'm not sure I would like to live in the hyper-cooperative Herland, and it's not just because I am a man; I'd rather not subjugate my own desires to those of a group quite that much. Second, even in these fictional accounts, the societies still have recognizable features (since we, the readers, must be able to relate to some part of what we are supposed to enjoy). Third, these examples—especially those far outside the region of the morphospace human societies actually occupy—can highlight how similar our societies are and how small a part of the range our species actually occupies. With ant colonies as an alternative, all human societies look extremely familiar.

### *Pre-wired for Society*

The cultural universals that concern us here—the social suite—are focused on traits related to social organization that are shaped by natural selection and partially encoded in our genes. We are not exploring traits that, whether universal or not, are strictly expressions of culture and not encoded genetically (there are no genes for specific types of personal adornment or particular gods to worship, for instance). But even given this focus, the relatively circumscribed extent to which each feature of the social suite is manifested across diverse circumstances and societies suggests that societies that fall outside those ranges may not be realistically feasible. If we were to make an eight-dimensional morphospace based on the eight components of the social suite, all the societies on the planet today would occupy a small part of it. We do not, for example, find a functional society without love, friendship, cooperation, or personal identity.

Why is this the case? Like shells, might humans be constrained to explore only a part of this theoretical space? What kind of society are we naturally constrained—or, more optimistically, permitted—

to make? What is the blueprint, the evolutionary baseline, for a human society? In what way is it natural for humans to transcend their individual selves?

In answering this set of related questions, we want to be careful—not everything that is universal about human affairs has to do with genetics, and not all diversity originates in culture. That is too simplistic. Consider the worldwide popularity of Coca-Cola. While this depends in part on our species' evolutionarily calibrated sweet tooth, the soft drink's universality is not a product of genes or biology but rather relates to a specific set of events rooted in the tastes of American soldiers during the Second World War and the workings of modern capitalism, globalization, and branding.[47]

Conversely, cultural and behavioral diversity can also result from humans' innate ability to flexibly respond to their environments, to engage in social learning, and to make culture (an ability which is itself a part of the social suite). The diversity might conceal an underlying universality that, paradoxically, might relate more to our genes than to cultural exigencies. Evolutionary psychologists John Tooby and Leda Cosmides provide a fanciful illustration of this idea. They suggest a thought experiment in which aliens replace humans with jukeboxes, each of which has a repertoire of thousands of songs and the ability to play a particular song according to where and when it is. We would then observe that jukeboxes in different parts of the world played different songs at different times, songs that were similar to those on the jukeboxes near them. But none of this intergroup variation and intragroup commonality would have anything to do with the workings of culture.[48] This is a way of illustrating that humans might have an inborn ability to respond flexibly—but also predictably—to their environment.

In fact, there may even be *less* diversity among humans living in different places than we would expect because, in one very important respect, all humans experience the same "environment." It's as if the aliens had somehow placed all the jukeboxes near one another in similar surroundings, and all the jukeboxes are playing very similar, if not identical, songs. Animal species may face a wide variety of threats, but the biggest threat to humans, more than predators or any environmental exigencies, is *other humans.* If the most important feature of the human environment is other humans, then that feature is a constant across all physical and biological environments, from the poles to the equator. And our species has adapted to *that.* Social organization may thus converge on remarkably similar social structures if they are adaptive in this way.[49] This is the adaptationist explanation for the social suite. The reason for our common humanity is that we have always lived among members of our own species and have evolved to cope with precisely this exigency.

From this perspective, we see that genes may have come to work outside our bodies, having their impact at some distance from their source—like fireworks exploding far from their origin—helping to shape the societies far above the genes themselves. They may do this by affecting the human tendency to cooperate with and befriend others, to care for others' children, to value other people's individuality, and to love one's partners. Because of this, in all the seemingly strikingly different human cultures around the world, in all the repeated opportunities to make new societies, we see the same core patterns again and again. Even the social organization and function of political units, like tribal chiefdoms and modern nation-states, are grafted onto this ancient heritage, and they must respect the principles guiding the organization of smaller groups. Rapidly invented, deliberately designed, or wholly novel social systems that seek to abrogate the social suite cannot be as functional as organically evolved ones.

## CHAPTER 5

# First Comes Love

Until a few years ago, I had always assumed—naively, as it turns out—that voluntary kissing in a romantic or sexual way was something all humans naturally enjoyed. But apparently not the Tsonga people of southern Africa, according to ethnographer Henri Junod:

> When they saw the custom adopted by the Europeans, they said laughingly: "Look at these people! They suck each other! They eat each other's saliva and dirt!" Even a husband never kissed his wife.[1]

I should have known better. I had fallen prey to the classic trap known to anthropologists as *ethnocentrism,* wrongly presuming that what makes sense in one's own culture must also make sense in others. Kissing is so normative in my culture that it never occurred to me that it could be otherwise in anyone else's. In fact, I do not think I have ever talked to anyone who has found kissing unusual, so I was surprised to learn that romantic-sexual kissing is not universal at all.[2]

The cultures I am most familiar with—in Europe, the Middle East, and India—happen to enjoy kissing. Some groups even see it as mystical. A Hindu text written more than three thousand years ago described kissing as "inhaling" a person's soul.[3] But a global, cross-cultural survey found that kissing was present in only 46 percent of one hundred and sixty-eight cultures studied.[4] Kissing is most common in the Middle East and Asia and least common in Africa and South and Central America. No ethnographers familiar with forager or horticultural groups in sub-Saharan Africa, Amazonia, or New Guinea have ever reported witnessing romantic-sexual kissing.[5]

This regional and cross-cultural variation is hard to understand because something that resembles kissing has been observed in chimpanzees and bonobos, and there is also evidence that kissing can communicate valuable biological information about a partner's health, genetic compatibility, and state of arousal, all of which would suggest an evolutionary origin and function to the practice.[6] So we need to consider the possibility that kissing is innate but that, in some cases, it has been culturally inhibited. But it also strains credulity to think that kissing has been repeatedly, widely, and independently suppressed. Perhaps, after all, kissing is just a cultural practice like any other, a custom found in some places but not in others?

Disbelief, amusement, and disgust toward romantic-sexual kissing are surprisingly common among members of non-kissing societies.[7] Anthropologist Charles Wagley spent time among the Tapirapé people of central Brazil in the 1940s, and he noted this:

> Though Tapirapé couples showed affection, kissing seems to have been unknown. When I described it to them, it struck them as a strange form of showing physical attraction... and, in a way, disgusting.... A couple might stand close to each other during a conversation with the man's arms over his wife's shoulders and she holding him around the hips.[8]

Still, it is important not to assume that the absence of evidence is evidence of absence. Wagley also failed to uncover any evidence for female orgasm; his (male) informants said there was not even a word for it in their language.[9] No doubt the gender of the anthropologist and his informants played a role in the outlines of the sexual behaviors described. It seems exceedingly unlikely that female orgasm did not exist in this society (and there is good reason to suppose that subjects would not want to discuss all intimate aspects of their lives with even the best anthropologist). And if that was overlooked, was kissing secretly happening too? Maybe.

Clearly, some aspects of romantic interactions can be inscrutable to those who are removed from them, and there is tremendous variation among individuals and among groups. But despite this variation, might there nevertheless be something universal about romantic-sexual love?

What is the basis for one of the great mysteries of the human condition, the drive to form a *loving relationship*, not just a *sexual connection*, with another person? From the perspective of evolution, it is easy to explain why humans lust for partners. But why do they feel a special attachment to them? Why do they feel love for them? To understand the competing human urges to love, to possess, and to copulate, we need to consider both the diversity of human romantic-sexual interactions and what, if anything, lies at their consistent core.

Many norms and practices regarding sex and marriage, well beyond kissing, vary around the world. But other features do not. The invariant features, such as the physiology of orgasm, which should be universal, arise from our species' evolved biology and psychology. A key such universal feature is the tendency to engage in *pair-bonding*. This is the biologically guided impulse to form an intense social attachment to a partner, an impulse driven by molecular and neural mechanisms that are increasingly well understood. Evolution provides culture with the raw material with which to work, and mating systems are built on this foundation. And as we will explore in chapter 11, mating systems can, in turn, shape evolution (for instance, some cultural rules prohibiting cousin marriage affect the survival of offspring).

Marriage norms build on evolved biology in another way, beyond the predilection to form pair-bonds. Humans have evolved to obey group norms, and so a different aspect of our evolution, one connected to social learning and cooperation, also plays a role in supporting marriage. It generally makes humans feel good to comply with the cultural rules of their in-group, no matter what those rules are.

Let's explore both the variation and the universality seen among humans when it comes to romantic-sexual love—or, as it is known more clinically, *mating behavior*, a locution we have to tolerate given my intention to explore commonalities between humans and other animal species.

### *Variation in Marital Unions*

The forms of marital unions vary across societies and include not only the type seen most commonly today, heterosexual monogamy, but also polygyny and polyandry. Many societies also have homosexual marriage, of course, and other forms of bonding. The famous anthropologist E. E. Evans-Pritchard described an arrangement among the Nuer of Sudan in which a woman who was unable to bear children was allowed to marry and have children with another woman (via a male who would inseminate the second woman) and assume further male roles, including being called "Father" by her children and being allowed to inherit and transmit property.[10]

Yet beneath all this variation, there is something truly universal: a special sense of connection between human beings who have a sexual relationship. Rooted in our species' evolutionary past, the human capacity for romantic love arose from an ancient tendency, shared with other animals, to form pair-bonds. To be clear, what I am primarily focused on here is not merely the sexual practices or even the marriage systems that humans have manifested over millennia, but rather the overall mating strategy our species has adopted. This strategy is the base upon which so many cultural practices and individual idiosyncrasies are built.

Most people take the emotional attachment they feel for their partners for granted. It's also tempting to focus on the many violations of monogamy in humans—the fact that people have several partners in sequence or are unfaithful to their supposedly exclusive partners or that some societies practice polygamy. But to focus on this variation is like standing on our 10,000-foot plateau and focusing on the differences between two mountains of 10,900 and 10,300 feet while neglecting their commonalities. The reality is that humans have the distinctive capacity to love and to keep their partners.

Understanding marriage practices is complicated because we have to distinguish between cultural norms (do cultures see kissing as absurd?) and evolutionary underpinnings (is there a biology to kissing?)—a distinction that is sometimes hard, if not impossible, to make. Another difficulty is the tremendous variation in marital practices across geography and time, measured on several scales. For example, we see a progressive ebb and flow of polygyny across our prehistory and history. Our hominid ancestors, such as *Australopithecus,* who lived four million years ago, were likely polygynous (one way we know this is that the males were bigger than the females, which suggests that the males competed with one another to secure multiple partners or defend them, or that females came to prefer bigger partners). However, over eons, our species evolved into *Homo sapiens* (perhaps three hundred thousand years ago) and adopted a hunter-gatherer lifestyle with mobile foraging units that were relatively egalitarian and free from stark status differentials, and these early humans generally practiced monogamy. The reasons for this shift are complex but are believed to stem from adaptations to different kinds of environmental pressures, such as changes in food resources, which we will discuss in more detail below. And these shifts in behavior were paralleled by physiological changes, including the lack of outward signs of ovulation, the further extension of childhood, and menopause.

Then another shift happened, beginning most likely with the

agricultural revolution roughly ten thousand years ago, and continuing through the rise of nation-states about five thousand years ago (along with the large-scale socioeconomic inequality that emerged with such states), such that polygyny once again became more common.[11] This shift was due to the pressure of historical and cultural forces, not evolutionary ones. And then, finally, in the more recent past, monogamy returned as a norm, again for cultural reasons, first in the West (beginning two thousand years ago), and then, in the past few hundred years, spreading around the world (although pockets of monogamy had previously existed). Each of these steps has shaped our contemporary biology and cultural practices, and disentangling them is not easy.

The degree of controversy and confusion among academic writers about human mating practices is reflected in the titles of their papers—for example, "The Mystery of Monogamy" and "The Puzzle of Monogamous Marriage."[12] From many biological perspectives, the mating strategy our species has adopted is hard to explain, and humans even appear to have more in common with birds, which are mostly monogamous, than with other primates (as we will see in chapter 6).

Over the course of evolution in many species, parents initially come to feel a special bond for their offspring. In our species, that included a feeling of love. This feeling toward offspring may have then been hijacked and repurposed for pair-bonding, which is the special feeling, beyond just lust, humans feel for their mates. This process of repurposing is sometimes described as *exaptation*. An exaptation is an evolutionary preadaptation in which a trait evolves initially for one purpose but then comes to serve another. A classic example is bird feathers, which probably evolved initially as a kind of thermal insulation and later were used to support flight.[13]

The ultimate and proximate origins of pair-bonding behavior in humans and other hominids are the focus of a vast research effort. Scientists' understanding of these matters is still evolving. Mating behaviors, including kissing, may be driven by a host of immediate biological forces, such as the desire to detect the odor of a prospective mate. But the ultimate causes of these phenomena have to do with why, evolutionarily speaking, humans prefer some mates over others and why people have evolved the capacity to discriminate among them or feel attached to one in particular.

It seems that, across evolutionary time, humans evolved to love their offspring first, then their mates, and then to feel affection for their biological kin, then their affinal kin (their in-laws), and then their friends and groups. I wonder sometimes if we are in the middle of a long-timescale transition to becoming a species that feels attachment to ever larger numbers of people. But to understand human relationships outside of sexual ones, we must begin with the sexual and romantic connections, which preceded other sorts of ties over the course of evolution. Love of our mates is a key element of the blueprint.

To summarize and give a very general time line, our ancestors were polygynous until about three hundred thousand years ago, primarily monogamous from then until about ten thousand years ago, primarily polygynous again until about two thousand years ago, and primarily monogamous since then. There have been many exceptions and these dates are necessarily rough, but this is the picture in broad terms. Let's work our way backward in time with respect to human mating behavior.

### *A Brief History of "Traditional" Western Marriage*

Marriage systems are the social norms, beliefs, and institutions that regulate partner choice, reproductive behavior, marital obligations, and spousal affection. These cultural practices can specify the permissible number and type of partners, how a new household is established, the expectations each spouse can have of the other, the division of property following death or separation, and even who is supposed to pay for the wedding celebration. Marriage entails social, economic, sexual, and normative expectations in all societies and some degree of patriarchy in most of them.[14]

From a global, cross-cultural, and historical (or prehistorical) perspective, there is nothing essential about monogamy. The currently ascendant European-style practice of monogamy is just one marriage system. Things like this are sometimes not easy to see. Many biases have historically underpinned the social sciences, biases that have resulted in the neglect of key features of family life and that have assumed, for instance, that the psychology of American college students (the classic experimental subjects) is applicable everywhere. To acknowledge this, contemporary social scientists have adopted the acronym WEIRD (which stands for "Western, educated, industrialized, rich, and democratic") to describe societies that, in fact, represent only a minority of human cultures and are composed of people quite distinct from a hypothetical "average" human.[15] Indeed, roughly 85 percent of human societies have permitted polygyny at some point, and polygyny remains legal or generally accepted for at least part of the population in forty-one countries worldwide, primarily in Africa and Asia.[16] And according to a survey conducted from 2000 to 2010, in twenty-six out of the thirty-five countries with polygamy data available, between 10 percent and 53 percent of women aged fifteen to forty-nine were in polygamous relationships.[17]

The minority of societies in the anthropological and historical record that have practiced monogamy fall into two broad categories at opposite extremes. At one end are small-scale societies in ecologically demanding environments with few status differentials among the men, and at the other are some of the largest and most successful ancient societies, such as the Greeks and Romans. "Ecologically imposed" monogamy is adopted when the environment does not easily permit an alternative, analogous to a person who loses weight because there is no available food. "Culturally imposed" monogamy, as in the Greco-Roman case, is adopted as a norm; the analogy here is to a person who chooses to lose weight because thinness is favored for aesthetic or health reasons. Culturally imposed monogamy is the form that prevails today.

The ascendancy of worldwide monogamy reflects a swinging of the pendulum. Our Pleistocene ancestors had relatively egalitarian societies, in part due to their forager lifestyle involving relatively few possessions (if everyone in the group has no possessions, this does impose equality). But the rise of agriculture roughly ten thousand years ago and the subsequent rise of cities led to inequality and substantial status differences. Significant levels of inequality were reached fairly rapidly, within a few thousand years, with the establishment of kingdoms whose rulers could assemble enormous wealth and harems. Certain men acquired the wealth and status to have two or more wives each, while other men might have no partner at all.

The Old Testament frequently mentions polygyny, and King Solomon supposedly had seven hundred wives and three hundred concubines.[18] The Hindu Lord Krishna was said to have 16,108 wives.[19] Polygyny was practiced in virtually all parts of the world, including in the Americas prior to European contact. Until quite recently, polygyny was the norm in much of East Asia and had been for thousands of years.

As part of a broad set of innovations directed at establishing a more egalitarian and democratic society, roughly around 1000 to 600 BCE, Greek city-states introduced laws regarding monogamy, paradoxically returning humans to their forager roots.[20] Rome adopted and expanded this practice, instituting a variety of laws against polygamy.[21] Concerned about moral decay and a perceived sapping of imperial strength, Emperor Augustus implemented legal reforms between 18 BCE and 9 CE that encouraged men to marry. The reforms included restricting married men to having extramarital sex with registered prostitutes only; limiting the inheritance unmarried men could receive; legally formalizing the divorce process (to disfavor serial monogamy); and prohibiting concubinage for married men and making any offspring of such unions ineligible to inherit wealth. Roman emperors after Augustus reinforced these laws. In short, the Greek and Roman view of polygyny as barbaric progressively established monogamy as a legal norm. Of course, this standard was violated in all sorts of ways by both men and women.

With the expansion of the Roman Empire, monogamy spread over Europe, and this norm was assumed by the Christian church after the fall of the empire. Once again, I emphasize that legal norms and actual practices often differed, and the extent of sexual polygyny was constrained but not eliminated altogether. Much later, the Industrial Revolution in Europe prompted a further evolution, from legal monogamy to cultural monogamy, meaning that a man made an exclusive reproductive commitment to one woman, and vice versa, and this norm was internalized by most Europeans.[22] Still later, laws prohibiting polygyny were enacted in Japan (1880), China (1953), and India (1955).

Why did monogamy laws and norms become so widespread if they were such a poor fit with our species' lengthy polygamous history and prehistory? For starters, even quite modest levels of polygyny can deprive many men of partners, which men do not like. Consider a society of one hundred men and one hundred women with substantial differences in status, as shown in table 5.1.[23] Assume that sixty men with the highest status marry sixty of the women. Then assume that a subset of these men, say the top twenty-five, each marry a second wife, that the top ten men each marry a third wife, and that the top five men each marry a fourth wife. This degree of polygyny is not atypical in societies that practice it. As a result of these marriages, thirty-five out of sixty of the men (58 percent) are monogamous. Only men in the top 10 percent of status are married to more than two women, and the most wives that any man has is four. Yet it means that 40 percent of the men are unable to get partners. Conversely, 65 percent of the women have to share their husbands (and their overall households) with one or more other women, and many find this as unsatisfactory as the unmarried men find their predicament.

**TABLE 5.1: ILLUSTRATION OF EFFECTS OF POLYGYNY ON RATE OF UNPARTNERED MEN**

| NUMBER OF MEN | NUMBER OF WIVES | NUMBER OF WOMEN |
|---|---|---|
| 40 | 0 | 0 |
| 35 | 1 | 35 |
| 15 | 2 | 30 |
| 5 | 3 | 15 |
| 5 | 4 | 20 |
| Total: 100 men | | Total: 100 women |

Physical anthropologist Joseph Henrich and his colleagues argue that cultural monogamy spread in part because it offered an advantage in competition between groups.[24] Men lacking spouses resorted to violence within their own group or went on raids of other groups, provoking conflict.[25] Political entities, nations, and religions that adopted monogamy had a reduced rate of this sort of violence and could deploy their resources more productively, internally and externally. From this perspective, modern norms and institutions regarding monogamous marriage have been shaped by a kind of evolutionary process in response to the forces of intergroup competition and intragroup benefits.

With respect to the intragroup benefits, Henrich argues that "unmarried, low-status men will heavily discount the future and more readily engage in risky status-elevating and sex-seeking behaviors. This will result in higher rates of murder, theft, rape, social disruption, kidnapping (especially of females), sexual slavery, and prostitution."[26] In fact, something similar is observed due to the preferential abortion of female fetuses in countries like India and China today—it skews the adult sex ratio, resulting in greater numbers of males and, ultimately, unwed men, which has been associated with more violence and shorter lives (for the men).[27] We saw a small-scale example of this on Pitcairn Island. Providing an opportunity for every man to have a spouse meant that low-status males could become more risk-averse and future-oriented, less violent *within* their group, and more focused on providing for their children. High-status males, instead of seeking additional wives, could make long-term investments directed at acquiring wealth and caring for their offspring.

There is evidence that the spread of cultural monogamy, as a kind of sexual egalitarianism, may even have contributed to the emergence of democracy and political equality.[28] Historically in Europe, universal monogamous marriage preceded the appearance of democratic institutions, and it seems also to have set the stage for the rise of ideas about equality of the sexes.[29] Analysis of cross-national data from the modern world also reveals an association between the strength of cultural monogamy and the extent of democratic rights and civil liberties.

The legal prescriptions and cultural normativity of monogamous marriage are not in themselves inevitable. This institution—while arguably beneficial—has arisen in particular places and times. Researchers can quantify how tenuous these norms are by looking at a large sample of cultures. One comprehensive study of 1,231 societies (and a focused subset of 176) based on historical and anthropological records found that 84 percent of the societies practiced polygyny, 1 percent practiced polyandry, and 15 percent practiced monogamy.[30]

Superficially, the variety of human marriage systems would seem to challenge the claim that there are any universal features underlying human mating. Let's take a closer look at some very different marriage systems and attendant cultural practices, using four peoples on three different continents—the Hadza, Turkana, Tapirapé, and Na. What key underlying traits with respect to relationships with reproductive partners have humans evolved, and what role do they play in social order? When it comes to mating behavior, what is it that we all have in common?

### *Monogamy in Hadza Foragers*

"Sexy." "Hard worker." "Only wants you." "Understanding and gentle." "Doesn't use bad words." "Cares for kids."

That might sound like a profile on a dating site, but those words are the partner preferences identified by a group of East African foragers, the ancient Hadza people of Tanzania (a group my lab has studied in collaboration with anthropologists Coren Apicella and Frank Marlowe). The Hadza live in savanna woodlands, where they forage for their food, as all humans did up until approximately ten thousand years ago. There are only about a thousand Hadza left who live in the traditional way—though, even for those few, it's inaccurate to say that they are living *exactly* as they did twenty thousand years ago, since they have acquired manufactured clothes and iron arrowheads and have contact with modernized people. Still, anthropologists believe they are a viable proxy for how humans lived in the past in most ways. And the Hadza provide powerful insights into what marriage behavior might have been like in humankind's ancestral setting.

The Hadza live in small, fluid camps of roughly thirty people, and they move camp every six to eight weeks as they exhaust the food supply in the area. Partly because of this mobility, they have no fixed shelters and often sleep under the stars. When there is a dispute in a camp, people cope by splitting up and joining or forming other camps. The Hadza hunt and gather their food, have few material possessions, and speak one of the most distinctive and oldest languages on the planet.[31] They are extremely egalitarian, with little status difference among adults (including between men and women). Food that is brought into camp by someone is shared by all. This sharing partly reflects a potent cultural value, but it also results from the fact that, as the Hadza have no refrigeration or preservation techniques, anything not consumed will spoil and be wasted. Sharing also decreases the risk that an unlucky forager's family will go hungry, and so the practice serves as a kind of culturally defined insurance policy.

Marriage among the Hadza is a straightforward affair, and the contours of Hadza love and marriage are recognizable to modern American eyes. When it comes time for a young Hadza woman, typically at the age of seventeen or eighteen, to choose a partner, usually someone two to four years older, the couple will have a discreet, brief courtship, culminating in sex, and then they will begin sleeping together at the same hearth. Marriages are not arranged, and young lovers are free to choose their partners, but both the young people will typically seek the approval of their parents. Although there is no wedding ceremony, they are considered married—by themselves and others. As anthropologist Frank Marlowe noted with some understatement, "Female choice seems to be the main factor leading to marriage, because young single men appear willing to marry a wide range of women."[32] Still, there can be problems if more than one man is competing for the same woman. "When this happens, it can lead to violent, even fatal, conflict. Because of the danger, others sometimes get involved and ask the woman to choose one suitor rather than keep stringing them both along, but young women appear to want to shop before marrying."[33]

In other words, the Hadza, in most respects, seem to value the same things in their spouses that contemporary humans in WEIRD societies do. Marlowe painstakingly interviewed eighty-five Hadza adults and asked, "If you were looking for a husband [or wife], what kind of man [or woman] would you want? What is important to you?" He then grouped all their answers into seven categories: character, looks, foraging ability, fidelity, fertility, intelligence, and youth, as shown in table 5.2. Hadza men and women had extremely similar preferences. The most frequently noted desirable trait was character, which included not hitting your partner (roughly 58 percent of men and 53 percent of women cited this as important). In other detailed work, Apicella asked respondents to make trade-offs. She asked a hundred and twelve Hadza men and women whether they would prefer a partner who was attractive or one who was a good forager; only 6.3 percent preferred the attractive mate. Similarly, when she asked them to directly choose between physical attractiveness and good parenting, only 9.1 percent chose attractiveness.[34]

Nonetheless, physical appearance is important to Hadza women and men, and the Hadza, like people everywhere, do indeed discriminate among individuals on measures of attractiveness (such as symmetry, voice pitch, and waist-to-hip ratio). Attractiveness was cited by 42 percent of men and 41 percent of women as a desirable trait.[35] Moreover, Hadza women were much more interested in men's ability to forage and much more likely to value intelligence in a man than vice versa. Men were much more interested in a woman's fertility.

**TABLE 5.2: TRAITS MENTIONED BY HADZA AS IMPORTANT IN A POTENTIAL SPOUSE**

| CHARACTER | LOOKS | FORAGING ABILITY | FIDELITY | FERTILITY | INTELLIGENCE | YOUTH |
|---|---|---|---|---|---|---|
| Good character | Shorter | Good hunter | Doesn't want others | Can have kids | Intelligence | Young |
| Nice | Thin | Can get food | Stays home | Wants to have kids | Think | |
| Won't hit | Good body | Hard worker | Good reputation | Will have kids | Smart | |
| Compatible | Big | Fetch water | Likes you | Lots of kids | | |
| Good heart | Big breasts | Fetch wood | Cares about home | | | |
| Understanding | Good looks | Will feed | Only wants you | | | |
| Gentle | Good teeth | Can walkabout | | | | |
| Goes slowly | Good genitals | Can help work | | | | |
| Share words | Good appearance | Cook | | | | |
| Won't fight | Sexy | | | | | |
| Good person | Good face | | | | | |
| Good soul | | | | | | |
| Cares for kids | | | | | | |
| Not bad words | | | | | | |
| Can live together | | | | | | |
| Won't sin | | | | | | |
| One's heart wants | | | | | | |

Among the Hadza, monogamy is the norm, and only about 4 percent of men have two wives at once. The two-wife arrangement is unstable, however, and a Hadza woman will typically leave her husband if he takes a second wife or even if he has an affair. Even though 65 percent of men say it is all right for a man to have two wives, and 38 percent of women say it is all right to do so, this situation is rarely realized. As Marlowe reported, "Many men have told me the reason for their divorce was that their wives left them. When I asked why their wives left them, many said with sincere puzzlement, and some sadness, they did not know why. But when I asked if they had started a relationship with their current wife before their previous wife left them, the answer was very often yes. Still puzzled, they would repeat, 'I don't know why she left me.' These men had no intention of divorcing; they merely wanted a second woman."[36]

But their wives—in part because they could gather their own food and share in what was brought into camp by others—had no interest in this arrangement, and they departed. Often, when a husband returned after a period with another woman (something his wife may have learned through gossip), he discovered that his wife had not only decided their marriage was over and left him but also had already taken up with another man. In fact, Hadza women were quite willing to consider having two husbands (19 percent of women thought this would be fine, though no men did). Still, polyandry was never observed.

Consequently, it was rare for Hadza men and women to be involved in a polygamous union. Overall, roughly 20 percent of Hadza stayed married to the same person for life, and most Hadza had two or three spouses over the course of their lives. Sexual jealousy was quite apparent in both Hadza men and women, and in situations involving infidelity, 38 percent of men said they would try to kill the other man, while 26 percent of women said they would fight the other woman. Extramarital affairs were the leading reason given for divorce among the Hadza, with disputes over children and "wife doesn't want sex" at the end of the list.

Among the Hadza, men who are better hunters are able to marry younger women, and they appear to have greater reproductive success as a result.[37] But they are not more likely to have more than one wife at the same time or to have more wives over the course of their lives. There is considerable debate among anthropologists about why hunting proficiency should matter at all to women, given that food is shared widely and therefore a woman should gain no benefit from her husband's provisioning.[38] One possible explanation is that being a good hunter is a marker for *other* qualities, such as health and intelligence, which women might independently value in a mate. In this light, being a good hunter is a form of costly signaling (something a person does that is not easy and not necessarily useful in itself—like skiing or playing guitar—but that sends a message of his or her underlying qualities).

Another explanation is that women do actually benefit from their husbands' provisioning when they are at their most vulnerable—namely, when they are pregnant or have very young children. In general, Hadza men spend 5.7 hours per day foraging, and women spend 4.2 hours. On average, women bring more calories overall back to the camp than men (57 percent versus 43 percent of the total).[39] But when a woman is pregnant or nursing, her husband appears to up his game, and the balance shifts. So, during a critical period such as reproduction, women could indeed benefit from the men finally shouldering the burden. In fact, pregnant women's and nursing mothers' need for food has been suggested as one reason that humans, unlike most animals, form pair-bonding attachments—both women and men can benefit from this provisioning behavior, and therefore the offspring of the union are more likely to survive. The preferential survival of offspring of parents who felt attached to each other would, over the course of evolution, favor the practice of such attachment and shape the evolved psychology of our species accordingly.

Meticulous examination of foods brought back to Hadza camps

by Marlowe demonstrated that, among couples with children under one year of age, men provide 69 percent of the calories consumed by the family. Moreover, careful study also revealed that if a man finds small food items that can be concealed—such as the Hadza's favorite, honey—he will sometimes reserve these for his family alone. To make these observations, Marlowe sat in the middle of a camp and insisted on weighing everything that anyone brought back. The Hadza were happy to let him, an outsider, do this. But he noted, "What others do not see, one does not have to share. In larger camps, people sometimes waited just outside camp until dark, then signaled me to come and weigh their food, but to be discreet about it. When Hadza men sneak food into their huts, they tell me that they are doing so because they want to feed their family."[40]

No living society is an exact proxy for our ancient past, but observations such as these about Hadza partnering practices shed light on the possible origins of the typically intense bond humans feel for their mates that is part of the social suite. The nineteenth- and twentieth-century communes seeking to counter this attachment might have saved themselves some disappointment had they been able to consult the Hadza first. It's clear that, even in an exceptionally egalitarian environment, nuclear families are a persistent feature and couples have strong preferences for each other and their own children.

An enduring sense of attachment among the Hadza makes sense. From the man's perspective, if the woman feels attached to him—feels love for him—he can retain her as a partner. It also makes sense from the woman's perspective because a man who feels attached to her will help with rearing children, especially when the woman is vulnerable. Hadza women live in an environment and have a way of life that would allow them to rear children independently of men. But even so, they can do so more efficiently and safely if they have husbands helping them.

These Hadza sensibilities and observations shed light on our evolutionary past. This provisioning behavior likely evolved jointly with the pair-bonding sensibility because it addressed an important evolutionary challenge: Why should a man help feed children who might not be his own? By offering a man high certainty of paternity—by pair-bonding with him—a woman is better able to secure her husband's investment in her children. In other words, she's saying, *If you see that I love you, you can be sure these are your children.* Mutual attachment solves an evolutionary conundrum. In essence, the Hadza research suggests that the evolutionary psychology of both men and women is to exchange love for support.

### *Polygyny in Turkana Pastoralists*

How did things change for our ancestors after animals and plants were domesticated and the agricultural revolution laid the groundwork for wealth accumulation and economic inequality? Just five hundred miles to the north of the Hadza, in the Rift Valley of northwestern Kenya, the Turkana, who number about two hundred and fifty thousand, practice polygyny. Unlike the Hadza, the Turkana treat marriage more as a contract between families than a private arrangement undertaken by individuals. Fairly rigid economic strata are respected in the choice of marriage partners, with rich women marrying rich men. To our WEIRD, twenty-first-century eyes, their marriage practices are a good bit more utilitarian and seemingly less romantic than those of the Hadza. The question is why.

Turkanaland is hot and dry, with so little water that it is used only for drinking, and the people bathe by rubbing themselves with animal fat or even dung. The Turkana are pastoralists, herding animals and living a highly nomadic existence. They move and rebuild their camps eight or more times per year, just as often as the Hadza, in search of natural forage and water. It takes an enormous amount of labor to manage herds in this tough environment where rainfall is minimal and uncertain. Both sexes contribute to Turkana livestock management beginning at the age of five. Members of a male-headed extended family live together in a mobile camp along with their livestock (cattle, camels, sheep, goats, and so on), who are corralled during the night and taken out to forage during the day. Men and women alike want as many children as possible to help with the labor.

Turkana marriage practices are connected to this demanding way of life.[41] The traditional Turkana marriage typically proceeds in stages. Men and women meet at dances, weddings, and watering holes, where they may flirt.[42] After a young woman has been identified as a suitable bride by the man or by his parents—or by his existing wives, if he has any—the man gets approval from his father and relatives. This approval is crucial since he will be relying on them to contribute to the assembly of bride-wealth. *Bride-wealth* or *bride-price* is a form of marital economic exchange in which the groom's family pays the bride's father an arranged sum (it is the opposite of a dowry).[43] Among the Turkana, this might be as much as a hundred and fifty head of livestock (which is not a trifling sum even in the United States today).

After a prospective match has been identified, the man must then negotiate with the bride's father to get permission to marry her, with complex rules governing the amount of the bride-price. Once a bride-price has been agreed on, the bride accompanies the man to his homestead, and the livestock are transferred. A marriage ceremony involving both clans takes place, culminating in an ox-killing ritual. The wedding ceremony itself is typically attended by all adults within a day's walk, and, like weddings all over the world, it includes dancing, feasting, gossiping, and flirting among unmarried guests.

These requirements can feel overwhelming. To assemble the bride-wealth, a suitor must seek familial contributions (which create obligations for him in the future). Assembling the livestock often takes months. It's so laborious that it is surprising any marriages take place at all. As one anthropologist put it:

> A father's brother may live 40 or 50 miles away in one direction, a paternal cousin or best-friend, godfather, etc., etc., equally far off in another. It is . . . a time of very great difficulty and emotional disturbance for the suitor, who in some cases will be practically exhausted physically and psychologically by the time he has all the bridewealth assembled. Even then his troubles are not at an end since he must supervise the handing over and indulge in further haggling with his new father-in-law to try and reduce the actual numbers.[44]

Contrast this operation, involving so many meetings, rituals, movements, and economic transactions, with the relatively informal and personalized approach of the Hadza. The presence of capital assets (livestock) in this culture seems to complicate things considerably.[45]

Traditionally, the Turkana practice exogamy, meaning that they marry outside their clan. Since marriage is a way to seal relationships between families, the Turkana believe there is no point in a man acquiring a wife, let alone a second wife, who comes from the same family he does. Consequently, marriage norms serve to broaden individuals' social connections. A man (and his family) whose herds have been wiped out due to chance events such as drought or disease can rely on a network of social interactions, often cemented by marital ties.

Since marriage among the Turkana is very much about joint production—and harder to dissolve than it is among the Hadza—men and women alike are very interested in having congenial relatives, and everyone gets involved in the couple's choices. A father will reject a son's choice of partner if, in his judgment, the prospective wife is "lazy," "talks back," or is from a family of "witches" (that is, people seen as antisocial or nonconforming). Even co-wives get into the mix and will carefully vet the proposed new wife, as one anthropologist described:

> She will be their constant companion and fellow-worker. If she is a poor worker, then she will be a liability in the network

> of family cooperation.... Thus, the criteria of a good wife which are always looked for by a man, and especially by his mother and wives, always include the following, as told to me by both men and women: she must not be a bad, lazy worker ("akale-nyana"), that is, she must be able to milk, water, and generally look after stock properly; she must be able to build good fences and huts; she must be able to make good vessels, do good bead-work, etc.; secondly she must not have a slanderous, malicious tongue, she must not be a witch, and she should be of a pleasant disposition and character.[46]

Moreover, fertility is prized, and a young woman will often conceive a child with her intended in order to demonstrate this and secure a marriage. However, infertility is not legitimate grounds for divorce, and an infertile wife will often be given the child of another wife of the husband to raise.[47]

In practice, young women can, and often do, refuse suitors, and the Turkana do not force their daughters or sisters to marry men they do not like. A young man and woman will often force their respective parents to accede to their marriage by having sex so that she gets pregnant. The man may even forcibly take a bride; sometimes, the woman may connive to be forcibly taken. In these cases, it appears, a bride-price is still paid, although possibly less than expected or after some delay.

Young people among the Turkana (and also in many forager populations such as the Hadza) face the problem of insufficient food to meet their bodies' high energy demands, which is, indeed, our ancestral condition. This typically delays the sexual maturity of both boys and girls to the late teenage years (whereas puberty has been accelerated in better-nourished populations in the developed world in recent decades). In part because of late puberty, in part because men must accumulate so much wealth, and in part because men are competing for fewer women (given the polygyny), the age at first marriage is fairly late; for women, the average age is 22.4 years, and for men, it is 32.6 years.[48]

That age gap between husbands and wives arises because it takes a man about ten years to accumulate the livestock to pay the required bride-price. The age gap increases by roughly another ten years for each subsequent wife, so that a fourth wife is usually about twenty-two while the husband is about sixty-two.[49] Older men who choose not to keep acquiring wives and instead employ their wealth to help their sons marry are seen as having "good hearts." Turkana norms also require younger brothers to marry after older brothers, which means younger brothers often postpone marriage until they are well past thirty-five. Sometimes this precludes their marrying altogether.[50] Having to wait for an older brother to marry, especially if he is consuming the family's wealth in the meantime, is a serious handicap.

One 1996 study found that, among Turkana males, 30 percent emigrated out of the traditional pastoral sector, and 8 percent died before marrying.[51] The percentage of males who emigrate resembles what is seen in polygynous Mormon sects in the United States today (where hundreds of boys and young men may be cast out of a community).[52] Finally, about 1 percent of Turkana men described themselves as "hating women" and were described by others as "men who think they are women," roughly corresponding to men having a homosexual orientation. Unsurprisingly, few of these men marry. This overall rate of nonmarriage among males makes sense, of course, since many of the potential brides have been taken by polygynous men.

How does this pattern affect *women's* reproductive opportunities? They pay a price. A twenty-year-old woman who marries a sixty-year-old ultimately has two or three fewer children than a twenty-year-old who marries a forty-year-old. This partly reflects the fact that older men die sooner, so their wives have fewer years of active reproduction (even if they eventually remarry), and partly reflects an age-related decline in male sexual performance.[53]

Co-wife conflict is ubiquitous in polygynous households, a fact that comes as no surprise to evolutionary psychologists and those who are practitioners of polygyny themselves. Because the Turkana often choose wives from different families in order to broaden their safety net, they typically do not practice sororal polygyny—the practice whereby men marry sisters or, sometimes, cousins. When co-wives are relatives, they can more easily share a household and cooperate, partly because of the evolved human tendency to help blood relatives and partly because of the obvious fact that the sisters may love each other and have prior experience living together. In sororal polygyny, wives also tend to be closer in age and fewer in number, which further restricts competition for resources and reduces conflict. But while sororal polygyny is especially common in cultures in the Americas, general polygyny tends to be the usual pattern in Africa. An examination of ethnographic data from sixty-nine nonsororal polygynous cultures fails to turn up a single society where co-wife relations could be described as harmonious.[54] Detailed ethnographic studies highlight the stresses and fears present in polygynous families, including, for example, wives' concern that other wives might try to poison their children so that their own children might inherit land or other property.[55] All these stresses suggest an additional rationale for the rise of cultural monogamy.[56]

Despite the differences in traditional marriage patterns in the Turkana and the Hadza, Turkana men and women do generally have affection for their partners, as the courtship rituals, jealousy, and accidentally-on-purpose pregnancies among the couples suggest. They choose who their hearts desire. In *A Wife Among Wives*, a 1982 documentary made by David and Judith MacDougall, prominent, long-term ethnographers of the Turkana, one man, Lorang, noted: "Girls don't follow the old customs these days. They don't listen to their parents' advice. Even if a man brings the bride-wealth, the girl refuses. Not even her parents can make her wait for a man with bride-wealth. Although he brings it, she refuses him. She suits herself. And that's to follow some boyfriend whom she's picked out for herself."[57]

A female interviewee, Yanal, said this when she was asked why a young woman might disobey her father: "Because of her feelings. Her insides rebelling.... She can refuse to marry either a senior man or a young one. Even an old man. She might go off with a young hunter, a man without a single animal, and refuse to marry."[58] The Turkana understand that a bad man, even if he is rich, will not be a good husband. In *The Wedding Camels*, another documentary by the MacDougalls, a woman named Akai said: "You can always refuse. Even if he's rich and handsome. If the girl chooses a poor man and won't change her mind, she can refuse her parents' choice. She can run away with her lover and live on a mountain. You can ignore your family... if it is not what you want in your heart."[59]

### *Multiple Fathers*

The contrast between the marriage practices of the Hadza and the Turkana provides clues about why the agricultural revolution, with the possibility of wealth accumulation and status inequality that it offered, might have inspired a comeback of polygyny. But there are other, rarer, marriage systems that not only shed light on the play of culture in determining romantic interactions in humans but also further illuminate key features of human mating behavior.

The rarer counterpart to polygyny is polyandry—more than one man for each woman. While there are some cultures practicing polyandry in Africa (for example, among the Lele and the Maasai), it is rare on that continent, and the usual examples come from the Himalayan region and parts of India and the Americas. As in the case of polygyny in the Turkana, the marriage practice appears to be a cultural adaptation to environmental exigencies that is also integrated with broader economic features.

When polyandry is practiced, it usually involves brothers marrying the same woman (known as fraternal polyandry, the mirror image of sororal polygyny) and co-occurs with other marriage types (including both monogamy and polygyny) in the same society. The youngest husband often leaves the marriage and marries his own wife when he can. Polyandry is especially likely in ecological situations where sustaining a household requires more than one man—for example, when one man must travel long distances to support the family and another is needed to guard the home. In the Himalayas, where land is scarce, the marriage of one woman to all the brothers of a family makes it possible for the brothers' plots of land to remain undivided and of sufficient size to be efficiently arable. In ecologically hardscrabble environments, where it can take many years of adult labor to raise a single child successfully, having three or more parents to help rear the children increases the odds that each of those children will thrive. In any case, children can be produced only as fast as a woman can give birth to them, so polyandry thus becomes a kind of cultural birth-control or birth-spacing strategy—ironically, from the *male* point of view. The solution to these various ecological constraints is for brothers to team up.

Some cultures practicing polyandry have a belief system about paternity that is hard to reconcile with the biological reality that a baby is conceived by sex between one man and one woman. This is known as the doctrine of single paternity, and an understanding of it stretches back to well before 500 BCE. Most cultures grasped this idea long before the biology was worked out by modern science. But across many cultures throughout Amazonia, and in a few scattered elsewhere, children are believed to have *multiple* fathers. In a book on the subject, anthropologists Stephen Beckerman and Paul Valentine described over a dozen cultures with beliefs about *partible paternity,* as it is known. These cultures are often from very different language groups and are widely separated, suggesting that the concepts did not develop due to isolated events reflecting some "pathological" occurrence, nor were they simply copied from neighboring groups.[60] The existence of these similar belief systems in so many successful cultures indicates that the idea of shared paternity is not incompatible with a functional society able to care for its children and avoid internecine conflicts.

In cultures that believe in partible paternity, the woman's role in producing babies is typically denied, and women are seen as mere receptacles. Some think the baby is actually made by an accretion of semen, like a snowball. Among some, the *men* are felt to pay a physical price for the pregnancy. Some men say that they must expend so much energy to make a baby through repeated intercourse that they become wasted from the effort.

Of course, for most human groups, knowing the identity of a child's biological father is a fundamental object of social organization. Paternity certainty and the social and biological processes that make it easier for males of our species to acquire such certainty (including, unfortunately, cultural and religious practices that sequester women) have played a crucial role in human evolution. Together with the gender-based division of labor, the sharing of food, and the lengthy period of caring for young, paternity certainty was a crucial facet in the transition from our hominid ancestors to our own species (as we will see in chapter 6).

The argument is that men provision women during their vulnerable perinatal period (as we saw with the Hadza) but *only* to the extent that they can be confident they are advancing the survival of their *own* offspring, not another man's. Evolutionarily speaking, that means that women provide men with paternity certainty in order to receive food and support. In the evolutionary tool kit, pair-bonded monogamy and female love may have served as authentic signals to increase the confidence of males that offspring were their own, thus fostering paternal investment in children.

Some scholars have wondered how provisioning could have played a role in the origins of pair-bonding if some members of our species had other ideas about how babies were made. Must humans always have understood the one-man-and-one-woman reproductive facts, and must men always have seen other men as a threat in regard to paternity certainty? That would seem logical. But it is not essential. Human culture is endlessly variable, and it should not surprise us that a few cultures believe that a single child can truly have multiple fathers. Men in such cultures are less troubled when women have sex with multiple men before and during pregnancy.

For example, among the Tapirapé (who found kissing disgusting), Charles Wagley observed:

> One act of copulation was not considered sufficient for conception. Intercourse had to continue. A male had to provide more and more semen "to build the flesh of the child." They laughed at me when I ventured to say that a single copulation was enough and that in my land women sometimes found themselves pregnant after one sex act. In any case, intercourse had to continue during pregnancy, but it did not need to be with the same male. All men, however, who had intercourse with a woman during her pregnancy were considered the genitors, not merely the sociological fathers of the child. It thus often happened that a child had two or three or more genitors. One was publicly known as the mother's mate at the time, so he was called *cheropu* (father). The others were known by gossip.... Matters could, however, become too complex. When it was known that a woman had had intercourse with several men (four or five or more), then the child had "too many fathers."[61]

"Too many fathers" is an interesting concept for those of us raised in WEIRD societies. Children in societies like the Tapirapé's could be seen as having multiple genitors (biological fathers) and one pater (social father), which is the reverse of our view that a child can have only one biological father but may sometimes have multiple social fathers (such as godfathers, foster parents, stepfathers, family friends referred to as "uncles," and guardians).

Among the Aché of Paraguay, any man who had sex with a woman for the whole year before she gave birth could count as a father, and these secondary fathers (other than the one deemed the primary father) could play important roles in the lives of the offspring.[62] Secondary paternity is often actually negotiated in these cultures, with women asserting or withholding the identity of possible secondary fathers, and the men accepting or rejecting the label. In most parts of the United States and other industrialized democracies, a woman also declares who the father of the child she gave birth to is, and these declarations are generally accepted. And yet, from genetic studies, we know that as often as 1 to 2 percent of the time, these are false-paternity events—meaning that the father was someone other than the woman's steady partner or the person she declared it was.[63]

Belief systems about multiple fathers make more sense when we recognize that women around the world do often have sex with more than one man; that it seems (given clues in the known biology of sperm) that, over human evolution, it was not uncommon for semen from multiple men to be present at the same time in a woman's reproductive system; and that it can often be advantageous to the survival of both women and children for beliefs about partible paternity to be prevalent (since a woman can benefit from support from more than one man).[64] In one sample of two hundred and twenty-seven Aché children followed over ten years, 70 percent of those with one father survived to age ten, but 85 percent of those with two or more fathers did.[65] While this advantage might relate to social factors such as provisioning of a woman by multiple fathers, a theory for which anthropologists have accumulated evidence, it's also clearly the case that a woman who attracts more partners might be fitter (for example more beautiful, better able to produce a healthy baby, and so forth) and that her good genes might also contribute to the superior survival of her children. We cannot be certain, but we can be relatively confident that partible paternity does not *harm* children's outcomes.[66]

So there seems to be a tension between, on the one hand, partible paternity, and, on the other hand, the male-provisioning hypothesis, which offers a foundation for the evolution of pair-bonding. Why do the majority of societies care very much about paternity—in most, a man will provide for a woman and her child only if he knows that the child is his genetic offspring—while a few cultures do not care much about biological paternity at all?

We can square these ideas with a more *female*-centric point of view.[67] Both practices (caring a lot and not caring much about genetic paternity) may simply be better for women, as we have seen. Researchers might have overlooked this possibility for methodological reasons. We must sometimes rely on old studies to get insight into the most traditional ways of life (of less contacted tribes and of those less affected by modernity), but most of those studies were done by men. And historically, male anthropologists were less interested in women's experiences, unable to access those experiences from their informants, or both. It wouldn't be the first time scientists were overly preoccupied with male subjects to the detriment of women's observations.

Another possible resolution for the tension between the existence of beliefs about partible paternity and the idea that male provisioning is crucial to pair-bonding is to adopt a more expansive perspective on the role of culture in human evolution. Humans may be genetically predisposed to produce myriad cultural forms, and if we see *culture* itself as something that our species is pre-wired to produce, we can be more forgiving of variation, even of those practices that seem to swim against our biology. We can see that many cultural practices might, from some points of view, enhance survival in difficult circumstances. Our ability to be cultural animals (with whatever beliefs, practices, or technology) and engage in teaching and social learning is an important part of the social suite.

Finally, in trying to square these unusual paternity beliefs with human biology, we cannot overlook that provisioning and pair-bonding evolved when early humans had much more limited cognitive and cultural capacities. Any cultural overlays regarding how babies are made that humans are consciously aware of were later additions to fundamental propensities laid down long ago.

### *No Fathers*

Finally, let's consider an example of one of the most radical and complicated types of formal human mating practices of which I am aware. The multiple fathers of the Tapirapé have nothing on the customs of the Na people of the Himalayas.

Most societies express at least some anxiety about young, unfettered love. The image of a hapless Juliet eloping with her lovelorn swain (or a Turkana girl doing the same) is a cliché that stirs us because it kindles both sympathy and concern. The image derives its power in part from the seemingly incontrovertible reality that teenage passion probably is not the best method around which to organize a community. But the Na people of the Himalayas have few such worries about unbridled teenage lust. On the contrary, what often worries Na adults is the possibility that Romeo might have actually stumbled on lasting love.

It's hard to imagine a society with less interest in what modern Westerners call marriage than the Na. They are a group of mountain farmers near Tibet numbering about thirty to forty thousand. Anthropologist Cai Hua's tellingly titled *A Society Without Fathers or Husbands* was written in 2001, but the Na's unusual sexual practices have been subjects of Chinese texts for over two thousand years. Even Marco Polo found the Na noteworthy, writing, "The natives [in this kingdom] do not consider themselves injured when others have connection with their wives, provided the act be voluntary on the woman's part; in this case, it is not considered a misfortune, but a jackpot."[68]

Na households are matrilineal. A woman lives with her mother, sisters, brothers, and maternal uncles. Therefore, unlike what we see in most cultures, it is standard for *all* members of a household to be consanguineal relations. Unrelated men from outside the family very rarely move into the household. However, men do frequently visit the household to have sex with women there; in fact, women are not supposed to have sex in others' homes, which the Na compare to a "sow in heat crashing through the fog."

People in Na society do not know or care who their own biological fathers are. Despite endless probing, Hua was unable to find any interest in the topic. Though the Na do realize that offspring can resemble the man who was the genitor of the child, most deem paternity irrelevant, since it involves no special rights or duties.[69]

The Na know that men, and sex, are required for a woman to give birth to a baby, but they believe that the babies are already in the woman, like seeds in the ground, and they are merely "watered" by the men's semen (they say: "If the rain does not fall from the sky, the grass will not grow on the ground").[70] They feel that it makes no difference who does the watering. These beliefs about how babies are made, of course, are the opposite of those of the Amazonian cultures we considered above, in which the people see semen as crucial to producing a baby. This also explains why men in Na society have no obligations to their biological offspring, whereas the perceived genitors in societies with partible paternity do.

The Na explain that "in mating, the aim of the woman is to have children, and the aim of the man is to have a good time and to do an act of charity"—though the ethnographic accounts make it clear that the women enjoy their sexual activity too and have total control over whom they choose to have sex with, since the Na insist on absolute consent of both parties.[71] Like all societies, the Na have an incest taboo. They also have strict norms against what Hua calls "sexual evocation"; it is completely taboo "to speak of sexual, emotional, or sentimental relationships in the presence of consanguineal relatives of the opposite sex and to make any allusions to sex."[72]

There are several types of sexual relationships (Hua calls them "modalities") among the Na. The most common is the practice of *nana sese*, or a "furtive visit," where a man will come to a woman's house after nightfall (and leave before daybreak). He does not eat in his partner's home and he tries to avoid any contact with members of her household. Many women (as well as men) in Hua's studies had dozens of partners, sometimes reaching one hundred over the course of a lifetime. Partners in this activity call each other *acia*, which roughly means "lover." The focus of this relationship is light and casual, and, as Hua notes, "arguments between two acia usually do not get very far, since they only discuss their sex life. When together, they never bicker over the worries of day-to-day life." And "a trifle is therefore all it takes to end the relationship."[73] *Acia* are lovers "who remain strangers to each other."[74] It was not uncommon for even a mildly attractive woman to have slept with all members of her age cohort in her village—and in nearby villages. This relationship type is almost exclusively about sex.

Just as in contemporary hookup culture or online dating services, specific cultural norms govern this activity. Young men will often go out after dark looking for partners they know, often climbing the walls of a woman's household compound. Sometimes, several men will show up at a woman's door at the same time, each trying to persuade the woman to pick him for the night. When men are rejected, there are generally no hard feelings because they can "always try tomorrow." Or they can just try another house.

Before a woman receives an *acia*, the two will usually have gotten to know each other through prior interactions and flirtation. The conversation for arranging a visit, if one is planned in advance, is quite straightforward and can be initiated by either the man or the woman. It goes something like this:

Woman: Come stay at my house tonight.
Man: Your mother is not easygoing.
Woman: She won't scold you. Come secretly in the middle of the night.
Man: Okay.[75]

According to Na women, the criteria for choosing a mate are "first and foremost, physical beauty, then a sense of humor, vivacity, roguishness, courage, and work capability, and, last of all, kindness and generosity." For men, the criteria for choosing an *acia* are "beauty and...physical charm, allure, a gift of conversation, good manners, and kindness to others."[76] The premium on beauty (and youth) means that once women reach thirty, they are much more rarely visited, and women who are unattractive or disabled might never receive visitors. But liaisons still continue into older age (men typically stop visiting women when the men are around sixty years old, and women stop accepting visitors when the women are about fifty years old). The number of regular partners an older woman has might settle at about three, and the men who visit often just knock on the household door rather than clamber over the walls.

A second sexual relationship type among the Na is the *gepie sese,* or "conspicuous visit." In this type of arrangement, a man visits a woman's household openly, is seen by other household members, and is permitted to join meals with the family, though he is supposed to be unobtrusive. Such couples always start with furtive visits. Then, "if the partners' feelings for each other have deepened, they exchange belts as a symbol of their desire that their feelings and *love* for each other will last," and the conspicuous visits begin.[77] Such relationships may last months, years, or decades, but they end instantly when either party so decides. People are typically in only one conspicuous relationship at a time. Unlike an *acia,* a man in a *gepie sese* can come before nightfall and leave after breakfast. Here, the members of the couple have an expectation of (public) sexual exclusivity, or sexual privilege. They state that they will be exclusive, but there are no public sanctions for violating this norm, and either member is allowed to practice furtive visits on the side unless these are discovered by his or her current partner, in which case the relationship is seen to have ended. Hua outlines cases in which sexual jealousy appears here, because if a man finds his partner in bed with one of her *acia*, he will occasionally (but not always) attack him. This relationship type is about both sex and love.

A third type of sexual relationship is *ti dzi*, or "cohabitation." (There is also a fourth that is quite similar to this, though it's exceedingly rare and involves payment of a sort of bride-price for a woman; Hua calls it "marriage.")[78] In *ti dzi*, the man and woman move in together, usually into the woman's household. This arrangement must be preceded by the furtive or conspicuous relationship type. The community considers the couple to be "intimate friends" and does not see them as *related* to each other or even married. Cohabitation typically occurs only when the existing household has few if any members of the opposite sex, who are required for economic productivity; it is more a response to situational exigencies than a reflection of personal preference. These relationships are also not exclusive and can end abruptly if one partner so decides. In many cases, the cohabitants will eventually come to act as if they were *siblings* rather than sexual partners. Hua concluded that the Na brand of cohabitation is not a type of marriage in any conventional sense. This relationship does involve sex, and sometimes love, but it's mainly about work.

In 1989, in a sample of several villages, Hua quantified the occurrence of these sexual modalities and found that over 85 percent of people were living lives involving only furtive or open visits.[79] The Na way of life has persisted despite repeated efforts to eradicate it. Several times since 1950, the Communist government in China has attempted reforms, out of concern that Na practices sap work effort, disrupt the "means of production," and foster immorality (mirroring the concerns of Emperor Augustus in ancient Rome, except that women have many male partners here, rather than the reverse). But the reforms have mostly failed.

Overall, sexual jealousy seems rare among the Na, though it is not entirely absent, as noted above. The Na revere tolerance of multiple liaisons, and jealousy is seen as unnecessary. If a man complains that his partner in an *acia* relationship is also sleeping with other men, the people in the village will mock his foolishness and even see his jealousy as shameful. "Why not just go sleep with someone else?" they will ask. "There are always more girls in the next village."[80]

But love isn't totally absent among the Na either. Partners do describe feeling love for each other, and stories are recounted of couples who flout the norms and flee together to form an exclusive relationship because they have fallen in love. Hua argues that fidelity and exclusivity are the "forbidden" desires here, which makes them more exciting.

Why do the Na have the institution of the visit rather than the institution of marriage? In the vast majority of societies, marriage confers some form of metaphoric or even literal possession of another person. Wedding vows incorporate this sense of exclusivity with variants of "to honor and keep" and "forsaking all others," and married couples generally live under the same roof and have expectations of each other sexually and economically. In some U.S. states, impotence is still officially grounds for divorce.[81] None of this pertains to the Na, however, where partners are independent and live separately with no exclusive sexual or economic claims.

Hua concludes that humans have several fundamental—and, I would argue, biologically rooted—desires, two of which are to possess one's partner and to have multiple partners. It's hard to square those seemingly contradictory impulses within the same group, and there really are only two options: possession without the pleasure of diversity or the pleasure of diversity without possession.[82] Across evolutionary time, attachment has proven the stronger force. Moreover, societies cannot satisfy both, institutionally speaking, and it seems that the Na—perhaps uniquely—have resolved this dilemma in favor of the latter.

But the choice is not so easy. An elaborate cultural edifice is required to suppress our deep, ancient desire to possess our partners and feel attachment and love for them. Like the Baining people's suppression of children's play, the Na have to work hard to maintain a functional culture where "to have and to hold" has none of the resonance seen in virtually all other societies.

This should not surprise us. Consider the recent resurgence of interest in polyamory in contemporary Western cultures. Advocates claim that having multiple sexual and romantic partners reflects humans' "natural" state (although, as we have seen, this belief is not in keeping with the cross-cultural evidence). But a similarly complex set of rules and regulations is required to govern so-called free love. As one observer told the *New York Times*, "There are so many more rules in non-monogamous relationships than in monogamous ones."[83]

Hence, to achieve this rare cultural form, the Na exploit one of our species' other great traits, the ability to cooperate and share, applying it here to the pursuit of partners. And they take advantage of the human ability to invent ways of living and find meaning and coherence in them. That is, the capacity for social learning and cultural transmission, which is itself part of the social suite, is a crucial aspect of our adaptation as a species and of our ability to manifest many marital patterns overlaid on an impulse to pair-bond.

Still, formal institutions cannot entirely eradicate either of these human desires (to both love and possess one's partner), originating as they do in the most fundamental aspects of human nature. People violate norms of all kinds in all societies. So, while the Na could function perfectly well solely with furtive visits, they also allow the institution of the conspicuous visit, providing some satisfaction of the desire for possession. Moreover, even among the Na, some couples who are "carried away by the burning fire of their love" run away together to possess each other completely, unsatisfied with mere visits and uninterested in multiple partners.[84] This is analogous to the accommodations that many societies have made to the institution of marriage to allow for a change in partners, accommodations such as permitting divorce or allowing men to have concubines.

Many have argued that the highly unusual sexual practices of the Na disprove the ubiquity of marriage, suggesting that monogamy cannot have any biological underpinning. But the existence of variation does not mean that there is not a central tendency in our species. As scientists, we can lump as well as split—that is, we can look for commonality as well as variation. Our human blueprint is the first draft, not the finished version, of our reality. The underlying motivation for the Na's relationship structure is the basic human desire for multiple partners, and the underlying motivation of the institution of marriage is the equally basic desire to possess one's partner. The exceptional Na case proves that an aspect of our humanity as deep and fundamental as a desire for attachment—a desire, in fact, to *bond* with one's partner—can never be fully suppressed or replaced, even by a highly elaborate set of cultural rules created for the purpose of cutting precisely that connection.

### *Love by Arrangement*

In many countries around the world, such as China, India, Indonesia, and Nigeria, a sizable fraction, and sometimes the majority, of marriages are arranged. Whereas romantic love is seen as a precondition for marriage in most Westernized settings, it is often seen as impractical, unnecessary, or even dangerous in many parts of Asia and Africa. However, most of these cultures do recognize the existence of romantic love (for example, at least 5 percent of marriages in India today are love matches).[85] And, more important, while romantic love *before* marriage is viewed with suspicion, love *after* a couple is united is seen as a natural and very-much-hoped-for consequence of marriage even in countries with arranged marriages.

Consider the testimony of Sandhya (age twenty-nine) and Ankur (thirty-one), an Indian couple whose engagement was arranged by their families through a newspaper ad in the matrimonial column. They had spent only a few chaperoned hours together before they were wed in a traditional ten-day Hindu ceremony featuring hundreds of their parents' guests.[86] The couple spoke with a columnist for a popular magazine in 2014 about the romance that emerged from this impersonal beginning:

> *How long did you guys talk before you met in person?*
>
> Sandhya: I think a couple of hours. He liked me and he told his parents that he was interested, so his parents called my parents.... [Later], we sat for, like, 15 minutes face-to-face. It's kind of embarrassing because it is so...
>
> Ankur: Because our entire families are there, so you are not really talking too much.
>
> Sandhya: He was, like, all shy and I was talking, and then he just went home, and the next day, his parents called and said that he wants to get married to me and my parents were like, "Is it OK with you?" and I said, "OK!" and then we got married! Now it's like falling in love each day with him.

Another woman in an arranged marriage described to the *Times of India* the exact moment she knew she was in love with the stranger who had become her husband one month earlier:

> My car broke down while returning home around 10 pm. I called my husband and informed him about the situation. I could sense the concern in his voice when he was talking, and to my surprise he reached the spot within 15 minutes. He was quite worried about my safety and hugged me as soon as he came near me. And that day, I fell in love with my husband.[87]

Academic studies of couples in arranged marriages support the idea that a key factor fostering the emergence of love is the sense that one's partner is *committed*.[88] A crucial facet of the hopeful anecdotes above is the focus on love and attachment, not just lust. Of course, other factors contribute to the deepening of love in arranged (and non-arranged) marriages, including self-disclosure, kindness, and physical intimacy. But commitment is key. One survey respondent who was in an arranged marriage put it this way: "Love starts from an absolute commitment, and that begins to break down barriers in my heart and head, so feelings of joy and contentment and peace can bubble up."

To be absolutely sure about the relative amount of love in arranged marriages, we would need to do an experiment in which people were randomly assigned to either find their own partners or have their spouses chosen by their parents. This is clearly impossible. But surveys in nonexperimental (that is, real-world) conditions are nevertheless informative. They have documented that satisfaction in arranged marriages generally is not lower, and is sometimes even higher, than it is in love matches.[89] One study of couples who had been married for an average of ten years found no statistical difference in self-reported ratings on a scale of passionate love between couples in arranged marriages compared to Western-style love matches.[90] Respondents in another small study of couples in arranged marriages from a variety of backgrounds rated their level of love on a scale of 1 to 10. It was 3.9 at the time of marriage and 8.5 twenty years later.[91] Love is a key part of marriage, no matter what kind. A worldwide survey of 9,474 respondents from thirty-three countries on six continents practicing many types of marriage—including both arranged and non-arranged—consistently found "mutual attraction/love" as the first or second most highly ranked attribute with respect to marriage partners.[92]

We have seen how culture and ecology can shape marital unions. The relatively egalitarian Hadza, for instance, are monogamous in part because women are less dependent on men for food. Among the polygynous Turkana, however, cattle ownership creates inequality and status differentials, in turn conferring advantages on certain men, and marriage functions to create labor units suited to the demands of nomadic pastoralism. Polygyny fits in with all these features. But we have also seen that even the strongest ecological or cultural forces rarely, if ever, overcome one core feature of human relationships, even among the marriage-phobic Na. That feature—a key characteristic of our species—is the pair-bond between sexual partners, a special kind of attachment to one's mate that we humans surely inherited from our primate ancestors, that is seen in cultures with all marriage types, and that people experience as love. The drive to love your partner is universal.

## CHAPTER 6

# Animal Attraction

In a standard laboratory test evaluating mating interest in rodents, a male rat will eagerly press a lever to get a female rat to drop out of the ceiling. The rat thinks, essentially, *A female; that's awesome!* But male prairie voles, a different type of rodent, are a bit more discerning and consider, *Who is that female?* It appears that the prairie vole derives a positive stimulus from a specific female, not all members of the opposite sex. Not only that, voles who are experimentally isolated from their particular partners display hormonal and behavioral changes consistent with depression. These symptoms are not relieved by the presence of just any replacement partner; only the vole's own partner will do. In another experiment, scientists suspended voles from their tails and noted their reactions. If a vole was pair-bonded with a mate, it struggled. If it was not pair-bonded, it also struggled. But a pair-bonded vole whose partner had been taken away acted as if it were depressed (I am tempted to say "grieving"), and it did not struggle.[1] Furthermore, "widowed" female voles rarely take a new mate.[2]

The terms *pair-bonding* and *social monogamy* are often used interchangeably, but they are not the same. *Pair-bonding* is an internal state reflecting perceptions and sentiments of attachment; *social monogamy* is an external practice or behavior.[3] In humans, this is akin to the distinction between being in love and living together. Among animals, if a male and a female merely wish to mate, they do not need to stick around after their gametes meet. But pair-bonded animals do stick around. Pair-bonds can even exist in polygamous species, such as gorillas, where one male and several females are connected to one another. It is the attachment, not the exclusivity, that is key.[4]

A pair-bond in any animal—not just in our species—is therefore a stable, mutually dependent, though not always exclusive, sexual relationship of some duration; it is accompanied by behavioral, physiological, sometimes cognitive, and (in humans) emotional attachment.[5] Humans are not alone in the animal kingdom in forming such bonds, though the practice is unusual even among primates. Stated most plainly, pair-bonding means that one is not indifferent to the identity of one's partner.

To understand why humans do not just lust for or merely reproduce with their partners but also love them, we need to go a lot farther into the past than a Roman emperor concerned with social order who legislated monogamous marriage. Natural selection has equipped us with the capacity for attachment and love, and this is built on our species' capacity for pair-bonding coupled with our capacity for conscious thought and emotional experience.

Exploring the biology of pair-bonds in animals and humans marks a shift in our consideration of the social suite. We will now move beyond the sociocultural descriptions of human groups—the shipwrecks and the communes and the couples in small-scale societies we have so far considered—to begin to understand how evolution has shaped social behavior and how genetics and physiology actually work in this regard.

### *Pair-Bonding in Animals*

The existence of an *enduring* and *sentimental* sexual association, whether monogamous, polygynous, polyandrous, or homosexual, is the core feature that differentiates the sexual practice of our species from most other animals.[6] To understand the ultimate evolutionary origin and purpose of pair-bonding in humans, we have to take a step back and look at the reproductive strategies of animals more generally.

For instance, social monogamy is extremely common in birds—90 percent of bird species are socially monogamous and mate roughly for life—but rare among mammals. A comprehensive study classified 2,545 mammalian species into three reproductive social systems. In the solitary system, females forage independently and interact with males only while mating; this is seen in 68 percent of mammalian species, including cheetahs and armadillos. In the group-living system, breeding females share a common territory with one or more males; this is seen in 23 percent of species, including deer and roosting bats. And in the monogamous system, a male and a female inhabit a common territory and are together for more than one breeding season; this is seen in 9 percent of species, including night monkeys and small-clawed otters. Only 29 percent of primate species are monogamous.[7]

By evaluating when and how mammalian species branched off from one another in the distant past, scientists can date the timing of the evolution of these mating patterns and determine how often the practice of social monogamy emerged independently. Biologists analyze such branching all the time to understand the origins of anatomical or physiological features in animals. The same approach can be used to study mating behaviors.

The common ancestor of all mammals appears to be a rodent-like animal that lived at least ninety million years ago; its females lived in a solitary fashion, with males occupying ranges that overlapped the ranges of several females.[8] This was also the pattern at the origin of each distinct branch of the mammalian tree. Social monogamy and pair-bonding between mates arose out of such solitary arrangements.

One school of thought holds that social monogamy developed in these various species as a consequence of natural selection for *paternal* care of offspring (in humans, we discussed this as an issue of male provisioning in chapter 5). Pair-bonding species, such as gibbons, wolves, voles, bald eagles, and humans, do something more than just stay together. They also often jointly raise their young. At some point, it must have become effective from an evolutionary standpoint for males to help raise their young rather than simply produce more of them, and since males would want to help only their *own* offspring survive, they needed to be sure that their female partners mated only with them. But when scientists took into account which species evolved from which others, they determined that paternal care was probably a *consequence* of social monogamy rather than a *cause.* In general, across evolutionary time, male animals came first to be bonded with particular females and thereafter to help raise their young.[9] So parental care is likely not the driver of attachment between partners, even if this seems logical.

What else might explain pair-bonding in these mammals? An interesting clue comes from the fact that monogamy virtually always appears in the evolutionary record in species that were previously solitary (rather than group-living). Scientists think that females of some mammalian species either gradually required larger ranges for foraging or grew intolerant of other females in their territory (possibly due to resource competition). In either case, the females spread out geographically. This would have made it much harder for the males to find and defend more than one at a time. It then became more efficient for both males and females to reproduce within long-term and relatively exclusive bonds with each other, and species evolved this behavioral pattern. Further support for this idea comes from the fact that pair-living species live at substantially lower density, with a median of only fifteen individuals per square kilometer, whereas solitary species have an order-of-magnitude-higher density, with one hundred and fifty-six individuals per square kilometer. Under these conditions of low population density, resource competition, and social intolerance among females, guarding individual females—and eventually developing a special attachment to them—likely represented the optimal reproductive strategy for males.

It turns out that monogamy has evolved from solitary ancestral species many times. We know that it has evolved at least sixty-one times in mammals alone. Since it serves a purpose, different branches of the animal tree have independently discovered it. This is an example of *convergent evolution*, where genetically distinct organisms evolve similar traits independently, often in widely separated locations. However, although pair-bonding does solve a mating challenge in many other species, the evolutionary mechanisms for the emergence of pair-bonding in humans in particular are still under investigation. Nevertheless, the key point is that this social practice related to how animals (including our own species) interact with one another is encoded in their genes, and it is shaped by natural selection.

### *Monogamy and Social Life*

Other than humans, there is no species of great apes that mates monogamously (though the so-called lesser apes, the gibbons, do).[10] And substantial paternal investment in offspring is also rare in primates, though it is normal for humans. All African apes live in groups and are polygynous, and it is likely that the common ancestors of hominids lived in a similar arrangement. That is, in the primate line leading to our species, the fundamental shift to social living occurred with the appearance of primate ancestors who first made the switch from solitary living to unstable groups. Eventually, they came to live in stable, polygynous groups. And pair-bonding arose from this arrangement.

Group living afforded males and females easy access to each other, but it also presented challenges for reproduction, such as the prospect of inbreeding. One way to get the benefits of living socially while avoiding inbreeding is for the species to evolve sex-based dispersal, in which members of one sex (usually the males) wander away from the ancestral home while the other sex sticks together (a practice widely seen in primates). In short, one sex leaves and the other stays put, resulting in the formation of relatively stationary groups of genetically related individuals (for instance, groups of females and their mostly female offspring of several generations). This pattern also sets the stage for kin selection, an evolutionary strategy in which individuals engage in considerable risks or sacrifices to help others in their local groups precisely because these others are actually related to them. Mutually beneficial cooperative interactions, another part of the social suite, are an obvious advantage of group living and help explain why group living emerged in primate species in the first place.

Pair-bonding appears to have arisen after group living had already been established. This is confirmed by a reconstructed phylogenetic tree for 217 primate species showing that this happened multiple times in different branches within the primate line and at widely disparate times, ranging from sixteen to four and a half million years ago. But pair-bonding in primates probably had different evolutionary origins than it did in other mammals, since it did not arise from previously solitary species.[11]

In humans, pair-bonding and monogamy likely arose out of polygynous, group-living social arrangements of our distant hominid ancestors. One piece of evidence for this polygynous past is human sexual dimorphism. On average, men are substantially bigger than women, consistent with the idea that males competed for females and that bigger males were able to mate with more females (or that females preferred bigger males for their own reasons), thus passing on their genes for size. But bear with me; the story is complicated. Humans have also become more *monomorphic* (physically similar across both sexes) over the more recent evolutionary past. This shift suggests increased evolutionary pressure for monogamy beginning with the origin of our particular hominid species, *Homo sapiens*, a few hundred thousand years ago.

The most sexually dimorphic trait in humans is upper-body strength, which is thought to provide males with an advantage during fighting and thus possibly to result in a greater number of partners and offspring, as we see with gorillas.[12] But there has been an overall decrease in sexual dimorphism in human traits, such as the size of canine teeth, with men's canines becoming less Dracula-like, which suggests that the fitness advantages of size and strength for fighting have been reduced over time. Some have proposed that the (very ancient) cultural invention of weapons mooted the advantage of big canines. Stones in men's hands superseded teeth in their mouths. This is known as the weapons-replacement hypothesis. But the reduction in male canine size *predates* the appearance of tools in the fossil record.[13] And larger size and bigger teeth could still be useful in fighting even after weapons became available (at least until the invention of gunpowder, which radically equalized all people regardless of size).[14]

These observations suggest that the maintenance of upper-body dimorphism in humans—despite the reduction in dimorphism in other traits, like teeth—may *not* be entirely explained by direct male competition. So what else might explain it? Physical anthropologist Coren Apicella has argued that some dimorphism in upper-body strength has been maintained, evolutionarily speaking, because of women themselves. For instance, if women preferred men who were good hunters, and if upper-body strength facilitated hunting, then the desires and choices of females rather than competition among males could lie at the root of this persistent dimorphism. This dimorphic trait, in other words, still provides an advantage—and may explain why, in most cultures, heterosexual women often find broad shoulders attractive. And upper-body strength, a trait that is related to hunting success in Hadza men, does have beneficial reproductive consequences, Apicella found.[15] In short, female choice, male provisioning, social monogamy, and body dimorphism are probably all linked.

So, regardless of the reason for the prior appearance of group living in primate species, and any advantages this offered, how is the transition to pair-bonding achieved if the evidence suggests that our ancestors lived in a more promiscuous, probably polygynous, and likely cooperative social structure already? The classic explanation of monogamy in mammals that we discussed earlier—that female dispersion obliges males to bond with individual females—is undermined by an arrangement in which animals live in groups. There must be some other mechanisms.[16]

### *Female Power*

Many of the ideas we have so far considered are focused on male actions related to things males do for themselves (such as guard females to prevent others from mating with them or exchange food for mating opportunities), for their mates (protect females from predators while the females tend their young or provision the females with food when they are doing so), or for their offspring (prevent their killing or starvation).[17] But females are not passive actors in our evolutionary story. To provide just one example, they exercise choice over their bodyguards (known as female "choosiness").[18] And males and females should, in general, affect the evolutionary trajectory equally.

Dominance hierarchies in groups would seem to subvert monogamy as particular males in groups of our hominid ancestors dominated others for mating opportunities with co-residential females, pushing evolution further in this direction. Yet, paradoxically, according to evolutionary biologist Sergey Gavrilets, these same hierarchies in our group-living ancestors may have contributed to pair-bonding, by *increasing* female choice.[19] Essentially, the male "losers" in this arrangement—the lower-ranked males shut out of the competition for female partners—had to come up with an alternative strategy to acquire mates. Given the arithmetic reality that low-ranking males were more numerous than high-ranking ones, natural selection might make it effective for such males to offer resources *to females* rather than fighting *other males* for dominance. High-ranked males could of course still fight with low-ranked males and claim mating opportunities, rendering any investment in females by low-ranked males wasteful. Nevertheless, when alpha males of some species (for example, gorillas, elephant seals, and red deer) are preoccupied with fighting, weaker bystander males can take advantage of the commotion to mate with otherwise inaccessible females. This is actually known among zoologists as the "sneaky-fucker strategy."[20]

But here is the catch. If females come to prefer male gifts to male jousting, then low-ranked males could ultimately outcompete high-ranked males across evolutionary time. To the extent that male success is based on nonphysical characteristics, female preferences modify what counts as being "on top" for males. In such a scenario, male provisioning and female faithfulness can coevolve in a self-reinforcing manner. The Gavrilets model suggests that, ultimately (except for a small fraction of the top-ranked males), males among our hominid ancestors came to secure mates by providing for them, and females evolved very high fidelity to their mates in order to elicit provisioning. As human brains increased in size and as gestation and lactation became costlier, it makes sense that provisioning became an attractive strategy. This shift from brawn to caring also fits our evidence of decreasing physical size and strength dimorphism between males and females, as noted above.

This line of analysis does not predict that females will be completely faithful but rather that the strength of female pair-bonding to male mates will depend on a balance of good genes (potentially supplied by top-ranked males) and better access to food and care (provided mostly by low-ranked males). Once this evolutionary process started, it would lead to a kind of *self-domestication*, as more and more females reproduced with less aggressive males (a topic we will return to in chapter 10). The result would be a group-living species of mostly faithful females pair-bonded to mostly good-provider males.[21] We would be set on the path to the evolution of attachment and love.

Just like human anatomical structures, human social arrangements—such as monogamy and group living—are subject to natural selection acting on our genes. And so our genes can be seen to affect not just our bodies, but also our societies. The transition to pair-bonding in our species was a breakthrough biological adaptation—an epochal shift in our species—and it is still with us today, everywhere, undergirding one of our most central social institutions, marriage.

As we have seen, however, a final distinctive feature of our species is the sheer *variation* in mating behavior, which includes polyandry, polygyny, homosexuality, and celibacy. All members of gorilla species can be said to practice polygyny (males with one partner are merely on their way to assembling their harems, and males without any partners are those who have been shut out by stronger males), but that is clearly not the case in our species. Variation in human mating systems is at least partly a consequence of ecological constraints, but it also reflects the way our social behaviors are pre-wired rather than hardwired, given our highly sophisticated brains. This ability to modify the basic human blueprint, to realize intimate attachments in diverse manners, no doubt also reflects our uniquely advanced, evolutionarily guided capacity for creating and maintaining culture (as we will see in chapter 11). Our many marriage systems reflect the human capacity for social learning, a key part of the social suite. We can flexibly apply a cultural overlay to the deep truth of our loving attachment to our partners.

### *Male and Female Strategies*

As we have seen, male and female mammals generally come at pair-bonding in different ways, some of which can seem annoyingly clichéd. But to recapitulate briefly: In terms of physiological cost, females invest more heavily in egg production than males do in sperm production. Furthermore, eggs are finite, whereas sperm are essentially limitless. This differential investment in the production of gametes is further accentuated over the subsequent course of reproduction, as females invest much more at every stage of rearing offspring. Females are more vulnerable to predation and food shortages while pregnant or nursing, and their lifetime reproductive capacity is limited to the number of babies that they themselves give birth to.

Even among a natural-fertility population such as the Hadza, where no birth control is practiced, a woman cannot produce many more than about a dozen offspring.[22] The number of offspring men can produce ranges up to many hundreds, as among imperial rulers (to the point where such men have been termed *sexual despots*). Genghis Khan and his brothers are estimated to have sired so many offspring that one out of every two hundred men alive today are their descendants (this number rises to nearly one out of every twelve men in parts of Central Asia).[23] In addition, most women capable of having offspring wind up doing so, whereas many men have historically been shut out of reproduction.

Given these differences, female mammals are usually choosier in selecting mates than males are, typically preferring males with better genes or more resources. Male mammals also desire good genes, but they can afford to be more interested in their mates' absolute fertility and perceived ability to raise offspring.

In some animal species in which there is high variation in status among males, both males and females may prefer polygynous unions, and this can set up a self-reinforcing feedback loop across evolutionary time, with more and more status differences in males, more status preferences in females, and rising polygyny. Males at risk of being shut out of reproduction may rationally respond to this state of affairs by taking riskier, potentially even violent, actions to secure a mate and by discounting their futures more steeply (this is sometimes known as the "crazy-bastard strategy," in contrast to the sneaky-fucker strategy noted above). Consequently, ecological (and, in humans, cultural) factors that flatten out status differences among males—meaning that males at the top do not differ very much from males at the bottom—can reshape male behavior by rendering males less violent across evolution.[24]

If this sounds confusing, it's because animals may have idiosyncratic responses to evolutionary challenges and because, if there is one thing we know about humans, it is that our species is extraordinarily variable in terms of behavior. Humans have evolved a flexible mating strategy in both males and females that is based, on the one hand, on long-term pair-bonds and the biological and psychological apparatus to support that and, on the other hand, on short-term copulations. Our evolved psychology reflects *both* these *competing* ancestral strategies based on our species' having faced different ecological and evolutionary pressures in the distant past.[25]

### *Behavior Genetics*

Over the past fifty years, scientists working in the field of behavior genetics have amassed increasing evidence that genes shape human behavior. The study of genetics and heredity began with how genotypes (genes and their variants) shape phenotypes (an organism's physical appearance and function). But phenotypes eventually came to be seen as any manifestations of genes, moving beyond physical appearance to include the way the brain works and, ultimately, human personalities and behaviors. Using a variety of techniques, such as studying twins and analyzing tiny variations in DNA, behavior geneticists have explored whether the genome as a whole and also certain genes in particular help to explain complex phenotypes like neuroticism, decision-making, and friendliness. So broad and powerful is the impact of genes on behavior that, in 2000, psychologist Eric Turkheimer formulated this "first law" of behavior genetics:

"All human behavioral traits are heritable."[26] One extraordinary study combined data from 2,748 publications involving over fourteen million twin pairs and 17,804 human traits. It concluded that virtually every behavioral domain had genetic determinants.[27] Roughly speaking, genes and environment are equally important in determining the extent to which people manifest numerous traits, from religiosity to risk aversion.

Still, figuring out which genes really matter for a given complex behavioral trait (or any phenotype) can be difficult. It can be like trying to learn what makes a car run if you have never seen one before. You might discover that cars without ignition keys do not run and that those with keys do run, but that does not mean that the ignition key determines whether the car runs or not or that the ignition key is the "cause" of the car's operation. There are many parts in a car that must work together to make it move.

Some phenotypes are indeed simple, however. You might remember from high-school biology that different variants of the hemoglobin gene encode the production of different kinds of hemoglobin, and some variants result in sickle-cell disease. This is the classic example of the way genes work, because a single gene encodes a single phenotype—in this case, whether the hemoglobin is normal or sickled. However, the vast majority of human phenotypes are much more complicated. For example, a person's height depends on the actions of dozens of genes working together. Imagine that there is a set of a hundred spinning rollers on a slot machine, all of which have to line up just right for the jackpot but other combinations of which yield smaller payoffs. Each combination of rollers produces a slightly different result, making it extremely hard to predict the outcome each time the machine is played.

The effects of genes are thus hard to discern—because there are many ways genes can influence phenotypes, especially in the case of behaviors. To complicate matters, just as many genes can affect a single phenotype (making that trait polygenic), a single gene can affect many phenotypes—for example, a gene that affects obesity might also affect the ability to process cholesterol. This is known as *pleiotropy*.

Given all this complexity, it is important to understand a bit of shorthand when speaking about genes. When scientists say that there is a gene or genes "for" a trait, it means that *variations* in a genotype correspond to *variations* in a phenotype, not necessarily that a specific gene is solely responsible for a specific phenotype. For example, there is a gene called *DRD4* that contains instructions for making a receptor for the neurotransmitter dopamine. Some studies have (controversially) suggested that people with different variations in this gene (variants of a gene are known as alleles) may be more or less likely to seek out novel experiences, migrate to new places, or develop ADHD.[28] We can then say that this gene is *a* cause (although not the *only* cause) of these behaviors. But that is very different than claiming that there is *one* particular gene "for" ADHD. The variation helps us to understand function (for example, what dopamine does), but all humans have dopamine systems. Another way to understand how genes work is to compare them across species, rather than within them. The appearance of new genes coding for a particular trait in one species can, when compared to another (perhaps similar) species lacking both the genes and the trait, indicate the crucial role of genetics in explaining the trait of interest.

Natural selection is essential to understanding all things human, including our societies. For behavioral traits, just as for physical traits, evolution works by the "survival of the fittest." In each generation, chance mutations mean that an organism's progeny will have small genetic changes that might increase or decrease the likelihood that an individual will survive and reproduce (the changes can also be neutral). This is where natural selection comes in—the environment in which an organism is located will naturally favor some individuals but not others, and those that are favored will have more descendants. The environment acts like an animal breeder, modifying selected traits over the course of a few generations by choosing who gets to reproduce. But evolution does not know where it is going. It has no overarching objective, no endgame. Genes, individuals, and species have no way of predicting what traits will be useful in the future or which traits might set the stage for subsequent modifications. Genes are just a way that biological systems store and transmit information.

It is also hard to discern the impact of specific genes because they work *probabilistically*. In other words, genetic code is not like a simple computer program that runs exactly the same in every computer. There is rarely a one-to-one relationship between genotype and phenotype. Instead, genes yield particular behaviors more than others *on average*—just like smoking cigarettes substantially increases your chance of getting lung cancer *on average* but does not make it a certainty.

The environment in which genes find themselves is not always the same either. Although genetic information is basically unchanging throughout an individual's lifetime, human physical and sociocultural environments are quite variable. As a consequence, genes can have very different effects on individuals depending on the context. Returning to our car analogy, turning the ignition and pressing on the gas pedal will not move a car if it is stuck deep in the mud.

Genes can be affected by their environments at many different levels. The biochemical environment in the nucleus of a cell can influence how genes get translated into proteins. The cellular environment outside the nucleus can influence how those proteins are transported or how they perform their functions. The environment inside the body can affect the availability of substances to interact with those molecules. And even the environment *outside* the body—both physical (like rain or sunshine) and social (like being surrounded by friends or foes)—can modify how genes are expressed or activated. Still, although a gene may lead to one phenotype in one environment and a different phenotype in another, scientists generally first look at the gene and the phenotype to see if there is a relationship. If there is, then they can explore whether and how this relationship depends on measurable environmental factors.[29]

It's increasingly clear that genes and the environment interact to shape our lives as individuals, not just our evolution as a species. When people ask whether human destinies are shaped more by genes or by the world humans live in, they implicitly assume that the world is independent of the genes. But this assumption is disproven in several ways. Genes can lead people to seek out particular environments; for instance, if you feel cold easily because of your genetic heritage, you might choose to live in a warm climate. Genes can even lead people to create and shape environments; for example, when it comes to social lives, genes help determine whether someone chooses to have many or few friends. And, as we will see in chapter 10, an individual's genes are a fundamental part of *other people's* environments, which means that genes in one person can affect outcomes in other people.

### *The Faithful Prairie Vole*

Our review of the emergence of pair-bonding in other animals across evolutionary time demonstrates that natural selection has had *something* to do with human mating behavior (and with group living). But sorting out the actual anatomical, physiological, and genetic bases for this reality is another matter. Digging deeper to explore this topic, investigators like neuroscientist Larry Young have turned to the humble vole. Prairie voles (*Microtus ochrogaster*) are naturally monogamous (or at least most of them are), whereas their closely related cousins meadow voles (*Microtus pennsylvanicus*) and montane voles (*Microtus montanus*) mate promiscuously.

In prairie voles, a number of neurotransmitters, including the hormones oxytocin and vasopressin, are known to regulate pair-bonding.[30] Monogamous prairie voles have higher numbers of receptors for vasopressin underneath the front of the brain (the forebrain) than their polygamous meadow-vole cousins. A similar pattern of vasopressin receptors is seen in monogamous species of mice and marmosets when comparing them to related polygamous species.[31] Moreover, even *within* the same species, some individual voles simply have more receptors, or receptors that are better situated, and this fosters pair-bonding too.[32]

Oxytocin and vasopressin have quite a few functions in mammals. Oxytocin is known to stimulate uterine contractions in childbirth and promote lactation immediately following birth, and it helps regulate a kind of reciprocal feedback loop of affection between mothers and their offspring. The more oxytocin released, the more emotionally connected a mother feels to her offspring and partner. Oxytocin levels are also partially regulated by the behaviors of others (this can even occur across species, as when dogs and their owners experience an increase in their oxytocin levels when they gaze into each other's eyes).[33] Vasopressin plays a role in several stereotypically male functions, including erection and ejaculation, aggression, territoriality, and scent-marking.[34] However, both hormones have roles in both sexes.

During pair-bond formation in voles, it appears that neural pathways involved in partner recognition (smell, for instance) and rewards (such as those related to sex) are activated concurrently with the oxytocin and vasopressin pathways. This leads to an association between the two stimuli and thus the development of a partner preference. This is how and why voles recognize their mates. The lack of vasopressin pathways and receptors in the forebrain of the nonmonogamous species means that the *selective* association of the sexual reward with a partner's odor is not made, and so the animals are unable to form pair-bonds. To be clear, many other hormones and neurotransmitters, like dopamine and opioids (involved in reward systems), also play a role, as do other neural circuits (for example, the pathways related to memory). In fact, some research in humans suggests that different hormonal cocktails may produce different relationship styles in our own species.[35]

We might assume that this hormonal physiology is immutable. But in an extraordinary experiment, Larry Young's group was able to show that a promiscuous species of vole could be rendered essentially monogamous by manipulating the expression of a single gene (*Avpr1a,* which codes for a vasopressin receptor).[36] This gene was transferred from monogamous prairie voles into promiscuous meadow voles in order to increase the expression of vasopressin receptors in the forebrain of the latter. Each genetically altered male vole with his new pair-bonding brain circuits was then subjected to a standard test of affection called the Partner Preference Test (PPT) to determine if he preferred his own partner or a "novel female of comparable stimulus value." The amount of time a vole spent huddling with each of these two females was measured. The genetically transformed voles much preferred their partners; control voles from this promiscuous species did not.

This line of research shows that one or a few genes can control the pair-bonding behavior of a species. An experimental change in the manner of expression of particular genes, like an evolutionary event, can profoundly alter social behavior and render the voles monogamous.[37]

But we should not forget that genes and groups of genes do not work in isolation. The genetic modification in this experiment had its effect in concert with numerous other genes and along with numerous other preexisting neural circuits and biological and socio-ecological factors. The effect that variants of one gene can exert on the action of variants of other genes is termed *epistasis.* If you inherit genes for going gray early in life (as I did), you may not express them if you also inherit genes for going completely bald even earlier (as I did not). In the case of the voles, while the change in the vasopressin-receptor gene was crucial to the effect, it was not the sole *cause* of the effect or even *sufficient* for the effect to occur. The *Avpr1a* gene is not working all by itself, even if it is critical. This is analogous to thinking that turning a key in a car's ignition is all that propels a vehicle forward. Rather, it is the action of the key in concert with all the other functions and structures of the car that makes the vehicle move.

Monogamy is also associated with enhanced parental care, especially among fathers. Research from the lab of evolutionary biologist Hopi Hoekstra involving two closely related species of mice (the deer mouse, *Peromyscus maniculatus,* and the oldfield mouse, *Peromyscus polionotus,* which are promiscuous and monogamous, respectively) has explored the genetic basis of parental behavior and its relationship to monogamy. By crossbreeding these species of mice and by using genetic sequencing, she was able to identify twelve genomic regions involved in parental care (with vasopressin genes again being important). Some acted just in males, some just in females, and others in both sexes. These results suggested that parental care evolved independently and via different genetic routes in the two sexes, even if it is manifested in similar behaviors in both sexes (like huddling and nest building). In short, the experiments indicated that, in these species, there is a genetic basis for parental behavior, that it differs between males and females, and that the attendant changes in genes and behaviors are seen more often in fathers in monogamous species than in promiscuous ones.[38]

Research on the genetics of human pair-bonding (and parenting behavior) is still in its infancy, and human pair-bonding is clearly much more complex than it is in voles or mice. I want to emphasize that the work with voles does not mean that precisely identical processes (neuroanatomically, physiologically, genetically) are occurring in humans—though the processes surely are similar.

Still, despite these caveats, one group of investigators went looking for genes associated with human pair-bonding, using a partner-bonding scale modeled on work with primates. In a study of twins and their spouses involving 2,186 people, they found that a particular variant in the vasopressin-gene receptor, known as allele 334, analogous to the one found in male voles, was associated with decreased pair-bonding behavior in men.[39] When surveyed, men who did not have this particular variant had stronger positive sentiments about their spouses and fewer perceived marital problems. Roughly 15 percent of the men carrying no copies of this allele experienced a marital crisis, whereas 34 percent of the men carrying two copies reported a marital crisis—a doubling of risk. In fact, the *spouses* of men with variants of this gene reported lower marital quality, as if the gene in one person affected the thoughts and feelings of another, a crucial idea regarding inter-individual genetic effects to which we shall return.

There may also be associations between vasopressin-receptor expression and other phenotypes in humans as varied as age at first sexual intercourse, autism, and even, intriguingly, altruism.[40] This further supports the idea that the evolution of pair-bonding was connected, in parallel or in sequence, to the evolution of other social behaviors.

### *Loving Mates After Loving Offspring*

As we have noted, oxytocin is relevant to the brainy parts of reproduction (such as bonding with offspring) as well as the bodily parts. For instance, rats cannot tell their own young from others'. But they do not need to; since rat pups are immobile, the mothers can identify them by their location. However, lambs walk immediately upon birth, and their mothers need to be able to identify them by smell in a large herd that includes others' offspring.[41] Oxytocin is relevant to this process. By giving a sheep oxytocin, one can actually induce it to bond with any particular lamb, even if the lamb is not its own.

These neurological functions of oxytocin appear to have been co-opted by evolution for purposes beyond the care and identification of offspring. Mechanisms associated with mother-infant bonding, shared across all mammalian species, were apparently modified among certain species (including our own) so that females came to feel about their partners as they did about their babies, taking advantage of aspects of sex to establish or maintain a bond. The neural circuits that light up in a woman's brain are similar whether she looks at her baby or her partner.[42]

When a female is giving birth, oxytocin is released to hasten delivery by, among other actions, enhancing the contraction of smooth muscle in the uterine wall. In fact, one of nature's marvelous tricks is that the baby's first sucking at the breast upon birth stimulates a release of oxytocin that can help uterine blood vessels clamp down as the placenta is delivered, staving off a hemorrhage that might kill the mother and thus dramatically lower the baby's own chances of survival. Oxytocin also relieves anxiety. But consider this: stimulation of human breasts, which happens more easily during face-to-face sex (almost nonexistent in other primates), also releases oxytocin. The possibility that, evolutionarily speaking, heterosexual face-to-face intercourse activates maternal-child bonding pathways would help explain why human breasts are large compared to other primates' as well as why they are large outside the period of lactation, which is also unusual.

Scientists speculate that stimulation of the vagina by the penis may prime similar bonding pathways. This in turn may help explain another mystery in evolutionary biology: why human penises are relatively large compared to those of our primate cousins. Gorilla penises, for instance, are only one and a half inches long.[43] Young argues that human penises may have evolved to be larger than necessary because they stimulate the vagina in a way that simulates the birth process. Coupled with the unusual fact that humans often have sex face-to-face, women may register the context in which they are having this hormonal experience and thus associate their sexual encounter with a set of physiological experiences that originally evolved to facilitate mother-child bonding.[44] This oxytocin reflex occurs while imprinting on a particular man, which supports pair-bonding.

The story of the genetic roots of pair-bonding in men is even more speculative. But Larry Young has argued that the bonding and love that men feel for their partners might be an exaptation of the neural pathways that originally evolved in men for the guarding of territory. Male animals often identify, mark, and defend territory, and this behavior requires a host of related adaptations, including the ability to remember the territory and feel attached to it. According to Young, in male brains, females may have become an extension of the concept of territory.[45] To be clear, Young is not even remotely suggesting that this is a conscious link in humans, or that women are possessions, or that this is the only factor at play in the evolution of love and pair-bonding in men. Still, the extent to which men's sense of connection to women is tied to territorial feelings is illustrated by certain aspects of violence in our species; wars have often been fought over women in small-scale societies, and propaganda posters encouraging warfare even in the modern era have invoked the trope of defending women. The mass rapes committed by Soviet troops in Germany in World War II and by the Pakistani army during the 1971 war of Bangladeshi independence might also be seen as grotesque perversions of this evolutionary conflation of male territoriality and sexual behavior.

Evolutionary explanations of human behavior are hard to pin down and can easily become what scientists call just-so stories. We do not yet know which, if any, of these theories best explains the physiology of human pair-bonding. But it's clear that many intricate genetic, hormonal, and anatomical changes have come together over the course of human evolution to shape the way we think, feel, and act regarding sex and love, providing a foundation for our cultural and moral practices to build on.

### *Partner Choice and Natural Selection*

Our genes play a role not only in our general sense of attachment to a partner, but also in the specific partners we choose. From the point of view of natural selection, this should not be surprising, since the choice of partners affects reproductive success and the type of genetic material each generation bequeaths to the next one. The fact that genes have been shown to play a role in partner choice further supports the claim that evolution has shaped this key part of the social suite.

Human spouses generally resemble each other in many respects, including attractiveness, health, religion, politics, and so on. Some of this similarity comes from the widely observed tendency of spouses to convert to each other's religions or adopt their food tastes. But most of this similarity comes about due to homogamy, or the tendency of like to marry like in the first place, which is known as assortative mating. Assortative mating can occur for traits that are mutable (such as religion) or immutable (such as a person's height or ethnicity).[46] Confusingly, humans sometimes manifest a tendency toward the practice of "opposites attract" with respect to certain traits as well.

A century ago, evolutionary biologists and statisticians Ronald A. Fisher and Sewall Wright separately proposed the idea that if spouses resembled each other superficially (that is, phenotypically), they would also resemble each other genetically.[47] Since most phenotypes arise from the actions of many genes working together, assortative (or disassortative) mating could result in genetic correlations between spouses at thousands of locations within the genome.

Evolution might actually have shaped people's preference to mate with others who are similar to them. There are two ways this could happen. First, the good-genes hypothesis posits that animals evolved to choose mates with versions of a gene or genes (alleles) that increase their Darwinian fitness.[48] Every individual is seeking mates with the "best" genes. In monogamous species with mutual mate choice, those with the "better" allelic variants of genes can choose to mate with each other, leaving those with the "worse" alleles to similarly assort with their own type.[49] Imagine that muscle strength is advantageous and is thus preferred in a mate. Strong people would therefore prefer to mate with other strong people, since it would maximize the fitness of their offspring. This would leave frail people to mate with other frail people. Both strong people and frail people would prefer strong mates, but, since choice is mutual, assortative mating by muscle strength would arise. We would observe a similarity in genes between partners as a result of such a process.

A second underlying method of mate choice is the genetic-similarity theory, which posits that people prefer genetically *similar* individuals as optimal mates.[50] Under this hypothesis, similarity in any given gene in a mate pair might confer an advantage that is not available to a solitary individual.[51] Imagine that there is no benefit to mating with strong people but that there is an efficiency benefit (and a fitness advantage) to reproducing with people of the *same* muscle strength. If you have genes that lead you to like certain foods or feel cold at certain temperatures, these challenges might be more efficiently confronted if you mate with an individual who has similar traits; the two of you can search for the same food or avoid arguments about the bedroom thermostat.[52]

There is, however, another way that genes might play a role in mate choice; it's known as the compatible-genes hypothesis, and it involves *disassortative* mating. Individuals might choose mates who have gene variants that increase fitness when paired with *contrasting* variants of those genes. This theory relates to the idea of heterozygote advantage. For instance, the blended state of genes for both tall and short stature might be optimal. Maybe being tall attracts the attention of predators, and being short is inefficient for hunting, so medium height, as dictated by a mix of genes, is best.

It would be absurd to suggest that partner choice is solely, or even primarily, driven by genetics or our evolutionary past. We cannot discount the importance of one's own conscious desires, interests, and prejudices honed both by culture and life experience. But there are nevertheless a set of inherited processes that influence those choices. Scientists have evidence for all three of the above mechanisms in diverse species, including in humans. Assortative mating has been well described for specific traits such as body size, personality, and other apparent features in many animals.[53] Disassortative mating has also been described, including for some nonapparent traits such as immune function.[54]

Let's consider an area where evolution may have shaped humans to be disassortative in partner choice. So-called human leukocyte antigen (HLA) proteins are on the surface of immune cells and are instrumental in combating infections; they also appear to play a role in odor, kin detection, and pregnancy. It is advantageous for an immune system to have a great variety of such proteins in order to combat a large and diverse array of pathogens. It is therefore best to be heterozygous for any given HLA gene; this way, the two copies of the gene code for two different variants of the resulting protein.

If close relatives mate, it is more likely that both copies of various HLA genes in their offspring will be the same, or homozygous, which is suboptimal. Consequently, it is to an individual's advantage to reproduce with someone who is not kin and who therefore has a greater likelihood of carrying dissimilar variants of all the HLA genes. And humans may somehow have the ability to choose partners who have different HLA genes.[55]

How is this possible when there is no readily visible indicator of someone's HLA genes? One way humans do this is by odor. Many animals use olfactory signals to choose mates with dissimilar HLA genes (known as *MHC genes* in nonhuman animals). There is some evidence in humans too, and this was first tested experimentally in 1995.[56] Forty-nine women were asked to rate the body odor of forty-four men by blindly smelling T-shirts that the men wore to bed for two consecutive nights. On average, women found the odors of men with dissimilar HLA genes to be more pleasant. The smell of dissimilar men's T-shirts was also more likely to remind them of their most recent mate's smell.[57] A few subsequent studies have found similar results. I emphasize that choosing mates based on odor arising as a result of HLA plays only a small part in partner choice, but there does appear to be some important role of biology here.[58]

As a result of such a process working behind the scenes, we would expect the HLA genotypes of couples to be somewhat dissimilar.[59] For a variety of methodological reasons, however, this is difficult to ascertain. For example, people tend to choose partners of similar ethnic background, so they would tend to have similar HLA gene variants just based on this shared heritage. To control for this possibility, we need to focus on unions *within* ethnic groups and see if they are similar or dissimilar beyond any background resemblance. One study of this kind in a Hutterite community found evidence of disassortative mating on five HLA genes, but another study of an isolated Amazonian group found that the extent of dissimilarity in HLA genes in couples was *not* different than what would arise from random mating.[60]

Preferences for particular body odors in partners might not relate solely to the HLA system. Smells might communicate information about partner quality and suitability for other reasons. And another pathway for odor preferences could be a kind of imprinting, whereby people prefer certain odors based on their exposure to their parents, especially those of the opposite sex.[61] This might provide one mechanism by which evolution shaped reproduction, supporting the choice of similar, not just dissimilar, mates. Since genes play a role in human attitudes (such as aggressiveness and risk aversion) and behaviors (such as alcoholism, altruism, and wanderlust), and since couples with similar values tend to be happier, it may be that partners with similarity in such genes might be more reproductively successful.[62] If people can somehow become aware of the genotypes of others via olfaction or other cues, then this might be a mechanism by which assortativity is sustained in our species. Women in particular often rate smell as a very important physical attribute in men, sometimes even more important than appearance.[63]

Political scientists Rose McDermott, Dustin Tingley, and Peter Hatemi have shown that assortative mating based on political ideology could operate partly through olfactory cues. They found that, to a small but discernible extent, people prefer the smell of mates with a similar political orientation! In this experiment, one hundred and twenty-five people blindly rated the odors of twenty-one targets who were chosen because they were on the extreme ends of the political spectrum. The targets provided body-odor samples (gauze pads that were worn in their armpits for twenty-four hours), and subjects smelled these and rated them in terms of how attractive the odor was. The subjects expressed a noticeable preference for the smell of the opposite sex, but, astoundingly, they also preferred the odor of people with a similar political ideology.

Some of the qualitative comments from this experiment are startling:

> [One] participant asked the experimenter if she could take one of the vials home with her because she thought it was "the best perfume I ever smelled"; the vial was from a male who shared an ideology similar to the evaluator. She was preceded by another respondent with an ideology opposite to the person who provided the exact same sample; this [previous] participant reported that the vial had "gone rancid" and suggested it needed to be replaced.[64]

To be clear, there are no genes (or odors) for membership in any particular political party. However, scientists believe that there are genetic predilections for more fundamental and biologically relevant traits that might affect survival; for example, one's "conservative" tendency to obey one's parents or to respect authority. In the modern world, such genes might come to play a role in party affiliations and political beliefs by affecting someone's interest in novelty, tradition, privacy, and the like. Preferential attraction to the smell of people with similar political beliefs would be a remnant of an ancient process designed to maximize reproductive success.

In my lab, using data from 1,683 unrelated heterosexual spousal pairs drawn from two populations and examining over one million genetic loci, we explored whether genotypic assortative or disassortative mating occurs across the whole genome rather than just with HLA genes.[65] We also constructed a comparison set of individuals artificially partnered in our data (that is, pairs of heterosexual strangers), which allowed us to measure the degree to which assortativity differed from random mating within the same population. We found hundreds of genetic loci across the genome that exhibited more assortativity or disassortativity than would be expected due to chance.[66]

We then did some further analyses to explore the possible role of assortativity in human evolution. We measured whether regions of the human genome that show a tendency to be similar across spouses have been evolving faster over the past thirty thousand years, as we would expect if assortative mating conferred fitness advantages. And we found that loci exhibiting even moderate assortativity among spousal pairs were evolving faster than loci exhibiting no assortative mating or those exhibiting disassortative mating. In other words, something about assortative mating may enhance the fitness of humans and thus increase the prevalence of the relevant gene variants.

Our analysis showed that, overall, people in our sample chose partners from the population at large who were genetically equivalent to fourth cousins (even though their partners were not *actually* related to them). A different study in a large Icelandic population also found that the greatest number of offspring was seen in couples who were related at the third- to fourth-cousin level. Couples who were more closely related (such as first cousins) had fewer surviving offspring, as did couples with too little similarity.[67]

These and other studies support the idea that genes affect an individual's attraction to, and choice of, particular partners. Not just anyone will do. It's likely that genes conspire to play a role in bringing people together because the partner you select affects your own chances of survival, the chances of survival of your offspring, and ultimately the chances that your genes are passed along.

### *Paving the Way for More Complex Social Behaviors*

Monogamy can be seen as a kind of fundamental egalitarianism: everyone has the same number of partners. And once pair-bonding and the related processes we have reviewed evolved in humans, a number of subsequent evolutionary developments in our species' social life were facilitated.

Pair-bonding likely served as a preadaptation for a style of parenting based on the sharing of effort to raise offspring and the division of labor between male and female mates. A *preadaptation* is a trait that eventually facilitates a different purpose than the one for which it evolved (for instance, limb-like fins in certain species of fish served as a preadaptation to terrestrial life). Joint parenting is very useful in humans, given the relatively high costs of raising children, who are born with larger brains and delayed maturity compared to other primates. Pair-bonding between parents also allowed offspring to reliably recognize their fathers (because they stayed around during their youth). And, symmetrically, this allowed fathers to reliably recognize their offspring. This too was a radical evolutionary development, and it is uncommon in other animals.

Mutual recognition and sustained attachment paved the way for a new type of family structure that involved individuals of both sexes and multiple generations. The recognition of kin in these larger groups facilitated the further evolution of within-group cooperation, including behaviors like alloparental care, in which adults cooperate and invest in caring for children who are not their own. All this set the stage for more equal relations and cooperation within groups more generally, since many members of hominid groups were likely genetically or reproductively related and also co-located.

Eventually, however, once cooperation became a part of the human repertoire, the necessity of being genetically related to others within one's group may have declined. Most of the social connections afforded within bands of foragers (such as the Hadza) are not among kin at all, and genetic studies suggest this arrangement in human groups has been in place for at least thirty thousand years.[68]

Indeed, a distinctive feature of our species' social organization, as compared to that of other primates, is that humans live with many unrelated individuals. Technically, humans live in multi-male, multi-female groups, and since humans are pair-bonded to their mates, strictly speaking, these are multifamily groups. Moreover, unlike those of other primates, human families do not have to stay just with maternal or paternal relatives but can move from one group to another in a so-called multi-local residence pattern. The origins of these features of our living arrangements are complex, but one pathway is that pair-bonding and joint investment in child-rearing by both parents resulted in greater equality between the sexes, in particular with respect to residential decision-making. Both mothers and fathers could choose—perhaps at different times—to live with their own kin. With many members of every group exercising this choice across time, the result would be a set of well-mixed groups of mostly non-kin. In short, the low within-camp relatedness seen in hunter-gatherer camps emerged naturally from men and women seeking to spend time with their own kin.[69] Thus, pair-bonding and joint child-rearing set the stage for cooperation and friendship with unrelated individuals.

Within such groups of mostly non-kin, people could come to have unrelated friends, as we will see in chapters 8 and 9. The circle of sentiment and attachment could expand. Consider food sharing, an important type of cooperation in human groups. In order to share food with others—not just eat it together on the spot where it was acquired—one must be able to carry it from place to place; and so the *gathering* of food for the purpose of sharing probably coevolved with bipedalism, which freed the hands to carry food back to one's partner and progeny.[70] When this behavior emerged against a background of pair-bonded primates, sharing more broadly with others could then follow. Excess food was given to or taken by other nearby individuals who were typically unrelated. Moreover, the preexisting practice of co-feeding among primates was another preadaptation; individuals were happy to eat near one another.[71]

In essence, the emergence of pair-bonding and proto-families formed a nucleus for the development of broader features related to group living and the emergence of other aspects of the social suite. Moving outward from a circle of attachment and affection to partners, offspring, and kin, we proceed to our attachment and affection for our friends and our groups.

## CHAPTER 7

# Animal Friends

Just as our capacity for love has antecedents and parallels in other species, so does our primal, unshakable capacity for friendship. Primatologist Jane Goodall provided one of the most extraordinary illustrations. After scrambling for hours in the dense forest to keep up with David Greybeard, the first chimpanzee ever to befriend her, Goodall describes their haunting encounter:

> Sometimes, I am sure, he waited for me... for when I emerged, panting and torn from a mass of thorny undergrowth, I often found him sitting, looking back in my direction; when I appeared, he got up and plodded on again.
>
> One day, as I sat near him at the bank of a tiny trickle of crystal-clear water, I saw a ripe red palm nut lying on the ground. I picked it up and held it out to him in my open palm. He turned his head away. When I moved my hand closer he looked at it, and then at me, and then he took the fruit, and at the same time held my hand firmly and gently with his own. As I sat motionless, he released his hand, looked down at the nut, and dropped it to the ground.
>
> At that moment there was no need of any scientific knowledge to understand his communication of reassurance. The soft pressure of his fingers spoke to me not through my intellect

> but through a more primitive emotional channel: the barrier of untold centuries which has grown up during the separate evolution of man and chimpanzee was, for those few seconds, broken down. It was a reward far beyond my greatest hopes.[1]

The desire for social connection is so strong that it can even reach beyond species prone to making friends.[2]

Consider the relationships that the brilliant scientist Nikola Tesla formed with pigeons. Tesla arranged his living quarters specifically to attract pigeons; he had little bird beds scattered across his desk, birdseed by the windowsill, and a window always left ajar. At one point, bird traffic through his apartment in the Hotel St. Regis in New York City became such a nuisance that the management told him he had to stop his pigeon feeding or leave. Tesla chose the latter.[3]

Although Tesla lived a life of celibacy and near solitude, his interactions with pigeons filled an emotional gap and a craving for connection, as he explained in a 1929 interview:

> Sometimes I feel that by not marrying I made too great a sacrifice to my work, so I have decided to lavish all the affection of a man no longer young on the feathery tribe. I am satisfied if anything I do will live for posterity. But to care for those homeless, hungry, or sick birds is the delight of my life. It is my only means of playing.[4]

Tesla poignantly described his misery when his favorite bird died: "Something went out of my life. I knew my life's work was finished."[5] And he died a few months later, at age eighty-six. Tesla's grief and the fact that he may have experienced an increased risk of death after the death of his pet is consistent with the phenomenon of "dying of a broken heart" widely seen in human couples (and in some animal species too).[6] Human attachments—even to pets—are so fundamental and beneficial that they can improve people's health or, when lost, sometimes cause their deaths.

### *Human-Animal Bonds*

Approximately two-thirds of American households have pets, spending over sixty billion dollars a year on their care.[7] But they pay their owners back. Pets often affect people in the same ways that their human friends do. In one experiment examining the human stress response, researchers measured how long study participants could keep a hand submerged in ice water before withdrawing it in pain. The presence of a pet during the experiment was found to be almost as good at decreasing the physiological stress as the presence of a spouse or a friend.[8] More generally, the more social connections someone has, the higher one's tolerance for pain—and even a connection to animals will do.[9]

Pets also strengthen relationships among the humans who share them. As anyone with a dog in the family knows, they are a constant source of shared interest, humor, and stories, and they do not carry the pressures or expectations of many other discussion topics. The presence of pets encourages human interactions and possibly even increases empathy. Children with autism benefit from play therapy with guinea pigs, and wounded veterans see a reduction in psychological symptoms from equine therapy.[10] At Serenity Park, a bird sanctuary in Los Angeles, American veterans suffering from addiction, mental illness, and aftereffects of trauma have formed intense bonds with abandoned parrots, macaws, and cockatoos.[11]

Iconic pairings of humans and their pets stretch back centuries, from Odysseus and his dog, Argos, to little Fern Zuckerman and her famous pig, Wilbur. A young Alexander the Great, in a key event of his upbringing, defied his father's instructions and attempted to subdue the steed Bucephalus. Noticing that the horse feared its own shadow, Alexander shepherded the animal out of the sun and consoled it. It was through an act of comfort and support, not violence or domination, that their relationship began, and they remained inseparable.[12] Many of the First Pets of American presidents have become household names. Franklin Delano Roosevelt's Scottish terrier, Fala, attended international conferences alongside the president. In remarks that historians and political commentators later branded the Fala speech, Roosevelt famously impugned Republicans for making "libelous statements about my dog."[13] Fala was buried beside Roosevelt and given a statue in the FDR Memorial in Washington, DC.

Our affection for animals and kindness to them says a lot about the human capacity for love, friendship, and altruism, and I would argue that this ability to connect with other living things is a marker of our own humanity. I will return to this claim, but first I'd like to focus on the animals. Our pets—birds, dogs, horses—are often very social beings and seem especially capable of reciprocating our attempts to connect with them. But, even more than the species we live with, it is wild animals such as chimpanzees, elephants, and whales that afford deep insights into our social selves. When humans form friendships and thus assemble the social networks that are so much bigger than family trees, they are behaving in a way that has instructive precedent in these other species and that has, accordingly, been shaped by natural selection.

In fact, elephants and whales, independently by convergent evolution, have come to resemble us in our capacity for friendship. Convergent evolution, as we have seen, occurs when unrelated species evolve the same features via completely separate evolutionary paths, like birds and bats both evolving the capacity for flight or octopuses and humans evolving eyes that are structurally similar. Such similarities indicate that these features (flight, vision, friendship) are so useful and so tailored to an environmental opportunity that they appear to be inevitable. Moreover, the existence of animal societies reinforces the centrality of various aspects of our own societies. By seeing what we have in common with animals, we humans can better recognize what we have in common with one another.

### *Making Friends for Science*

Developing relationships with animals can help scientists understand the building blocks of a social species. Creative and costly procedures are typically required for researchers to observe animals in their natural habitats and insert themselves into their groups. In fact, one of the reasons scientists have taken so long to confirm what animal enthusiasts have long known is that it is often exceptionally difficult for humans to interact naturally with animals in the wild while also remaining safe and unobtrusive.

The challenges are especially great with animals that fly or swim. Some scientists have dived deep to track the social interactions of aquatic mammals. Others, like activist and meteorologist Christian Moullec, have taken to the air, in one instance flying with flocks of geese in a tiny, slow-moving, open aircraft twelve hundred miles from Sweden to Germany in an effort to understand their behavior.[14]

But scientific progress in understanding the social lives of animals began on the ground, in particular with chimpanzees. Jane Goodall was just twenty-six when she arrived with her mother in July of 1960 on the east shore of Lake Tanganyika. Her campsite, reachable only by boat, was nestled within the thick forests of the Gombe Stream National Park in Tanzania. Although she would eventually become one of the world's most respected primatologists, Goodall held no college degree, carried few supplies, and had no experience living in remote forests. She was, however, very driven. She lived among the chimpanzees until 1986, and over the course of those twenty-six years, Goodall built close relationships with approximately fifty members of the species, even titling her first book *My Friends the Wild Chimpanzees.*

During her first few weeks in Gombe, Goodall caught only fleeting glimpses of the primates before they scuttled off into the dense foliage. As she recalled, "As the weeks passed, the chimpanzees continued to flee and I often despaired."[15] All of this changed, however, after she crossed paths with David Greybeard, as she called him, the large male chimpanzee whom she befriended (and who, more important, befriended her). Goodall first observed Greybeard near her campsite for nearly an hour using blades of grass as tools to pluck termites from their nests (this observation of a primate using tools was in itself remarkable).[16] Greybeard began frequenting Goodall's camp on a nearly daily basis.

Years later, she described what made him special:

> Well, first of all, he was the very first chimpanzee who let me come close, who lost his fear. And he helped introduce me to this magic world out in the forest. The other chimps would see David sitting there, not running away, and so gradually they'd think, "Well, she can't be so scary, after all." He had a wonderful, gentle disposition. He was really loved by other chimps; the low-ranking ones would go to him for protection. He wasn't terribly high-ranking, but he had a very high-ranking friend, Goliath. And there was just something about him. He had a very handsome face, his eyes wide apart, and this beautiful gray beard.[17]

Greybeard continued his visits to Goodall. While he seemed to enjoy being in her general vicinity, he never initiated any direct contact. This all changed one day when Goodall reached over and groomed him "for at least a minute."[18] The relationship subsequently deepened, with Goodall and Greybeard not only interacting but also communicating with each other through their behavior.

Goodall, contemplating his inevitable death, wrote: "It will be a sad day too when my first chimpanzee acquaintance, David Greybeard, is no more. To me, he is not just a chimpanzee—he is, quite truly, a friend. My close contact with him led to the establishment of a bond between man and ape, a bond based on mutual trust and respect—in a sense, a friendship."[19] Later, Goodall wrote that, when David Greybeard died of pneumonia in 1968, "I mourned for him as I have for no other chimpanzee."[20]

Among Goodall's greatest contributions to the field of ethology was her avant-garde method of observing and interacting with her animal subjects (figure 7.1). Where other early primatologists assigned their research subjects sterile numbers, Goodall gave them names. Where others explained animal behavior in terms of simple behavioral laws, Goodall allowed room for complexity and individual personality. Above all, where others observed their subjects from afar, often avoiding interaction, Goodall welcomed and even sought interaction, embedding herself within the chimpanzee troop.

**Figure 7.1: Jane Goodall Grooming a Chimp**

Failures and frustration marred her first year in Gombe, but eventually, she was able to wander peacefully among the chimpanzees, mimicking their sounds and behaviors.[21] Her awareness of the potential triggers of aggressive behavior helped ensure safer and more amicable interspecies relationships. Beyond her knowledge of individual animals and their personalities, Goodall began to decode the broad social dynamics of chimpanzee communities. Anyone who has seen the famous pictures of Goodall grooming the chimps instantly grasps the extent to which she was successful in befriending them and understanding their social interactions.[22]

### *Friendships Between Animals*

A key challenge for ethologists evaluating animal behavior with respect to friendship is pinning down what counts as a friend.[23] One simple approach is to calculate the association index of two animals, which is based on the amount of time they spend together. If, over the course of a week of intermittent observation, I note that you and your spouse are together for four hours, that you are alone for six hours, and that your spouse is alone for ten hours, then the association index is 4/(4+6+10) = 0.20. In other words, the two of you spend roughly 20 percent of your time together. We can also compare this number across pairs of animals. This strategy resembles one of the most widely used methods of ascertaining friendships in human groups, which is to determine who people spend their free time with, as we saw in the study of crews at the Antarctic station.

Many animal friends are also relatives—often siblings, maternal aunts, cousins, or even grandmothers. In fact, among nonhuman animals, especially long-lived ones, a good predictor of persistent friendship bonds is matrilineal kinship (being related on the mother's side). This is true even among chimpanzees and dolphins, where females tend to leave the groups they were born in, making it harder to maintain such ties.[24] Nevertheless, observational field studies document that, especially when maternal kin are unavailable, animals such as chimpanzees and baboons form at least one enduring friendship with an *unrelated* individual.[25] And relationships between unrelated male dolphins can last for decades.[26]

Increasing evidence suggests that natural selection may have favored the creation of social bonds in general, not just social bonds to kin. Although humans are unusual in the extent of our friendships to non-kin, we are not alone. Let's look at primates, elephants, and whales to see how they make friends in order to set the stage for a closer examination of how and why humans do.

### *Primate Friendships*

Primates have bodies that originally evolved for life in trees; they have forward-facing eyes that afford good depth perception and very flexible arms, legs, and digits. The more than two hundred living primate species are divided into two suborders: the prosimians (for example, lemurs) and the more intelligent anthropoids. The anthropoids are in turn divided into monkeys and apes, and hominids are a kind of ape.

On the evolutionary line leading to humans, old-world monkeys split off about twenty million years before apes did. Monkeys thus more closely resemble other nonprimate mammals; for instance, while most monkey species have tails, no apes (or hominids) do. The nonhuman apes (gorillas, chimpanzees, bonobos, orangutans, and gibbons) are closer to humans, with the same basic body structure, substantial intelligence, and similar behaviors (such as tool use). They may also have distinct and repeated patterns of learned behavior specific to local groups, which is a fancy way of saying that they may have culture. Gorillas, chimpanzees, and orangutans have the capacity for some form of language, and individuals from all three species have been taught sign language and then gone on to invent their own words.[27] Of course, humans (and their hominid ancestors), whose bipedalism left their hands free to use tools and carry food, have even larger brains and more advanced intelligence, language, and culture.

One of the reasons it is so exciting to study other primates is that we can recognize ourselves in them—our emotions, our cognition, our morality, and our social lives, as we saw in our consideration of pair-bonding. But, of course, the species vary. Chimpanzees are relatively egalitarian; gorillas are more despotic. Male baboons disperse; female chimpanzees roam. Gibbons establish pair-bonds; bonobos do not. As with all animals, many of these differences relate to the ecological circumstances faced by the species over the course of evolution.

Since anatomical and behavioral traits vary across primate species, it's not surprising to see differences in social organization too. Even the typical group size varies across species, ranging from four to thirty-five, with a median size of nine individuals. Within these groups of different size, we can also quantify the extent of social interactions through grooming or play. In a group of four individuals, for example, there are six possible ties connecting pairs of individuals (4 x 3/2 = 6). If all of these ties are observed, we would say that the group has 100 percent *density* of its ties and that the network is fully saturated. In one study of thirty primate species with groups of various sizes, the density averaged 75 percent, but it ranged from 49 percent to 93 percent across the species.[28]

With such high density coupled with generally small group size in primates, from the point of view of each individual, all other individuals in the group are either friends (one hop) or friends of friends (two hops) away in the network. Nevertheless, as group size increases, it becomes inevitable that some individuals will be more connected than others. The existence of such popular hubs is a key feature of primate social organization. Finally, the strength of ties, not only their number, varies across individuals; in no species do all individuals bestow their attention equally on all members of their group.

Chimpanzees live in populations as big as one hundred and sixty, but they tend to spend most of their time in smaller, flexible parties of around ten individuals, splitting up and reforming in what is known as fission-fusion dynamics. Males stay with their natal group and females disperse to join other groups, typically around age eleven. Primatologist John Mitani has documented long-term bonds between unrelated males in a population of about one hundred and fifty chimpanzees in Uganda by examining who the chimps shared meat with, who they groomed, and who participated in border patrols together (figure 7.2).[29] As happens in humans, bonds were more likely to form among relatives. But bonds between non-kin pairs of males were actually more numerous. Three-quarters of the males formed their closest and most enduring friendships with unrelated individuals, with some friendships lasting a decade or more.[30]

**Figure 7.2: Male Chimpanzees in Ngogo, Uganda, Engaging in Social Behaviors**

Three behaviors of male chimpanzees are shown from left to right: grooming, sharing meat, and patrolling the territorial border.

Female chimps generally spend the majority of their lives, perhaps forty of their fifty years, living among non-kin. A detailed ten-year study of a group of nineteen adult female chimpanzees in Côte d'Ivoire found that, while they are less likely to form friendships than males, the friendships they do form may be stronger, and their friends are less likely to be kin than males' friends are. The majority of females (84 percent) had at least one preferred friendship lasting several years and ending only when the friend died or disappeared (see figure 7.3).[31] For instance, a chimp named Fos had three close friends—Cas, Gom, and

**Figure 7.3: Friendships in Adult Female Chimpanzees**

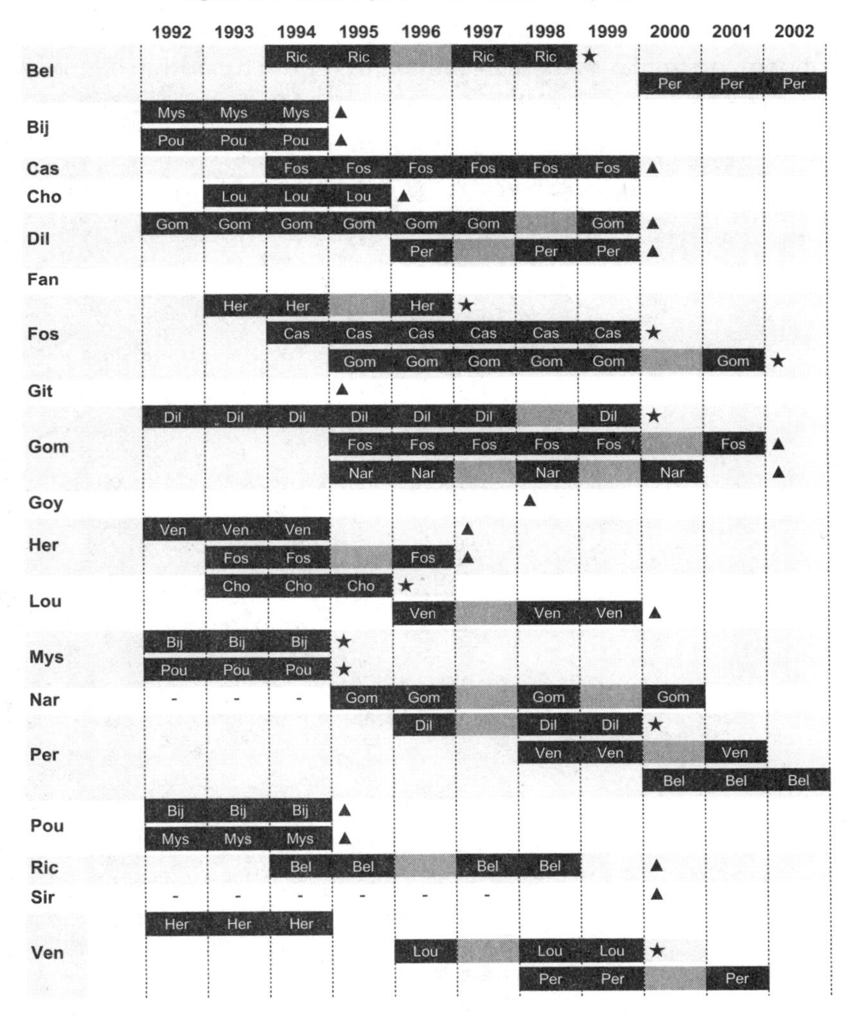


The beginning and end of primary friendships for nineteen female chimpanzees (rows) are shown. Three-letter names indicate individual adult females, and black bars indicate friendships in the respective years. Each chimpanzee can have one or more friends across time. For instance, Gom is friends with Dil, Fos, and Nar. But Goy and Sir have no observed friendships. Dark gray segments indicate years during which pairs of friends did not necessarily spend a lot of time together. A star indicates the death or emigration of the friend; a triangle indicates the death or emigration of the individual chimpanzee.

Her. She lost all three of them to death. The social network arising as a result of these friendships is shown in figure 7.4. Friendship patterns in baboons are similar to those in chimpanzees, and friendship bonds are even stronger among bonobos than among chimpanzees.[32]

**Figure 7.4: Network of Female Chimpanzee Friendships**

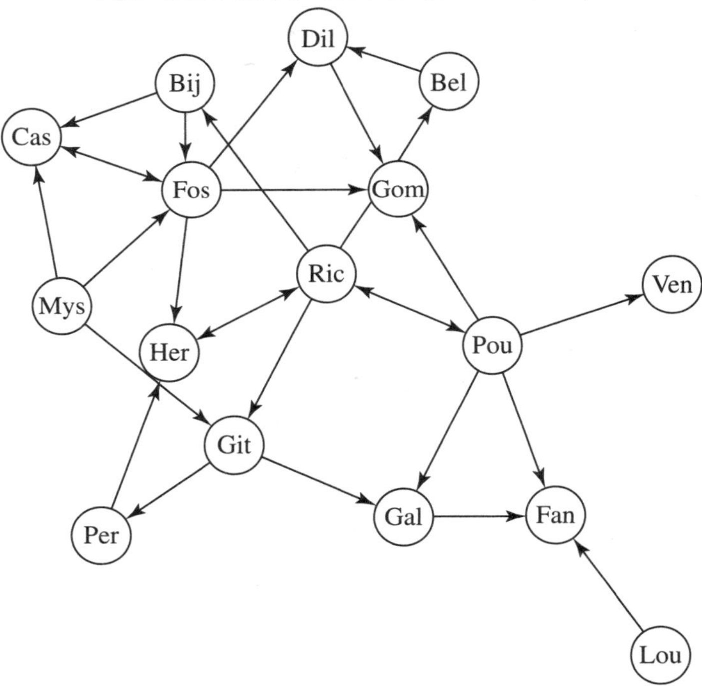


A network of sixteen female chimpanzees arising from sustained friendships between pairs of chimps. Ric is centrally located and has many inbound and outbound friendships (e.g., chimps who groom her or whom she grooms, as indicated by the direction of the arrowheads). Lou is on the periphery and has one friend, Fan.

### *Popular Individuals and Their Popular Friends*

Through their choice of friends, individual chimps assemble themselves into social networks. But individuals come to occupy different positions within the network that they collectively make, as we noted above with respect to hubs. A kind of leadership can arise from having many social ties and being in the middle of the network, as can be seen by comparing Fos and Ric to Ven and Lou in figure 7.4.

Anyone who has watched children on a playground knows that not all individuals are equally capable of making friends. Some desire friendship more than others or are better able to act on this desire. Some are better able to attract friends too. Social mammals are *status seeking*, that is, interested in befriending powerful, appealing, or popular individuals. Those desirable partners tend to be connected to other desirable partners because they get to choose who they are friends with. Partly as a result, less popular individuals wind up being friends with less popular individuals. This type of sorting results in a status-based society in which individuals associate with others who have similar social status and characteristics.

In social-network analysis, this is called *degree assortativity*. The word *degree* here refers to the *number* of social connections, and the word *assortativity* refers to preferential sorting that connects individuals who resemble one another. The tendency of individuals with similar numbers of ties to be connected is a general feature of the social networks of primates, which includes humans. One analysis of primate species (chimpanzees, baboons, bonnet macaques, wedge-capped capuchin monkeys, and others) found that sixty-eight of seventy groups showed degree assortativity.[33]

What's particularly remarkable about this tendency is that it is an atypical mathematical property of networks. In networks of neurons, genes, or computers, for example, we find just the opposite—popular nodes in these networks tend to be connected to many unpopular nodes. This is called *degree disassortativity*. Airport hubs like Chicago and Denver are connected to all the large airports, but most of their connections are to numerous smaller airports, and small airports are preferentially connected to hubs, not other little airports. You cannot fly directly from New Haven, Connecticut, to Lebanon, New Hampshire.

We might assume that popular individuals in leadership positions foster this degree-assortativity hierarchy for purely selfish reasons, but in fact leaders can function as a kind of social police. A little bit of hierarchy can level the playing field, creating greater opportunities for all members of the group to cooperate, coordinate on beneficial activities, and survive.

Evolutionary biologist Jessica Flack and her colleagues manipulated the network structure of a group of eighty-four pigtailed macaques at the Yerkes National Primate Research Center near Lawrenceville, Georgia.[34] They first measured connections based on whom the monkeys groomed or played with. Group leaders were identified by counting the number of times others silently bared their teeth at them during peaceful moments, an act of deference in this species, resembling a human smile (see figure 7.5). The scientists then strategically removed the highest-ranked individuals (they were *knocked out* of the group, as the scientists termed it) and then compared the resulting social networks to an unperturbed control condition.

When the high-ranking individuals were knocked out, chaos ensued. Conflict and aggression skyrocketed. This analysis sheds light on the impact of leaders on interactions within the group as a whole. First, after the leaders were removed, the group had fewer grooming and play interactions overall (this is visible by comparing panel 2c to 2a and 2f to 2d in color plate 2). That is, the remaining macaques became less connected to one another. This suggests that stable leadership promotes peaceful interactions not only between leaders and followers, but also between followers and other followers. The existence of popular leaders seems to facilitate social order throughout the group.[35] And given that these interactions create opportunities for beneficial mutual support, it is easy to see why nat-

Figure 7.5: Macaque "Smiling"

ural selection might favor the evolution of an individual interest in, and respect for, hierarchy.

When the group leaders were removed, those at the edges and the center of the networks jockeyed for power, which resulted in a lot of conflict. When the leaders were in place, however, there was less conflict and more connectivity between high-ranking and low-ranking monkeys. The degree assortativity was actually lower because the leaders affirmatively intervened in conflict between status-seekers and regulated social connections. Equally important, individuals adapted to the presence of the leader and knew that intervention was likely if conflict arose. The presence of leaders therefore allowed socially isolated followers greater access to the more popular members of the group without threat of reprisal from other status-seekers. This study showed that leaders pushed against this type of status seeking. Finally, without social hubs in the form of leaders, the macaques not only had fewer friends, but also fewer friends of friends. Hence, removing leaders made it harder for information and behaviors to spread via these indirect connections. This applied to both good and bad behaviors, but another feature of leadership was that the leaders tended to affirmatively prevent the spread of bad behaviors such as violence.

It's likely that natural selection played a role in the near-universal emergence of degree assortativity. But besides reducing conflict, what else is beneficial about this arrangement? One possibility is that degree assortativity slows the spread of infectious disease. Mathematically, having people connect to one another according to popularity reduces the risk of epidemics. The best way to grasp this point is to imagine that the network was degree *disassortative*, like an airport network mentioned above. Here, we would have a hub-and-spoke system, with one or two very popular individuals connected to all of the others in the group. In this situation, if any individual in the network became infected, the disease could reach *all* the others in the group in just two hops—from the initial individual directly to the hub individual, then through that hub to all the others. The principle applies in the converse too. Organizing the same set of individuals with degree *assortativity* generally helps to keep outbreaks more confined—for instance, on the periphery of the social network.[36] This collective immunity is wholly distinct from individual-level immunity, meaning the strength of each person's immune system.

The way friendships are organized within a group really does matter to the group as a whole *and* to every individual in it. To return to our diamond and graphite metaphor, a different arrangement of the same people yields very different properties. Most animals, including humans, would clearly benefit from a social organization that minimized the prevalence of disease and discord. Our social networks provide just such benefits, and these properties are seen universally in human populations around the world, and not just across primate species.

***Elephant Friendships***

It should not be too surprising to find precursors or analogues to our social behavior and friendships in other primates, given our phylogenetic relationship to them, but elephants also form some of the most significant and long-lasting friendships in the animal kingdom—a capacity that they, like David Greybeard, can also extend to humans.

Ethologist Joyce Poole spent years observing African elephant societies. Once, after a prolonged absence, she eagerly returned to them, accompanied by her family and her new baby daughter, Selengei:

> The elephants were less than two meters away from us when Vee stopped dead in her tracks and, with her mouth wide, gave a deep and loud, throaty rumble. The rest of the family rushed to her side, gathering next to our window, and with their trunks outstretched, deafened us with a cacophony of rumbles, trumpets, and screams until our bodies vibrated with the sound. They pressed against one another, urinating and defecating, their faces streaming with the fresh black stain of temporal gland secretions.
>
> Who can know what goes on in the hearts and minds of elephants but the elephants themselves? What we had experienced was an intense greeting ceremony usually reserved only for family...members who have been separated for a long time. I could only guess that I had been remembered and now, after all the time away, I had returned with other new, but familiar, smells: those of my brother, my mother, and my tiny daughter, held out to them in my arms.[37]

Much of what we know about the social organization of elephants comes from decades-long studies of two wild species, African elephants

(*Loxodonta africana*) and Asian elephants (*Elephas maximus*). One population is a group of over twelve hundred African elephants residing in Amboseli National Park in Kenya; they have been studied since 1972.[38] Another population of African elephants is under study in Samburu National Reserve in Kenya. The Asian elephants include a population of roughly one thousand elephants residing in Udawalawe National Park in Sri Lanka.[39]

Like many primates, elephants express emotion, empathy, and altruism. They often provide assistance to others who are prostrate or injured or facing predation or threats. Elephants will assist not only members of their own kind, but also, sometimes, injured humans—just like humans might care for injured animals. As Poole explains:

> Elephants seem to have categories in which they classify other animals. They have a particular dislike for species that are either predators or scavengers, even if those animals are not a threat to elephants or are scavengers that are not feeding on an elephant carcass. For example, if an elephant comes across a group of jackals and vultures feeding on a zebra carcass, typically it chases them off or at least shakes its head in what I would characterize as annoyance.[40]

Poole and other researchers have observed countless displays of compassion and friendship:

> On a number of occasions, I have watched elephants standing on either side of an immobilized elephant, propping it up between them.... Vladimir [an elephant] had contracted some kind of disease that had left him almost crippled... [and] several young males seemed to be looking after him. Albert [another elephant] was seen with him day after day, walking

> along at Vladimir's slow pace, accompanying him to the swamp, and then another young male took up the job. It was as if they understood that Vladimir was sick and needed help.[41]

Poole recognizes the likely evolutionary explanation for such friendly and altruistic behavior: elephants may be hardwired to assist members of their clan simply to increase their own genes' chances of survival. "But if that is the case," she writes, "why do elephants assist injured elephants *and* injured humans, but not, as far as we know, other species?"[42] That point suggests a set of evolutionarily guided abilities that transcend simple kin selection. Elephants seem to form opinions and make judgments about altruism. They recognize when a human is actually in *need of help.*

It could be argued that elephants' altruism and compassion does not come from any conscious empathy or understanding but rather from an instinct that simply became overgeneralized and was extended to other species. In our own species, we see a similar emotional extension that can prompt a child to tenderly anthropomorphize a captured ladybug, decorating its container so it will not be bored and refusing to turn it out into the cold. We also see this in other social animals, such as dolphins, who have famously been known to aid humans in distress.[43] Nonetheless, the overapplication of an evolved tendency to perform good deeds should hardly, in my view, be taken as proof of *imperfect* social inklings.

Elephant society is sex-segregated, matriarchal, and multitiered (meaning smaller groups are nested in larger ones). Males leave their natal groups and spend much of their time either solitary or in small groups of other adult males, like Vladimir and Albert above, joining female groups during periods of sexual activity. Male elephants form small cliques that have dense interactions within their own group and few interactions between groups. This type of organization—isolated, roaming males and stable groups of females—is notably different than the societies of primate species (though, interestingly, elephants share this structure with whales).

Core groups of females and periodically visiting, solitary males constitutes an ancient social structure. Seven-million-year-old fossilized tracks in the desert of what is now the United Arab Emirates indicate that this was already a distinctive elephant pattern in the Miocene epoch (see color plate 3). The tracks clearly show the movements of fourteen elephants; thirteen are females and their young walking together in a herd, stepping into one another's paths, and one is a large male whose solitary wanderings intersect theirs.[44] Even the size of this ancient group is the same as modern herds. This pattern of social organization also appears to result in a much larger fraction of preserved mammoth fossils from the past sixty thousand years being male rather than female. Moving about on their own and lacking the guidance of matriarchs with useful knowledge, solitary male mammoths more often wandered into dangerous places like crevices and sinkholes; moreover, they had no companions to help save them once they got into a fix.[45]

Each matrilineal family group (also known as a core group) typically numbers around ten and consists of closely related females and their offspring.[46] Members of family groups do almost everything—migrating, eating, drinking, and resting—together, often in unison. For instance, Poole would routinely sing "Amazing Grace" to her favorite elephant, Virginia, who, along with her entire family, would stop to listen for up to ten minutes at a time, "slowly opening and closing her amber eyes and moving the tip of her trunk."[47] Members of these groups often touch and smell one another in comforting and greeting actions. Family groups wait together while young calves rest during migrations, and, if a calf makes a distress call, several members of the group will vocalize and rush to the calf's aid. In fact, adults will sometimes even discipline calves that are not their own, and calves can suckle from mature females who are not their mothers. Elephants engage in the cooperative alloparental care we see in human aunts, older sisters, and grandmothers.

Scientists define a coefficient of relatedness, *r*, to describe the genetic relatedness of two individual animals; it is roughly the fraction of gene variants they have in common (ordinarily because they share a common ancestor). Identical twins have an *r* of 1.0 and wholly unrelated individuals have an *r* closer to 0. Partly because most females remain with their natal group, the average pairwise genetic relatedness among elephant core group members is about 0.2, which is around the level of the relationship of an aunt and niece.[48] Naturally occurring human groups of similar size, as in the case of the Aché foragers of Paraguay, have a value of 0.05 (roughly the relatedness level of second cousins).[49] Female elephants prefer to associate with those to whom they are genetically related, and both observational and genetic data suggest that most of the stronger ties are among matrilineal kin.[50] But elephants also form friendships with entirely unrelated individuals, just as chimpanzees and humans do, highlighting the fact that social interactions and assistance are not solely between kin.

Heartbreaking levels of poaching in recent decades have created social disarray, especially when matriarchs are slaughtered and can no longer guide the community. Consequently, roughly a fifth of the core groups in the Samburu Reserve in Kenya are composed of individuals who are not significantly related to each other. This depressing "natural" experiment suggests that while elephants may prefer the company of their kin, their social organization does not necessarily require it.[51]

Because elephant groups break up and reunite very frequently—for instance, in response to variation in food availability—reunions are more important in elephant society than among primates. And the species has evolved elaborate greeting behaviors, the form of which reflects the strength of the social bond between the individuals (much like how you might merely shake hands with a longstanding acquaintance but hug a close friend you have not seen in a while, and maybe even tear up). Elephants may greet each other simply by reaching their trunks into each other's mouths, possibly equivalent to a human peck on the cheek. However, after long absences, members of family and bond groups greet one another with incredibly theatrical displays, like the moving ceremony Joyce Poole described earlier.[52] The fact that the intensity reflects the duration of the separation as well as the level of intimacy suggests that elephants have a sense of time as well. To human eyes, these greetings (minus the urination and defecation Poole mentioned) strike a familiar chord. I'm reminded of the joyous reunions so visible in the arrivals area of an international airport terminal.

In Amboseli, the elephants live in fifty-five core groups, some as large as seventeen females and offspring. Among these African savanna elephants, groups periodically associate to form higher-level groupings known as tiers. At the lowest tier, level one, there is a single female and her own dependent calves. A group of related females may form the family group or core group at level two, the kind that we have been discussing so far. Multiple level-two groups may associate to form what is known as a bond group at level three, typically involving substantial kinship relationships. A bond group may coordinate its foraging activity over an area of several square miles for many weeks. Individual elephants may belong to the same core and bond groups for decades, in part because of the underlying kinship ties these groups represent.[53] Bond groups may associate at the clan-group level, level four, which also appears to serve an important function.[54]

Evolutionarily speaking, what advantages could have led elephants to evolve the cognitive capacity for a hierarchical organization like this, especially one that includes the clan level? As we have seen, the benefits at lower social tiers may include predator defense, foraging efficiency, and alloparental care, all of which are apparent up to the core and bond group, but those behaviors are very uncommon at the clan level.[55] Some scientists hypothesize that elephants make the effort to unite at such a large level in order to share information and improve learning opportunities.[56] This may give elephants a chance to exchange and preserve a kind of culture. Another explanation is that it helps elephants identify suitable mates. The capacity for, and respect of, such a form of social organization likely has some benefits.

However, it's also possible that the clan level has no explicit purpose and that it instead represents a kind of "runaway" propensity to social behavior (like people who get overexcited about their upcoming high-school reunions). Given the rough calculus of evolution, perhaps being overly social is better than being too antisocial. There might not be any evolutionary *disadvantages* to large clan gatherings.

In humans, the purpose of higher-order levels is more apparent. We have military and trade alliances between villages, tribes, and nations, for example, although we clearly did not evolve to cope with social complexity on a national level. Thousands of people contribute to the cooperative maintenance of the online encyclopedia Wikipedia, not because cooperation on such a large scale was something our species needed in its evolutionary past, but because it's a runaway form of cooperation originally relevant to the sharing of information and food in much smaller groups.

As always, the environment also affects social organization. In Sri Lanka, Asian elephants (a species that diverged from their African brethren six million years ago) live in more heavily forested areas with more consistent food supplies, and they have no natural (nonhuman) predators. Asian elephants form smaller groups than African elephants and prefer maternal kin to an even greater extent. They also may not form tiered social groupings quite like African

**Figure 7.6: Snapshots of Elephant Social-Network Fragmentation**

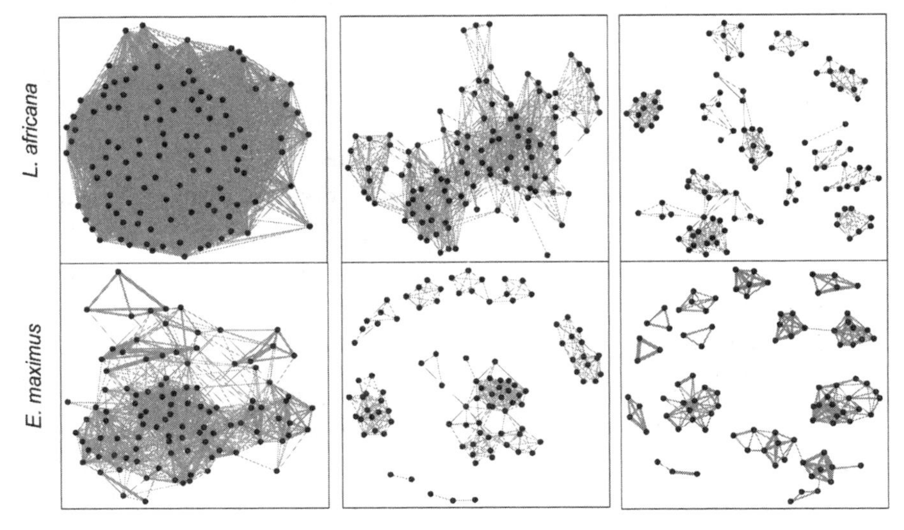


Elephant social networks are shown. Each dot (node) is an elephant and the lines represent relationships of varying intensity. The Asian-elephant network (bottom row) is less interconnected than the African-elephant network (top row). The network also contains fewer strong ties and thus fragments more quickly when ties are sequentially removed by increasing the association-index threshold for what actually counts as a social tie. This is shown moving from left to right by increasing the association index (i.e., requiring more time together before indicating a tie).

elephants do.[57] Mapping of the social networks of both species (see figure 7.6) illustrates these phenomena. African elephants have more direct and interconnected social contacts than Asian elephants, and any given elephant is no more than two steps from any other, an arrangement that is conducive to the maintenance of cultural knowledge. By contrast, some Asian elephants are up to four steps away from other elephants in the same group (composed of about one hundred individuals).

Still, if you compare the upper-left panel of color plate 2 to the mid-upper panel of figure 7.6, you will notice the similarity in the elephant and primate networks. Human social networks are also similar, as we will see in chapter 8. Once the concept of friendship takes root in a species, there might be just one basic way to become optimally socially organized in this regard.

### *Cetacean Friendships*

Whenever I feel overwhelmed by how hard it is to collect data about human networks, I think of whale researchers Hal Whitehead and David Lusseau and their colleagues. They collect data similar to mine, but they have to do it underwater and in frigid and dangerous conditions. And I think of the old line about Ginger Rogers, that she did everything Fred Astaire did except backward and in high heels.

Elephants are enormous terrestrial herbivores, and sperm whales are enormous aquatic carnivores, yet similarities abound. Female sperm whales form stable units of perhaps twelve genetically related members. Whales from the same unit will often stop diving in the afternoons and spend a few hours together, touching one another in what scientists term *caresses* and uttering characteristic vocalizations called *codas*. Again like elephants, two or more family units of sperm whales may travel together and forage for a time in a coordinated fashion, and they use sophisticated acoustic signals to coordinate their movements and activities over many miles. Adults in a family unit will deliberately stagger their deep dives to minimize the amount of time calves stay unattended near the surface, a form of communal child care.[58] Adult whales also attempt to bear up injured family members. In fact, hunters have historically exploited this tendency by deliberately injuring animals, especially young ones, and then killing the older females who came to help. When a whale is harpooned, others have been seen trying deliberately to bite the cable.

As is often the case with animal ethology, much less is known about males' social organization. Like male elephants, male whales live relatively solitary lives, either alone or in small bachelor groups, roaming far from their ancestral family units except at breeding time.

Also like elephants and many primates, whales live in complex environments that they can fill to capacity. All these species tend to be long-lived and to care intensively for a small number of offspring. These two phenomena are related in that older members of the species can be valuable past their reproductive years as repositories of ecological and social knowledge. For instance, menopausal whales are known to show leadership in the search for food when food abundance is low.[59] Humans may also owe their longevity to the requirements of social learning. The ability to care for younger kin and to transmit knowledge to them and to others in the group makes those who are no longer reproducing still useful.

Other cetaceans, such as orcas and bottlenose dolphins, also form long-standing friendships with unrelated members of their species. Ongoing work with dolphins is revealing the sophistication of their social networks, learning, tool use, and culture. Dolphin networks look like those of apes and elephants and have similar mathematical properties. Measurement of social interactions over many years among sixty-four adult bottlenose dolphins in Doubtful Sound, New Zealand, yielded the network shown in figure 7.7. On average, each dolphin was connected to five other dolphins, a value similar to humans'. The transitivity of dolphin groups is also very similar to humans'; on average, any two of a given dolphin's friends had a 30 percent chance of also being friends with each other (human values typically range from 0.15 to 0.40, depending on the population).[60] Moreover, dolphins in this network varied in their centrality, with some serving as hubs (the most connected dolphin had twelve connections).[61] Lasting friendships with nonrelatives were common.[62] For example, within dolphin cliques of ten to twenty adults found in a population of bottlenose dolphins in Port Stephen, Australia, the average pairwise genetic relatedness was on the order of 0.06 to 0.09 (roughly between first and second cousins), again similar to elephants and human forager groups.[63]

In most species, unrelated males and females rarely form meaningful ties devoid of immediate sexual implications. But we do sometimes find this type of platonic cross-sex interaction in dolphins, pilot whales, orcas, bonobos, baboons, and, of course, humans. Among the Port Stephen dolphins, females may preview mating partners who are connected via friendship ties to their male relatives. This is not so different from an older sibling vetting a teen's dating choices. And useful information can certainly be transmitted between pairs of animals whatever their sex. Orca whales also share many of these features, including gathering in large groups, apparently to promote social learning or mate-finding.[64]

**Figure 7.7: Network of Dolphin Friendships**

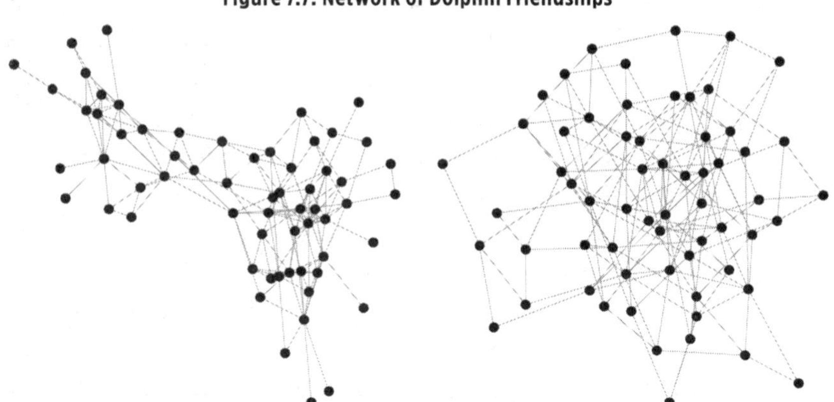

A real network (left) with sixty-four dolphins, wherein each animal is connected to five others, on average, and a random network (right) constructed with the same total number of nodes and ties, to which it can be visually and mathematically compared. In the real network, a few dolphins have a large number of ties (and are social hubs) and others have only one or two ties.

### *Anthropomorphosis*

In our quest to understand the human blueprint, the conversation about traits like friendship, cooperation, and social learning in other animals can quickly devolve into one of semantics. Is such a trait really present, or are we simply anthropomorphizing? Do chimpanzees really console or greet one another? Are elephants really happy to see a friend? Do whales really cooperate to care for their young? The same issue comes up when we speak of animal personalities, societies, or cultures. Are these phenomena objective or are they projections of our own minds, like the way people see a human face in a rock formation?

Some critics argue—wrongly, in my view—that the term *friendship* in animal settings reflects the experience of human observers more than the experience of animal subjects or the actual function of animal connections. Some argue that the whole concept of animal friendship is mistaken because even primates do not have a concept of the future. They do not have expectations for future social interactions with specific others and cannot comprehend a future time at which reciprocation by a friend would be needed or desired. They also do not possess a sophisticated understanding of the concept of *friend,* let alone the ability to declare (in some explicit way) to one another that they are friends.[65]

But are these capacities really required for friendship? That is a long list of caveats and judgments we place on our animal brethren, one that we would not dream of foisting on those humans who are similarly incapable of answering questions verbally, such as toddlers, the developmentally disabled, or the memory-impaired. We see virtually all humans as capable of friendship, even those who have only a tenuous grasp on the idea of the future or who do not have an explicit understanding of friendship. It seems to me that people's squeamishness about animal friendship reflects a certain degree of arrogance.

Still, this is not just a conceptual problem. The methodological issues are also significant. Researchers cannot simply ask animals if they are friends. They must instead rely on other ways of identifying nonreproductive social connections. Bonds among animals can be hard to pin down because they occur over time and because they can involve diverse categories of behavior (do they groom each other, share food, form defensive coalitions, or simply co-appear at the same place and time?). This has led to what psychologist David Premack refers to as the "Russian-novel problem"—that is, when looking at the history of two animals (or humans), it is often hard to know what caused what, since the two may have interacted in so many ways over such a long time, and they may also remember events differently and act according to this subjective experience.[66]

Along with primatologists Robert Seyfarth and Dorothy Cheney, I believe that our reluctance to acknowledge the rich social lives of animals is mistaken.[67] In my view, people's desire to distance themselves from animals may be driven by shame at how some members of our species abuse them. As social critic Matthew Scully argued in his haunting book *Dominion*:

> It is easy to look down upon animals as utterly alien to us, driven on by need and instinct in their grubbier, less rational way, slavering for food and attention like our pets, jostling at the trough on our farms, battling one another in the wild over mates and territory and status in their group. But the person who thinks himself entirely above and apart from this world need only take a closer look at his or her own daily existence, at the struggle and hurts and yearnings of body that still mark each and every human life.[68]

Just like us, animals serve philosopher Jeremy Bentham's "two sovereign masters," pleasure and pain. We humans are not so special.

It's actually not difficult to dispense with critics' reservations. It is true that we appear to be the only species that says "good-bye" and that other species seem to lack any special leave-taking rituals.[69] But we are not the only species that says "hello" and that has specific greeting behaviors. Indeed, anticipation of the future is not absolutely necessary for the formation and maintenance of friendship ties. Only the cognitive ability to remember the *past* is needed. Even if the ability of other animals to anticipate the future is unclear, there is no doubt that many species have memory and even a sense of time. I can see this in my own home, given how our dachshund, Rudy, will greet my wife's return to our house mildly when she has just gone out to fetch the mail but with berserk enthusiasm when she comes back after having been away all afternoon. We saw this in the case of the variable elephant greetings that were contingent on the duration of separation as well.

Second, the impetus to form friendships within various animal species can be seen as a kind of implicit knowledge or instinct that influences their behavior. Everything from the caching or singing behavior of birds to—in our species—the capacity of six-month-olds to form moral judgments and the ability of eight-month-olds to avoid crawling off a cliff can be understood as a kind of knowledge that its possessor cannot explicitly describe.[70] Animals can have friendship instincts just as humans do.

Additional powerful evidence of animal friendship comes from so-called third-party knowledge of the relationships between *other* pairs of animals. That is, animal A knows about its own relationship to animal B and its own relationship to animal C, but it also knows something about the relationship between B and C. In its most evolved sense, seen in ourselves and our closest primate relations, this recognition reflects what is termed a theory of mind, meaning an individual's ability to imagine the mental states of others and recognize that those states may be different from one's own.

In fact, some species have evolved not only a theory of mind but also, distinctly, a theory of relationships—which is evolutionarily advantageous, because recognizing relationships between other individuals helps predict their social behavior. The most basic type of such knowledge is when one animal knows the relative dominance rank of two other animals, not just its own rank with respect to the others. This important ability is widespread, seen in hyenas, lions, horses, dolphins, and, of course, primates, but also in fish and birds. Capuchin monkeys in conflict preferentially seek out allies that they know to be higher ranked than their opponents, and they also seek out allies that they know have closer relationships with themselves than with their opponents.[71] If two chimpanzees have a fight and a bystander offers consolation to the loser, this can reconcile the two combatants, but only if the bystander has a friendship with the aggressor.[72] All three animals understand what it means for two of them to have a special bond.

The requirement of group cooperation may actually be the ultimate evolutionary impetus for friendship. Animals in the wild can help one another through the exchange of goods (such as food) and services (such as grooming, support, and protection) over time. In the lab, unfortunately, experiments are typically more transactional, with goods used as a medium of immediate exchange ("Solve this puzzle and I'll give you a banana"), so we may not have a full picture of animal friendships in natural settings. Nevertheless, we know that many animals, such as chimpanzees, engage in contingent cooperation—immediate, tit-for-tat reciprocity—exchanging both goods and services with particular individuals. This explanation of animal friendships is known as the current-needs hypothesis.

Still, this cannot account for the existence of long-term, stable friendships. Chimpanzees that groom each other today are much more likely to form a protective alliance in a few weeks and be sharing food in a few years. That's hard to show in a lab. Primatologists theorize that emotions may serve as a kind of cognitive accounting for such exchanges over longer-term interactions in the wild. We feel good about other people in proportion to how nice they are to us, and we keep balance sheets of those sentiments and remember past interactions. This longer perspective allows for short-term imbalances, moving beyond mere "What have you done for me lately?" exchanges. This also facilitates the larger alliances common in primate societies.

Clearly, we could define the term *friendship* sufficiently narrowly so that it applied only to humans. But this is scientifically uninformative and sets up a separation between humans and other animals that obscures more than it clarifies. Speaking of animals as having friendships does not strike me as anthropomorphic at all. Other scientists have also taken up this banner. Primatologist Joan Silk argues:

> Friendship is the F-word; a word that many primatologists have been reluctant to use in print though we may use it freely when we chat with our colleagues about animals that we study. When we do use the term in academic venues, we feel compelled to cloak it in italics, as if this gives us some indemnity against charges of anthropomorphosis or lack of rigor.[73]

There has been a growing backlash in the past few years against what seems to be an overwhelming interest in negative aspects of animal societies—"red in tooth and claw"—with an extraordinary amount of attention lavished on topics like competition, conflict, manipulation, coercion, deception, and even (in other anthropomorphic terms) kidnapping, rape, murder, and cannibalism. But love, friendship, altruism, cooperation, and teaching surely also deserve their due. These are the traits at the core of the social suite, the ones that make it possible for humans to be a successful species and to keep the darkness at bay. Understanding this in other animals sheds light on the evolutionary origins and purpose of these

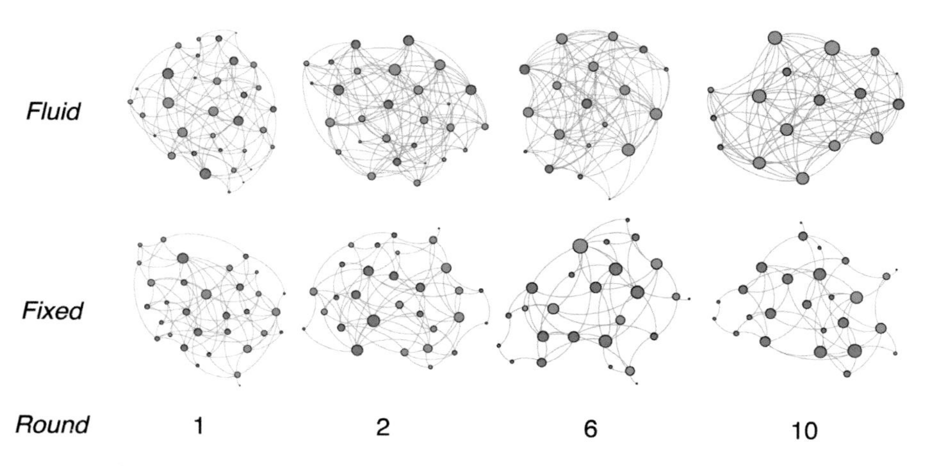


**Plate 1:** Manipulating the Rules of Social Interaction Affects Cooperation in Groups

The images show snapshots at four rounds (out of ten sequential rounds) of a network cooperation game involving two groups of roughly thirty people in a set of experiments. Blue nodes represent people who choose to be kind and cooperative; red nodes represent people who choose to be exploitative or uncooperative (and who defect). Node size is proportional to number of connections a subject has. In the fixed condition (bottom row), people are stuck with their neighbors (assigned at the first round by the experimenters) and cooperation declines because the subjects have no alternative but to defect when their neighbors take advantage of them (by not cooperating). Note that subjects with many connections are mostly defectors (since it's especially hard for well-connected people to stay cooperative in the face of exploitation). At the end of the game, there are just a few holdouts who continue to cooperate (on the far right), sticking together on the edge of the group. By contrast, in the fluid condition (top row), where subjects can also choose with whom to connect at each round (in addition to choosing whether to cooperate or defect), cooperation is higher (there are more blue nodes at the end). Moreover, cooperators eventually come to have more connections than defectors, as they are sought-after partners. By specifying the rules governing social connections, we can make a group of people mean or kind to one another.

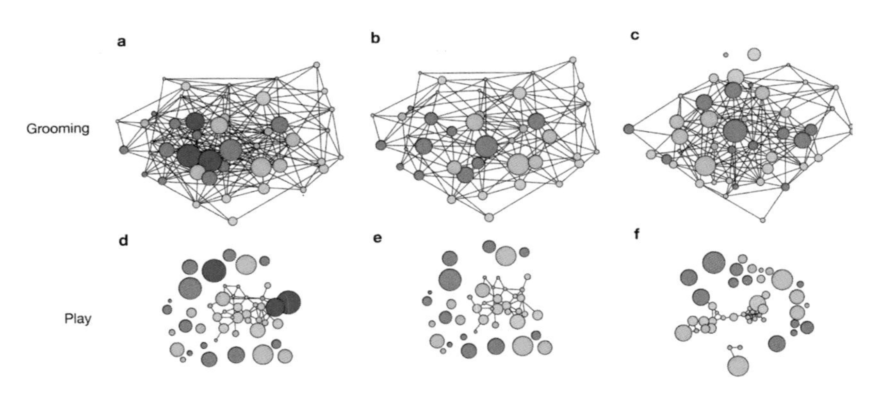


**Plate 2:** Changes in the Structure of Social Networks in Pigtailed Macaques When Leaders Are Removed

Each node is a macaque, with colors denoting roles: purple for leaders, bright pink for alpha females, red for matriarchs, light pink for other adults, and gray for socially mature but not fully grown individuals. Lines in the top graphs indicate grooming relations, and in the bottom graphs they indicate play relations. The left column is the original condition. The far-right column is the experimental condition in which the leaders (three in this case) were actually physically removed (knocked out) from their groups and then the network of social interactions was measured again by observation. The middle column is the original network with the leaders removed simply from the network map and the network redrawn; this is what the network of ties would look like if the three leaders and their connections were removed from the data, but not in reality. These networks look more like the original ones than the knockout networks, which means that the actual removal was more destabilizing than merely the loss of the three individuals and their ties.

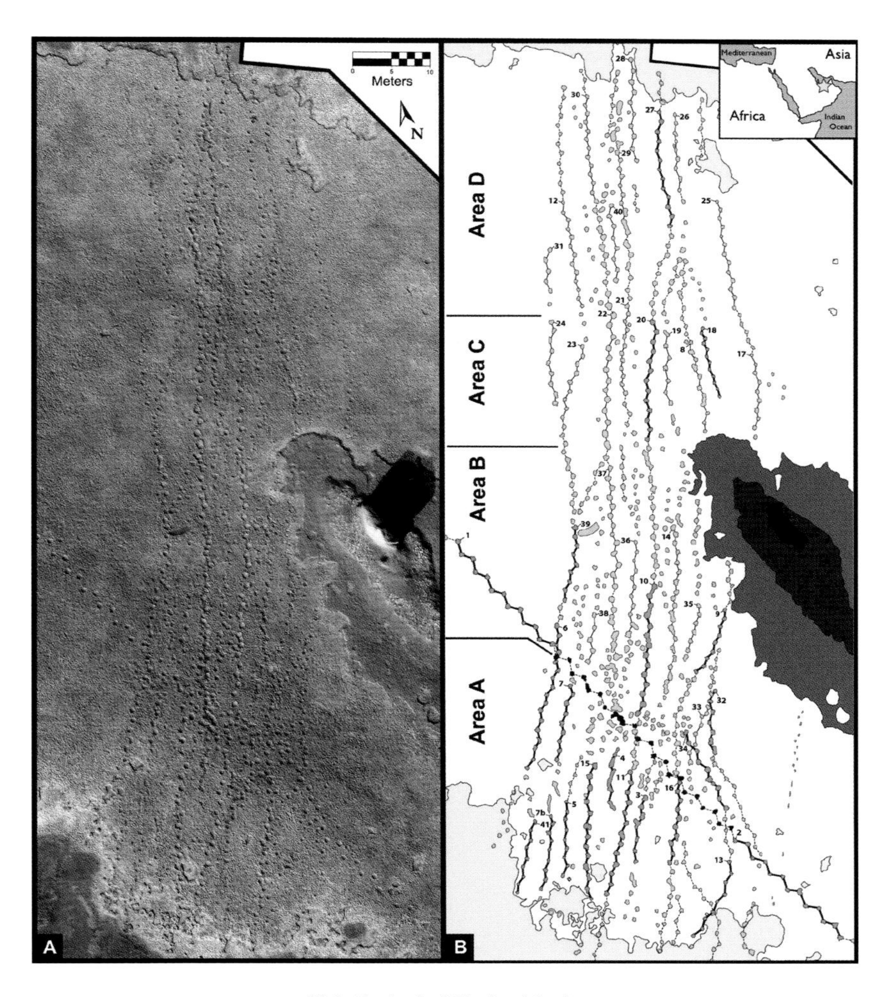


**Plate 3:** Ancient Elephant Tracks

The left panel shows an aerial photo of seven-million-year-old fossilized elephant tracks over a 260-meter distance in the desert in Mleisa, U.A.E., and the right panel shows a diagram in which paths of particular individuals in the female group are indicated; the path of the solitary male can be seen crossing the other tracks diagonally. The social organization reflected in these ancient tracks is the same as that found in elephants today.

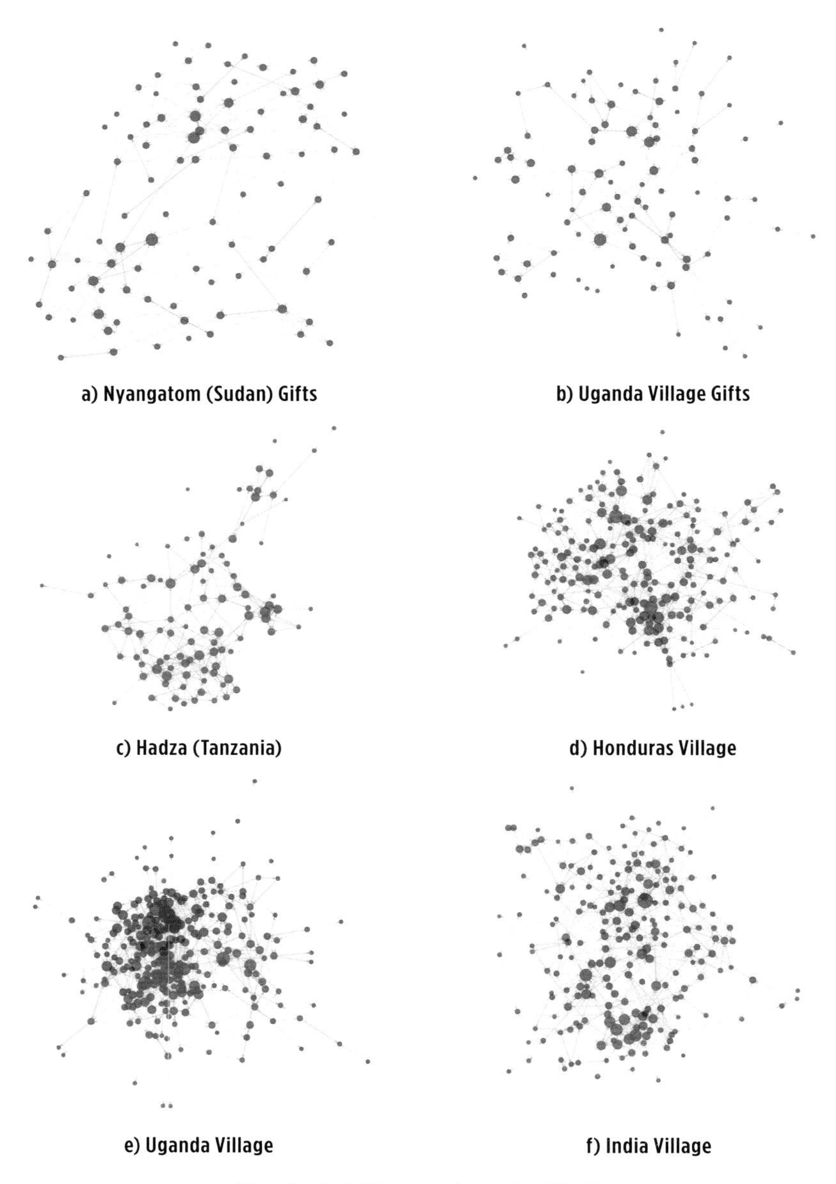


**Plate 4:** Social Networks Around the World

Social networks involving face-to-face interactions in small groups and villages that we have mapped from countries around the world are shown. Blue nodes are men and red nodes are women. Circle size indicates the number of connections each person has (bigger nodes mean the person has more social connections). Orange lines indicate close familial relationships (parents, children, spouses, and siblings); gray lines indicate all other sorts of relationships, principally to unrelated friends. These networks, which range in size from 91 to 261 people, show remarkable structural consistency around the world (but also some interesting differences, e.g., in gender segregation and in the presence of cliques of especially interconnected people within the broader network).

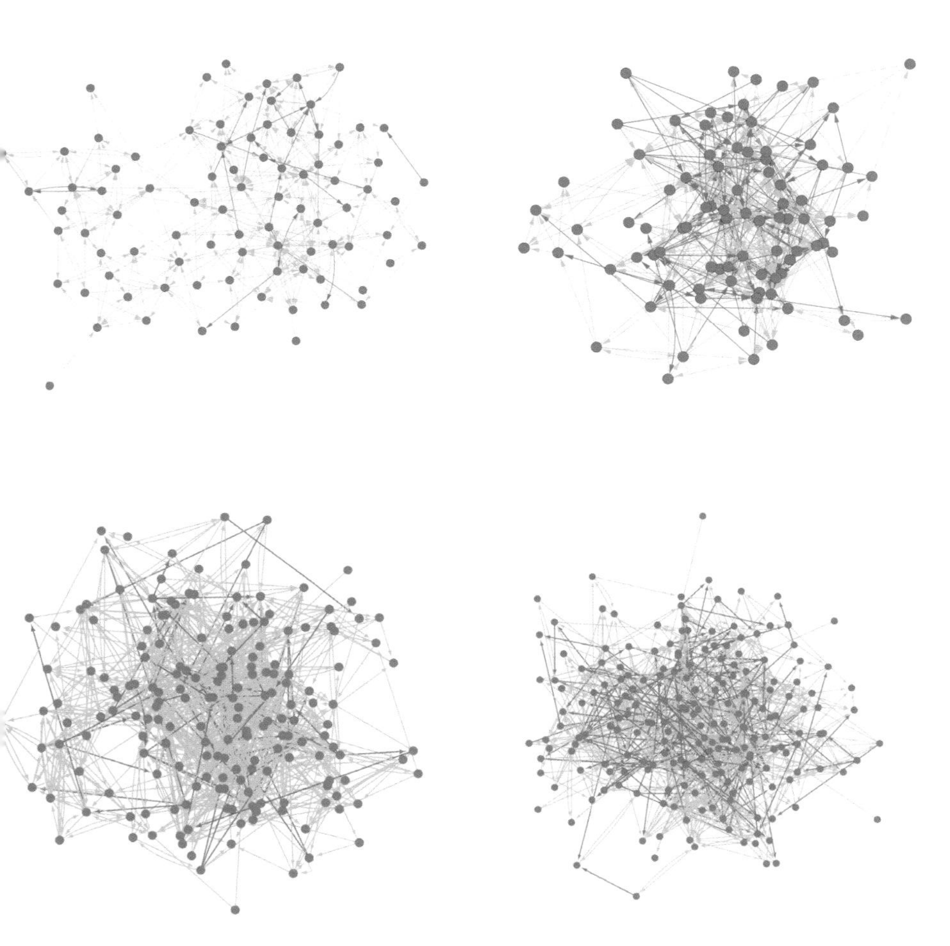

**Plate 5:** Levels of Animosity in Four Villages in Honduras

These four images are network graphs of villages in Honduras. Blue nodes are people; the gray lines indicate friendship (positive) ties and the red lines indicate antagonistic (negative) ties. The top row represents two smaller villages (N = 86 and 87) and the bottom row represents two larger villages (N = 204 and 184) from left to right. The first column is for villages with low animosity (8.5 percent, 9.6 percent) and the second column is high animosity (40.0 percent, 32.2 percent), as measured by the fraction of negative ties to positive ties in the village.

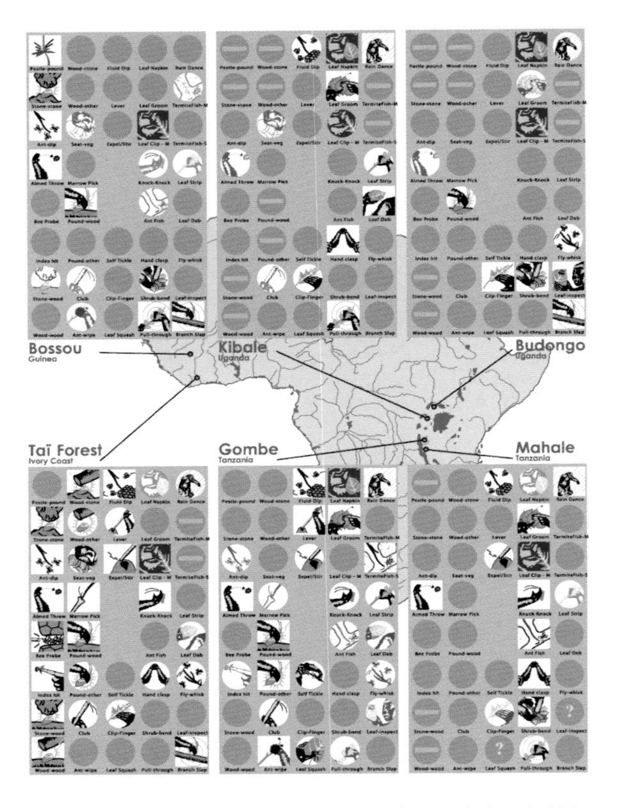


**Plate 6:** Elements of Chimpanzee Culture in Six Populations in Africa

The image shows six chimpanzee field sites in Africa and a broad set of cultural practices in this species. The practices are shown in the 5-x-8 arrays that indicate those behaviors which are customary or habitual at each site. Color icons show customary behaviors; circular icons show habitual behaviors; monochrome icons indicate behaviors that are merely present; and clear positions indicate that the behavior is absent. A horizontal bar indicates that the behavior is absent but that there is an ecological explanation (for example, the absence of algae fishing in a setting where there is no algae). A question mark indicates that the situation is uncertain. Overall, each field site shows a distinctive cultural pattern of behaviors.

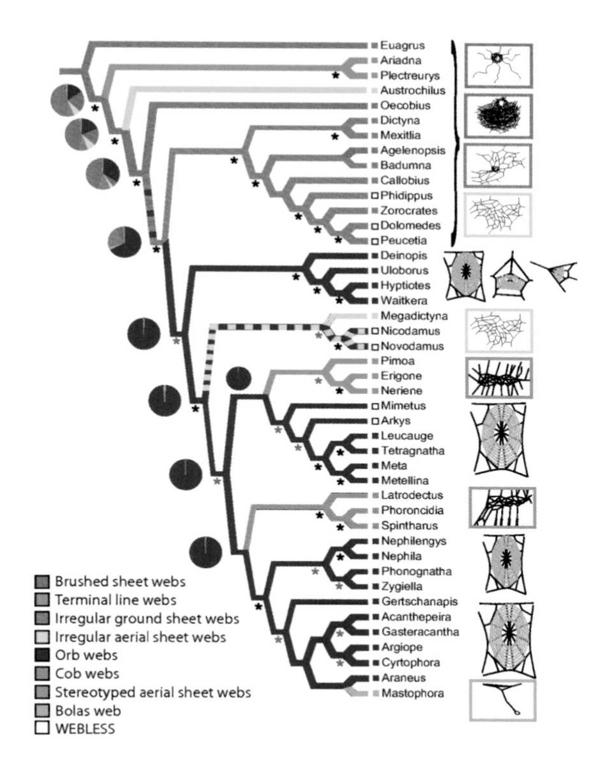


**Plate 7:** The Evolution of Spider Webs

Branch colors represent a reconstruction of the likely evolution of web morphology across the spider order across evolutionary time. Colors to the left of taxon names indicate web types. Black and gray stars and pie charts indicate the strength of the scientific evidence for the relevant branch points (and are not pertinent here). There is a tremendous diversity of spider web types that varies by species and that has been shaped by evolution. This is an example of a physical object made by an animal that has a structure specified by the animal's genes.

**Plate 8:** A "Zombie" Ant

The extended phenotype of the fungus *Ophiocordyceps unilateralis* is manifested in the Asian ant *Polyrhachis armata*. The fungus controls the ant's behavior, prompting it to climb a plant to a certain height and then bite into a leaf vein. The fungus then kills the ant and grows the large mushroom stalk from its head (visible here) from which spores can rain down on other ants. This photograph is reversed in that this is the underside of a leaf and the ant would ordinarily be seen hanging from below it, rather than right-side up.

traits in our own species. If even elephants and dolphins can have friends, this highlights how much all humans have in common.

### *Kinship Detection*

So far, we have taken it for granted that animals can tell the difference between kith and kin when they make social connections. But how and why do they do that? And what role could this ability play in the evolution of friendship within species and the formation of friendship ties between individuals?

Many animals are able to sort members of their own species into three categories: kin (of varying degrees of relatedness), familiar non-kin, and unfamiliar strangers.[74] *Kin detection* (or *kin recognition*) originated because it was evolutionarily useful to be able to avoid breeding with kin and to know whom to treat favorably. A preference for kin can be reflected in other domains too, including playing, assembling, attacking, and defending. Kin detection requires that animals express cues regarding their kinship status and that there be neural mechanisms and cognitive algorithms for detecting these cues and for computing a measure of genetic relatedness. The subsequent process of *kin discrimination* describes the further behavior of differential treatment of kin and non-kin.[75]

From an evolutionary point of view, one fundamental reason that kinship is important to behavior relates to a classic idea called *inclusive fitness* introduced by evolutionary biologist W. D. Hamilton in 1964. He tackled a puzzle that was apparent back to the time of Darwin: Why would animals ever risk helping others if there was no benefit to themselves? Such actions are widespread in the animal kingdom, but they were hard to explain with then-existing models of Darwinian fitness, which were based solely on the reproduction or survival of an individual. For instance, why would a prey item bother to expend the energy to make extra chemicals rendering itself bitter to a predator if the prey does not benefit from having tasted lousy (because it has already been eaten)? Or what about animals that sound alarms when they detect incoming predators? Doing so calls attention to themselves, which seems counterproductive to their individual survival. For that matter, why would elephants care for orphan members of their species (evidence of which we can see on any number of tear-jerking wildlife videos)? As Hamilton observed, "If natural selection followed the classical models exclusively, species would not show any behavior more positively social than the coming together of the sexes and parental care."[76] But there is much more to social interactions than sex and parenting.

Hamilton proposed a theoretical solution by expanding the concept of individual Darwinian fitness to *inclusive fitness*—an evolutionary benefit that accrues beyond the individual. He argued that, since related individuals have many genes in common, an individual could pass on its genes *indirectly* via its relatives. An individual advances its Darwinian interests with actions that enhance its own survival and reproduction *or* the survival and reproduction of its relatives.[77]

Hamilton's rule can be expressed in the form of an inequality: $rB - C > 0$. Here, C is the cost to an individual (meaning the risk to its survival and reproduction) in performing an altruistic act; B is the benefit to the recipient; and $r$ measures the coefficient of relatedness between the two individuals (ranging from 0 to 1). The idea is that genes for altruistic behavior will evolve, even if harmful to the individual, if the benefit to relatives is sufficiently high. In a passage about sacrificing oneself for one's relatives, Hamilton famously summarized this as follows:

> In the world of our model organisms, whose behaviour is determined strictly by genotype, we expect to find that no one is prepared to sacrifice his life for any single person but that everyone will sacrifice it when he can thereby save more

> than two brothers, or four half-brothers, or eight first cousins.[78]

This fundamental idea has given rise to many embellishments.[79] One modification has to do with the reproductive value of the giver and the recipient of the help, meaning that an older person sacrificing his or her life for a younger person would be more rational than the reverse. In fact, we saw this poignantly in the Fukushima nuclear disaster in 2011 during which elderly Japanese citizens volunteered to perform duties that would expose them to lethal radiation in order to benefit others.[80]

A further evolutionary reason for the emergence of kin recognition is the need to avoid breeding with too-close relatives, given the well-known survival disadvantages of such mating.[81] The twin objectives (helping kin and avoiding inbreeding) have long suggested that there must be *some* way that animals can tell kin from non-kin. But the mechanisms by which this is accomplished across diverse taxa are still not fully understood.[82]

The most straightforward mechanism would be spatial distribution. For animals that do not range widely, a good approach would be for them to care for nearby animals of their species. They could simply follow the rule "Be nice to animals located in this spot." Those animals are likely kin. But if descendants stay too close to one another, inbreeding will result. A solution to this tension between objectives would then be for a species to adopt *sex-based dispersal.* Females could stay put while males moved away, as we have seen in elephants and whales, or females could move away and leave the males behind, as happens with chimpanzees. As a result, all animals mate with members of the opposite sex who are not from their home turf (and are thus largely unrelated), and all animals are nice to other animals on their home turf (who, by felicitous design, are indeed largely related). This practice is widespread in the animal kingdom because, in this scenario, animals need recognize only places, not individuals. Many birds recognize their nest sites rather than their offspring, as sneaky ornithologists moving eggs and chicks around have discovered.[83] This also implies that species that range more widely should be less altruistic, as they can be less certain that the individuals of their species they encounter are related to them.

But what about animals that do not have fixed locations, such as migrating herbivores and non-nesting birds? Other means of identifying individuals would be needed. Some bird species use their characteristic songs to identify themselves not only as members of a species, but also as individuals. In a colony of seabirds who roost by the thousands, often on a featureless expanse, partners can face extreme challenges in finding each other (or their nests). For two species of penguin (the king and emperor penguins), the problem is even worse, because they do not have nests at all. They hold their eggs and young on their feet as they move around, huddling together in groups of ten birds per square meter in order to stay warm and avoid the fierce Antarctic winds. So the family has no choice but to find a way to uniquely identify one another—which they achieve through individual-level acoustic recognition.[84]

Of course, it's not enough to be able to send or detect individual signals. There must also be a way to recognize the signal, *linking* it with a particular individual. Physical association is a key way this linkage is accomplished, and mothers and offspring in species as diverse as penguins, elephants, and sheep recognize each other from the time of birth using a variety of sensory cues, including smell, sight, touch, and sound. Even a human baby can recognize his or her mother's voice at birth.[85] Animals who are in close association for long periods or who begin associating early in their lives are generally more confident that they are related. Animal siblings avoid mating with each other by, roughly speaking, developing an aversion to mating with those they recognize from birth. This means that latter-born offspring will be better able to distinguish kin than earlier offspring; latter-born offspring will have had more experience identifying their siblings, since they will have had the older ones present throughout their lives (but not vice versa). Among humans, mutual exposure during childhood similarly weakens the sexual attraction between individuals when they are adults (as we saw with the kibbutzim). Conversely, there is some evidence that biologically related people reared *apart* sometimes report an overwhelming (and usually terribly unwelcome) physical attraction when they meet as adults.[86]

In addition to location and individual identity, phenotype matching is another and even more complex possible mechanism for kin detection. Here, an organism evaluates how similar others are to *itself*. For instance, it might use the smell or sound of its siblings or mother as a template: *If you smell or sound like my mother, you must be related to me, at least a little; therefore, I should not mate with you, but I should help you.* In some species, organisms might actually engage in self-matching, whereby they recognize the extent to which members of the same species resemble themselves.[87]

In humans, the precise neurological and genetic bases for kin detection are not well understood. But we clearly have a biologically based capacity to detect kin. The great majority of people are disturbed at the thought of having sex with their close relatives. And humans can typically identify pairs of siblings on sight, even if they are not related to them.[88] Plus, humans think a lot about kinship. Mistaken kinship is the subject of countless myths, plays, operas, novels, and stories from Sophocles to Shakespeare.

### *From Mates to Friends to Societies*

An enduring, sentimental tie between individuals is what defines friendship, analogous to pair-bonding in mates. This individual constancy is mirrored by a collective constancy. Individuals in the social species we have considered come and go; they are born and die; and friendships begin and end. But *the overall social organization of the species stays the same.* Some turnover in social ties within groups may even be necessary for networks to endure.[89] It's like replacing planks in a boat. This is required to keep the boat seaworthy. But the plan of the boat, like the topology of the social network, remains fixed, even if all the individual boards are eventually replaced.[90] That is, the structure of social networks—arising from all the dyadic friendship ties and the genes within us—is a feature of our species itself. Remarkably, humans share this emergent structure with other social mammals, which highlights how fundamental this property is to all humans, regardless of their culture.

If we can share the structure of our eyes with octopuses, we can share the capacity for friendship with elephants. Primates, elephants, and whales manifest elements of the social suite because, despite the fact that they branched off from a common ancestor more than seventy-five million years ago, they have independently and convergently evolved these traits to cope with the challenges imposed by their environment. Initially, those environmental challenges were external. But eventually, the social groups these animals made became features of their environment too, further shaping and reinforcing their social behavior. Animals evolve to be better at living socially as they find themselves more often in social groups.

Still, friendship is rare in the animal kingdom. The human propensity to do this has been shaped by natural selection and is written in our DNA. Friendship in animal species serves the stunningly useful purposes of mutual aid and social learning. And it's the foundation of the capacity for an enduring culture that transcends individuals and transmits information across time and space. Many further aspects of human psychology are also related to friendship, like the joy and warm feelings we experience in the company of friends and the sense of duty we feel toward them.

This in turn suggests an even deeper observation about the role of friendship in our species. Our assembly into networks of friendship ties sets the stage for the emergence of moral sentiments. At their core, moral compunctions relate to how people interact with others, especially those who are not kin and for whom the bonds of kinship and the inexorable workings of inclusive fitness are not enough of a guide.

Most human virtues, I would argue, are *social* virtues. To the extent that we care about love, justice, or kindness, we care about how people enact these virtues with respect to *other* people. No one is interested in whether you love yourself, whether you are just to yourself, or whether you are kind to yourself. People care about whether you show these qualities to others. And so friendship lays the foundation for morality.

# CHAPTER 8

# Friends and Networks

Among the twelve victims of the July 12, 2012, mass shooting in a movie theater in Aurora, Colorado, were three young men in their twenties who used their bodies as shields to protect others from bullets.[1] As soon as the shooting began, Jon Blunk pushed Jansen Young, his girlfriend of nine months, to the ground and threw his body on top of hers. Alex Teves did the same for his girlfriend of a year, Amanda Lindgren. And Matt McQuinn interposed his body between the killer and his girlfriend, Samantha Yowler, as the shooting continued. All three women survived.

Lindgren described the experience: "I was really, really confused at first about what was going on.... But, it's like Alex didn't even hesitate. Because I sat there for a minute, not knowing what was going on, and he held me down and he covered my head and he said, 'Shh. Stay down. It's OK. Shh, just stay down.' So I did."[2] Similarly, Blunk pushed Young under a seat. "He laid up against me and had the other side of my body against the concrete seating, and I was pretty much boxed in," Young later recalled.[3] When the shooting was over, Young was the last to leave the theater. She was getting up when she realized that Blunk was "really wet." She could not believe what had happened, and "she tried to convince herself that someone must have thrown a water balloon."[4]

People sacrifice their lives for their partners and for their children and other kin.[5] From an evolutionary perspective, given kin selection and the other processes we have reviewed, this should not surprise us—although cases like Aurora are deeply moving. But people also give their lives for their friends, which is much harder to explain. Sacrificing one's life for unrelated individuals happens on the battlefield, of course, but soldiers are trained for mutual sacrifice against a common enemy. The amazing thing is that people also sometimes make such heroic sacrifices for their friends, people to whom they are *not* related and whom they have not been trained to protect.

The sentiment is beautifully captured in the New Testament: "Greater love has no one than this: to lay down one's life for one's friends."[6] President Obama quoted that Bible verse in January 2016 while remembering Zaevion Dobson, a fifteen-year-old boy in Knoxville, Tennessee, who threw himself over three of his friends to shield them from gunfire. Dobson and his friends had been sitting on their porch when they were shot at randomly. The friends, all girls, lived; Zaevion died.[7] While such acts of sacrifice are more commonly made by men, in July 2015, Rebecca Townsend, a seventeen-year-old girl, died after shoving her friend Ben Arne out of the way of an oncoming vehicle. Ben later said, "The last thing I remember is Rebecca pushing me and telling me to hurry up."[8] Later, her family would discover a bucket list of aspirations Rebecca had written two years earlier: "Kiss in the rain; fly to Spain; save a life."[9]

Friends make other meaningful sacrifices, such as donating a kidney or sharing scarce food in wartime prison camps.[10] We can see that the extreme attachment people can feel for one another is not limited to sexual partners or kin. Friendship is powerful. In fact, my friend Dan Gilbert, a professor of psychology at Harvard and the author of *Stumbling on Happiness*, argues that friendship is a key determinant of happiness, even more important than marriage.[11]

### *Making Friends*

We have focused a lot so far on kinship and kin-based altruism, mating and pair-bonding, and strict exchange relationships. Friendship is a fourth kind of social interaction between individuals. We can formally define *friendship* as a typically volitional, long-term relationship, ordinarily between unrelated individuals, that involves mutual affection and support, possibly asymmetric, especially in times of need. Close friends in most societies violate many of the customs regarding exchange-based relationships (what we might call tit-for-tat behavior) between unrelated individuals. Explicitly conditional or reciprocal exchanges ("I'll scratch your back if you scratch mine") are the types of cooperation and kindness that are seen when trust is low and friendship relations are weak or nonexistent.

We are supposed to respond to our friends because they have a need, not because of what they have done for us in the past or what we might expect from them in the future. Moreover, real friendships are based not on what each party can *do* for the other (mutual aid or usefulness), but on how each party *feels* about the other (mutual goodwill or sentiment).

Relations among friends are characterized by a number of primary sentiments, including closeness, affection, and trust.[12] A feeling of closeness to the other person is crucial to whether you deem him or her a friend and whether you will offer help, but such a feeling is not necessary when it comes to identifying kin. And people do not have to have sex with their friends or raise children with them in order to feel a special way about them.[13] Feeling close to someone involves some inclusion of the other in one's self-identification. When it comes to a friend, you perceive that things that benefit the other person also benefit yourself. You take joy in your friend's well-being. Scientists can measure this aspect of friendship by asking a subject to evaluate depictions of the subject and a friend as two circles with variable overlap (figure 8.1).[14]

**Figure 8.1: The Inclusion of Other in the Self Scale**

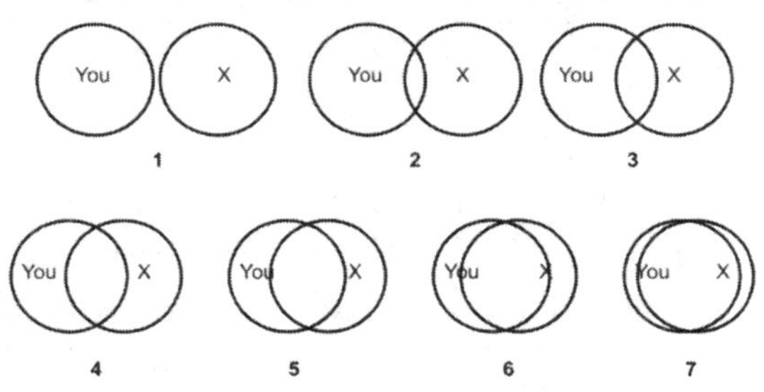


In using this scale to measure social intimacy between friends, subjects are invited to circle the picture that best describes the relationship between themselves (you) and another person (X). The greater the overlap, the stronger the friendship.

People demonstrate their friendship with others or find evidence of such friendship through time-intensive, exclusive behaviors, honest expressions of emotion, and accepting vulnerability (for instance, tolerating a friend's teasing because it indicates that one understands that the friend's intentions are good). Exclusive behaviors are ones that cannot easily be expressed to large numbers of people. Spending time with a friend or writing personal notes are good examples, because they are not scalable and are hard to fake.

### *Friendships Around the World*

Using a sample of sixty representative cultures from around the world, anthropologist Daniel Hruschka showed that core features of friendship are widespread.[15] Not one of these cultures lacked a notion or practice of friendship. In a broader sample of four hundred cultures, there were just seven that might possibly have lacked friendships. In five of these, the societies were extremely collectivist, and intimate friendships were strongly discouraged because they were seen as a threat to social unity. The discouragement of friendship is reminiscent of the approach to romantic relationships in some of the communes described in chapter 3, where, in order to foster group cohesion, people were allowed to have sex with everyone or with no one but nothing in between. Similarly, in totalitarian regimes, everyone is a comrade or referred to as a friend. To be fair, even in the (more admirable) kibbutzim, residents often use the Hebrew word *haverim*, meaning "friends," to refer to other members of the collective. Moreover, friendships are common even in totalitarian societies. But it is noteworthy that a powerful cultural overlay is required to suppress a natural impulse to have a few close friends, just like the Baining must make huge efforts to keep children from playing. Friendship is universal.

Friendship also tends to look very similar across cultures. In the sample of sixty cultures, as table 8.1 shows, mutual aid and positive affect are seen in 93 percent and 78 percent of societies, respectively, and this is not disconfirmed in any society in the sample (that is, in no society was it reported that mutual aid or affection should *not* be part of friendship). In 71 percent of the societies, it was explicitly noted that friends did not engage in tit-for-tat accounting of their exchanges. Of course, just as in the case of monogamy—which is for some people more of an aspirational goal than a strict reality—there is a difference between the ideal friendship type and the actual experience or practice of friendship.

Despite the similarities, certain features of friendship occur in other societies but are uncommon in the United States, such as sustained physical touching and formal commitment rituals. When President George W. Bush was seen publicly holding hands with Crown Prince Abdullah of Saudi Arabia in 2005, it struck many Americans as odd, but they were sending the signal that they were friends.[16] By contrast, some aspects of friendship seen as crucial in Western settings are uncommon in other cultures. Self-disclosure of personal information, which is typical in the United States, was seen in only 33 percent of the sample; and 10 percent of the time, it was noted that this was not a feature of friendship. Frequent socializing, a typical aspect of American friendship, was

**TABLE 8.1: CHARACTERISTICS OF FRIENDSHIPS IN SIXTY SOCIETIES**

| CHARACTERISTIC | PERCENT DESCRIBED | PERCENT DISCONFIRMED |
|---|---|---|
| *BEHAVIORS* | | |
| Mutual aid | 93 | 0 |
| Gift-giving | 60 | 0 |
| Ritual initiation | 40 | 4 |
| Self-disclosure | 33 | 10 |
| Informality | 28 | 0 |
| Frequent socializing | 18 | 55 |
| Touching | 18 | 0 |
| *FEELINGS* | | |
| Positive affect | 78 | 0 |
| Jealousy | 0 | 0 |
| *ACCOUNTING* | | |
| Tit-for-tat | 12 | 71 |
| Need | 53 | 0 |
| *FORMATION AND MAINTENANCE* | | |
| Equality | 30 | 78 |
| Voluntariness | 18 | 64 |
| Privateness | 5 | 66 |

noted in only 18 percent of the societies studied in this sample, and 55 percent explicitly noted that friends need not spend a lot of time together.

There are other cultural differences in how friendship is expressed. In 28 percent of societies, friends may publicly display their intimacy by being informal or taking liberties with one another in front of other people. Among Bozo fishermen in Mali, friends make lewd comments about the genitals of their friends' parents.[17] In Greece, male friends endlessly and publicly (and tiresomely, if you ask me) accuse one another of being masturbators. Some of these human behaviors, which can be seen as indicators of vulnerability or trust, may have parallels in behaviors seen in other primates. For instance, some groups of capuchin monkeys play a finger-in-the-mouth game in which one monkey places a finger into a friend's mouth, and the friend clamps down on it so that it cannot be taken out but not so much that it hurts the finger-poker.[18]

Friendship, Hruschka proposes, may have evolved as a system of social relationships that encouraged cooperation and mutual aid even in situations fraught with uncertainty and in environments characterized by variability.[19] In the relative absence of formal institutions, friendship is a kind of insurance against unexpected setbacks. My lab is evaluating this idea. We are conducting a public-health project, which we began in 2015, involving 24,812 people in western Honduras; our primary target is to improve maternal and child health. But we are also evaluating whether the introduction of formal institutions (like visiting nurses, health clinics, police stations, or credit unions) could replace friendships or at least modify their meaning and role. Providing a permanent or official source of health advice or medical care, for instance, means that social ties might change and the importance of friendship might be attenuated. While one's core group of friends might stay intact, some other friends could be lost. In fact, this may be an unacknowledged side effect of public-health interventions in the developing world—and a side effect of modernization in general. The introduction of formal institutions may weaken traditional friendship ties. This is the *Gesellschaft* and *Gemeinschaft* that provided the push and pull for the founding of communes.

Such a phenomenon does not occur only in developing-world settings. Working-class Americans report relying on their friends and neighbors for practical help such as child care, spiritual advice, car and home repairs, and cash gifts or loans more often than middle-class Americans do. This reliance makes them more rooted in their communities, with stronger ties there, whereas more affluent Americans tend to get support from formal institutions such as therapists, work colleagues and mentors, and legal and financial advisers. Being less reliant on the friends in their community, they can afford to be more geographically mobile.[20] In this regard, the long sweep of modernity may be diminishing our natural inclination to make friends.

Further evidence for the innate propensity for friendship in humans is its similar developmental course across cultures. In the first stage of friendship, from five to nine years of age, children focus on shared activities or superficial rewards ("I like playing with her Lego collection").[21] By about nine years old, children transition to more abstract notions of friendship, such as loyalty and trust, and recognize the duties and transgressions of friends. In the third stage, during adolescence, expectations can become still more abstract, focusing on commitment, empathy, and affection. Adolescents describe nurturing their friendships not only because of their friends but because of the relationships themselves. Of course, there is a strong cultural overlay to all this, and institutions such as sports teams, fraternities, and initiation rites (such as those seen in some traditional societies) may reinforce or facilitate friendship in culturally specific ways.

The impulse to make friends is often so intense that many young children have imaginary friends with whom they interact happily in complex ways. Children even report attempting to avoid hurting an imaginary friend's feelings. By the age of seven, 63 percent of children in one study had an imaginary companion.[22] These friends ran the gamut, as described by their creators, from "Derek, a 91-year-old man who is only 2 feet tall but can 'hit bears,'" to "Baintor, an invisible boy who 'lives in the light,'" to "Hekka, a 3-year-old invisible boy who is very small but 'talks so much' and is 'mean' sometimes."

My wife's imaginary childhood friends (Toctoe and R) were so important to her that, decades later, she will jokingly invite their intercession when we have a disagreement.

Friendship can be just as strong a bond as kinship. For instance, in one experiment, people were willing to spend 140 seconds in a painful posture to benefit themselves, 132 seconds for immediate kin (such as siblings or parents), and 107 seconds, on average, for cousins. They were willing to spend 103 seconds in a painful posture to benefit a children's charity. And for their best friends? They put in 123 seconds.[23]

### *The Banker's Paradox*

The diversity of goods and services that two people might exchange, the uncertainty in the timing of when help might be needed, and the prospect of rendering the items over a long period make tit-for-tat accounting difficult. But this is exactly what friendship is built for and what makes it valuable, evolutionarily speaking. For people around the world, the test of a real friend is that he or she gives you something *without* the expectation of a quid pro quo. In fact, an explicit statement by a supposed friend that a favor is expected in return is taken as a mark of a *lack* of friendship. Once again, the extent to which this is true likely varies somewhat across individuals, cultures, and circumstances. But friendship always involves relaxation of expectations of even exchange.[24]

In my lab's work mapping social networks in Tanzania, Sudan, Uganda, and elsewhere, we have taken advantage of this to identify friendship relationships via anonymous gift-giving. Among the Hadza, for example, honey is a very prized food, obtained with difficulty by climbing trees and knocking hives to the ground while being stung by bees. We wanted to see to whom Hadza respondents would give anonymous gifts, so we provided them with honey and asked them to choose recipients (this allowed us to map their important social relationships, most of which were with non-kin). Rather than scaling bee-infested trees, though, we just bought honey-filled straws at Costco and took them to Tanzania in suitcases.

The very fact that humans have friends and also the emotional apparatus to support these relationships (similar to the feelings of attachment and love that accompany sexual relationships in our species) is evidence that simple models of tit-for-tat reciprocation do not, and cannot, fully account for altruism and friendship in social life. But why did humans evolve this emotional apparatus? Evolutionary psychologists John Tooby and Leda Cosmides believe that this capacity for friendship reflects our species' response over evolutionary time to a kind of banker's paradox—a notion that captures the irony that people who need resources the most are precisely those to whom bankers have the least interest in lending money.[25] Analogously, when our hunter-gatherer ancestors needed help the most, others might have been the least likely to offer it for fear of never being repaid. Friendship might have evolved to cope with just such circumstances.

The cognitive adaptations needed to make decisions about whom to help in difficult circumstances would surely be attuned to whether people would be able or willing to repay the debt in the future. In other words, you would need to be able to establish whether a person was a good credit risk. Was that person *really* your friend? If the object of one's assistance becomes permanently disabled, emigrates, or dies, then one's investment in that person would be lost. But if the problems are temporary—requiring, for example, that you extend a branch to a drowning person while safely standing on the shore—then that person would be a good investment and would make a good friend. A system that allowed a sort of fluid mental accounting for these sorts of exchanges would be very valuable. Evolving the capacity to help others when they need it and when it does not cost much, as well as the capacity to track such connections, would be useful, evolutionarily speaking, for all involved.

Tooby and Cosmides argue that, in foraging societies, infection, injuries, food shortages, bad weather, bad luck, and attacks from other groups were constant threats with major evolutionary impact. As they note:

> The ability to attract assistance during such threatening reversals in welfare, where the absence of help might be deadly, may well have had far more significant selective consequences than the ability to cultivate social exchange relationships...when one is healthy, safe, and well-fed. Yet selection would seem to favor decision rules that caused others to desert you exactly when your need for help was greatest. This recurrent predicament constituted a grave adaptive problem for our ancestors—a problem whose solution would be strongly favored if one could be found.[26]

The way out of this predicament was for individuals *to have friends*, friends devoted to them *in particular*. The ability of friendship to be useful during times of reversals, when an even exchange is not possible, is precisely what makes friendship so valuable as an adaptation in our species. Moreover, to facilitate the accounting over time, in the case of friendship, our evolved psychology equips each of us with the sense that we are not substitutable. Individuality is crucial to this. The fact that you are irreplaceable to your friends even though you are unremarkable to strangers suggests that there is a deep connection between individuality—another key part of the social suite—and friendship.

But why couldn't people simply rely on kin for support in tough times? There are several reasons. Kin are sometimes competitors for family resources (such as parental attention and inheritances) in a way that friends are not. More important, collective tasks, such as hunting large prey and moving safely across terrain, can require group sizes that exceed what might be feasible if restricted to kin. Immediate kin might also lack the range of skills, knowledge, and abilities necessary to cope with a particular circumstance, especially if they share very similar genes and features. Connections with non-kin may be the only way to access new ideas or other resources. Non-kin ties may thus have been especially important to the emergence of our species' ability to create and maintain culture, an ability that is crucial to our survival, as we will see in chapter 11, and that is also part of the social suite.

If banker's-paradox dilemmas were features of our species' evolutionary landscape, then we would expect to see the evolution of diverse adaptations to make more credit available. In tandem with friendship, humans might have a desire to be recognized and valued for their individuality and even to feel jealous (and perhaps act vindictively) when developments in their social lives threatened their special status. Referring to someone as "irreplaceable" is a common form of praise. And many psychological phenomena in our species reflect the threatening nature of social replaceability, including the fact that we seem to like to form groups that are small enough for individuality to be appreciated. Ironically, then, individuality is crucial to the formation of social groups and to how the whole emerges from the parts.

From this perspective, the sense of alienation felt by many people who live in modern market economies makes sense. People are often dissatisfied with the feeling of anonymity engendered by formal institutions and bureaucracies. If your life is full of explicit and conditional exchanges with strangers that occur at a frequency and volume absent in our species' evolutionary past, it might well make you miserable. Our species' evolved psychology reads these exchanges as markers of how superficially, even meaninglessly, we are engaged with the people around us and how vulnerable we might be to a sudden reversal of fortune. Without friends, we feel naked.

### *The Genetics of Friendship*

A final line of evidence regarding the innate propensity for friendship—beyond its cultural universality, its developmental consistency, and its evolutionary rationality—is provided by recent work on the genetics of social interactions. Genes play a role in the formation, attributes, and structures of friendship ties.[27] In one study, my lab examined the friendship patterns of 307 pairs of identical twins and compared these to patterns in 248 pairs of fraternal twins. As shown in figure 8.2, the networks of identical twins are more similar in their structure than the networks of fraternal twins, documenting a role for our genes in the social networks we make for ourselves.[28]

This is not too surprising if you think about it. Some people are born shy and some people are born gregarious. We found that 46 percent of the variation in how many friends people have can be explained by their genes. But we further found that 47 percent of the variation in people's transitivity can also be explained by their genes. For example, in a group that consisted of Betty, Sue, and Jane, whether Sue was friends with Jane depended not just on Sue's genes and Jane's genes, but also on *Betty's* genes. How could this be? How could whether or not two people are friends with each other depend on a third person's genes? We think the reason is that people vary in their tendency to introduce their friends to one another, and Betty may be the sort of person who knits the social fabric around her together.

This was an exciting finding, and it got me and a colleague of mine, political scientist James Fowler, thinking about how people might act on the social environment around them to construct social niches. There was a long-standing literature in biology on such niche construction that considered, for example, rabbits digging burrows, birds building nests, earthworms modifying the soil around them, and bacteria secreting chemicals into their environment to make it

**Figure 8.2: The Social Network Structure of Identical and Fraternal Twins**

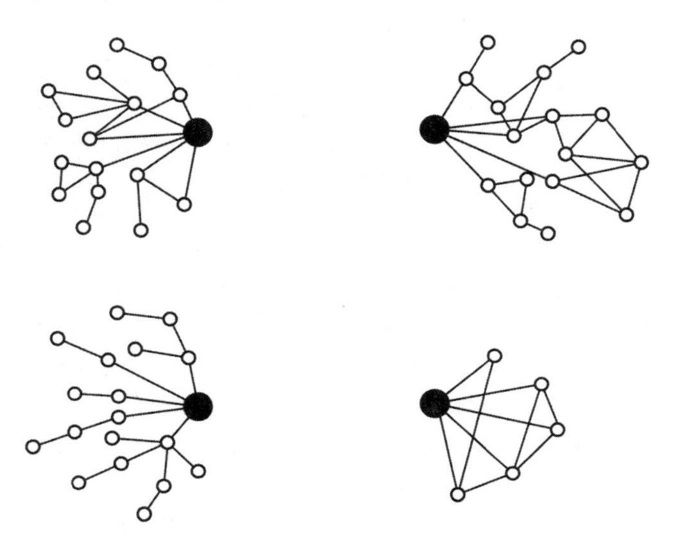

The top two panels show diagrams of the social networks of a pair of identical twins, and the bottom two panels show the networks of a pair of (same-sex) fraternal twins. The pairs of twins are indicated by the solid black circles, their friends are shown in open circles, and the friendship relationships among the various people are indicated by the lines. The fact that the networks are more similar (visually and quantitatively) among pairs of identical twins (top) than among pairs of fraternal twins (bottom) is apparent, consistent with the partially genetic basis of friendship networks. If genes played no role, the networks for identical-twin pairs would be no more alike than the networks of fraternal-twin pairs.

more convivial. But we began to consider if people might also engage in a kind of *social* niche construction, reshaping their social environments in ways that benefit them and that, in the past, might have affected our species' evolution (an idea we will explore in chapter 10).

### *Birds of a Feather*

Genes may affect not only network structure but also how people choose their particular friends—just as evolution has shaped our species' taste for particular mates. From an evolutionary perspective, this choosiness would be shaped by the friends one already has as well as other factors. For instance, you might pick your friends based on their ability to assess your needs and intentions or based on their capacity to see you as a special individual. You might also select friends who want the same things you want or who are able to incidentally provide benefits to you—like being a good way finder or food locator. Tooby and Cosmides argue:

> A person who values the same things you do will continually be acting to transform the local world into a form that benefits you, as a by-product of their acting to make the world suitable for themselves. Trivial modern cases are easy to see: e.g., a roommate who likes the same music or who doesn't keep setting the thermostat to a temperature you dislike. Ancestrally...a person who your enemies fear, or a person who attracts more suitors than she can handle, may be a more valuable associate than a reliable reciprocator whose tastes differ widely from your own.[29]

Choosing similar friends can have a further rationale. Assessing the survivability of an environment can be risky (if an environment turns out to be deadly, for instance, it might be too late by the time you found out), so humans have evolved the desire to associate with similar individuals as a way to perform this function efficiently. This is especially useful to a species that lives in so many different sorts of environments. However, the carrying capacity of a given environment places a limit on this strategy. If resources are very limited, the individuals who live in a particular place cannot all do the exact same thing (for example, if there are few trees, people cannot all live in tree houses, or if mangoes are in short supply, people cannot all live solely on a diet of mangoes). A rational strategy would therefore sometimes be to *avoid* similar members of one's species.

For these reasons, while humans have a general preference for homophily (which is Greek for "love of like"), desiring the company of people they resemble, this tendency may coexist with a preference for heterophily (essentially meaning "opposites attract") in some attributes.[30] You may want your roommate to have your tastes in music and ambient temperature, but it might be better to have someone very different from you to help you with your math homework.

Friends may resemble one another in subtle ways short of overt phenotypes. An important study by psychologist Thalia Wheatley and her colleagues Carolyn Parkinson and Adam Kleinbaum examined the brain responses of a group of forty-two friends embedded in a naturally formed social network of two hundred and seventy-nine graduate students. The subjects were each placed in a functional MRI scanner and shown the same set of fourteen video clips (including a sentimental music video, a bit of slapstick comedy, a political debate), and the blood flow to various brain regions was individually measured as they viewed the clips.[31] Pairs of friends had exceptionally similar responses to the stimuli. Neurologically speaking, therefore, pairs of friends are similar in how they perceive, interpret, and react to the world around them. It was even possible to predict who would be friends with whom based on the pattern of blood flow in their brains in response to this sample of clips!

Friends may resemble one another on a genotypic, not just phenotypic, level.[32] There are four reasons for this. First, similar genotypes may simply reflect that friends often come from the same place. If Greeks choose other Greeks as their friends, we should not be surprised to see that pairs of these Greek friends share gene variants, since they are from the same small corner of the world. Second, people may actively choose friends of a similar genotype. If athletic people prefer to hang out with athletic people, we should not be surprised to find that they have variants of certain "athletic genes" in common, such as fast-twitch muscle fibers or gene variants of the *ACE* gene associated with endurance.[33] Third, people with genetic similarities may prefer the same environments. For instance, if you like to run marathons at high altitudes and have the sort of body that is capable of that, you will likely encounter others on your mountain enjoying the same thing. It would not be surprising to find that you and your new friends carry gene variants that increase the binding of scarce oxygen in the blood. Finally, people with similar characteristics may be chosen by third parties to reside in the same place. For example, music conservatories admit people who are exceptionally musical, and students make friends with other students, so it would not be surprising if these students had similar variants of genes associated with musical ability, such as the *SLC6A4* gene, which has been linked with musical memory and being in a choir.[34] These four reasons are not mutually exclusive, of course.

There are fewer reasons that friends would be *dissimilar* in their genotypes. It's possible that certain environments might foster interactions between individuals with dissimilar traits. Leaders might affirmatively assemble teams of people with heterogenous skills; think of Jason choosing his crew of Argonauts to ensure that at least one of them had exceptional vision to pilot the boat while the others were good rowers. A second possibility is that people might actively choose to befriend those who are different types than themselves. In prehistoric times, hunting a large herbivore might have required a group of people with various different skills—fast runners, strong spearmen, good trackers, and so on.

Importantly, all these processes can work simultaneously, and humans may select friends and environments based on a wide variety of traits that produce different degrees of synergy and specialization, of homophily and heterophily. But overall, the observed balance skews toward homophily, or birds of a feather flocking together.[35] My lab quantified this. We evaluated the extent to which friends resemble each other genetically by analyzing 466,608 genetic loci in 1,932 unique subjects who were in one or more of 1,367 friendship pairs.[36] As with our work analyzing spousal pairs, discussed in chapter 6, we found that friends tended to be significantly more genetically similar than strangers drawn from within the same population. As a benchmark, the size of the effect roughly corresponds to the coefficient of relatedness we would expect for fourth cousins. That is, when people are left to choose friends among a group of people to whom they are not actually related, they have a discernible, if slight, preference for people who resemble them genetically. We can even use measures of similarity of the genotypes of any two people to create a friendship score that can predict whether they are likely to be friends in the first place.[37]

What do these processes tell us? They may reflect the extended workings of a kinship detector in humans.[38] Friends, in other words, may be *functionally related* rather than *actually related* (as in the case of kin). Forming social ties to such functionally related others who might perceive or cope with the environment in a similar way to oneself can result in *both* individuals benefiting from each other's deliberately or accidentally created benefits. If one individual builds a fire because he feels cold in the same circumstances as the other, then both benefit. This gives natural selection a lever with which to influence friendship choices since the friends we choose affect our prospects for survival.

It's possible that genetically guided friendship preferences originated in how humans evolved, long ago, to prefer certain kin over others. Imagine that you have ten cousins but can spend your time with only some of them. Furthermore, imagine that forming a group was beneficial to your survival. It would make sense for you to choose the cousins with whom you share the most genes. Humans would have evolved a way to identify the cousins to whom you were more similar on a genetic level. Then, the same procedures that evolved to allow you to prefer some of your cousins more than others might, by the process of exaptation we discussed before, come to be extended to people who are *not* your kin at all.[39] From among a population of people, you would naturally pick as your friends those people who were sort of like distant cousins.

Our analysis yielded a further supportive insight regarding the survival advantages of friendship. Using genomic techniques based on rates of mutation, we found that, overall, across the whole genome, the genotypes humans tend to share in common with their friends are more likely to be under recent natural selection (over the past thirty thousand years) than other genotypes. Consider a hypothetical example of a gene associated with the capacity for speech (such as *FOXP2*, a gene that may play a role in it). Imagine the first early hominid with a mutation that conferred the ability to make actual utterances rather than just grunts. The evolutionary advantages of speech—which are considerable—would be enhanced if the person befriended another person with a similar mutation so that they could talk to each other. Without that, the mutation might be of little use and would disappear in the population. The fitness advantage of variants of other genes (for example, variants related to warding off infections or reciprocating cooperation) might be influenced by their parallel presence (or absence) in other individuals to whom one is connected.

This fact that correlated genotypes between pairs of friends appear to be under positive selection (that is, they are increasing in prevalence) suggests the further observation that the genes of *other* people might modify the fitness advantages of an individual's genes. The human evolutionary environment is not limited to physical (sunshine, altitude) and biological (predators, pathogens) circumstances; it also includes humans' social circumstances.[40] These results are still more evidence that social interactions direct our species' evolution.

It is possible that humans paradoxically evolved a predilection for homophily in friendships precisely when they started to frequently interact socially with unrelated individuals—for example, when bands became large enough to involve fewer direct relatives or when humans began to disperse more widely, leaving their natal groups and interacting more often with strangers.[41] Once humans could not count on the people with whom they interacted necessarily being their kin, they would have had to develop other ways of connecting to genetically similar individuals who might offer benefits or whose fitness they might enhance from the point of view of evolution.

Friendship and interpersonal genetic preferences and effects can extend Hamilton's ideas about kin selection and provide natural selection with a way to work on groups of individuals based on friendships.[42] If you would sacrifice your life for two siblings who share 50 percent of their genes with you, why not for twenty friends who share 5 percent of their genes with you? The principle of kin selection that we discussed earlier simply had to do with the extent of genetic similarity between individuals, not whether this similarity specifically arose because people were actually relatives.

### *Social Networks Around the World*

So far in our discussion of friendship, our attention has primarily been on how pairs of people choose each other, how people interact when they are friends, and how natural selection has played a role in the universality of friendship. But when each person in a group chooses his or her friends, the group then assembles itself into a social network. All human groups do this. What's remarkable is how similar the resulting social networks are the world over. It's not just the predilection for friendship that is universal; it's also the way people go about naturally organizing themselves into broader networks. Natural networks have a complexity and sheer beauty not found in artificial networks (such as military chains of command or corporate-organization charts), and their existence raises questions about how they develop, what rules they obey, and what purpose they serve.

As we do with animal networks, when we map human social networks, we need to specify how the relevant social connections are identified. Who counts as a relevant social contact? Someone an individual has sex with? Lends money to? Gets advice from? Admires? Often, we can learn this simply by asking people so-called name-generator questions. Some of the standard name generators are as follows: (1) "Who do you trust to talk to about something personal or private?" (2) "With whom do you spend free time?" (3) "Besides your partner, parents, or siblings, who do you consider to be your closest friends?"[43]

In 2009, we asked a national sample of households two key name generators (the first two questions mentioned above), and we found that Americans identify an average of 4.4 close social contacts, with most having between 2.6 and 6.2. The average respondent lists 2.2 friends, 0.76 spouses, 0.28 siblings, 0.44 co-workers, and 0.30 neighbors in response to these questions. These numbers have not changed appreciably in decades, and we see similar results around the world.[44] People have roughly four to five close social ties on average, typically including a spouse, perhaps a sibling or two, and usually one or two close friends. These numbers can change somewhat over the course of a person's life (for instance, as people become widowed).

Over the past few years, my lab has developed tablet-based software called Trellis to collect network data in settings around the world, mapping face-to-face social interactions within defined populations in countries including the United States, India, Honduras, Tanzania, Uganda, and Sudan. In color plate 4, I show a selection of some of the networks we have mapped.[45] The visual and mathematical structure of the social networks that people make worldwide is strikingly similar, which, given that our networks resemble those of other social animals and that our genes play a role in friendships, should not be surprising.[46] Modernized networks resemble those found in forager societies, which further supports their innate nature.[47] Social networks are a fundamental part of the social suite and they play a crucial role in the blueprint.

### *On Enemies*

The capacity for making friends comes with a capacity for making enemies. Philosophers and scientists have thought about our obvious facility for animosity and hatred for a long time, but the actual mapping of antagonistic ties in parallel with friendship ties is recent and still quite rare. One study explored the antagonistic ties in 129 people in the urban communes discussed in chapter 3.[48] A classic 1969 study of eighteen novitiate monks collected information about members of the group who were disliked.[49] Other work has mapped connections between bullies and their targets in schools, and between supportive and toxic co-workers in workplaces.[50] Still other work has examined social interactions in a massive multiplayer online game by, for instance, using the placement of bounties on virtual enemies as a measure of negative ties.[51]

In light of the paucity of data on antagonistic networks and given the possibly important role that negative ties might play in social structure, in 2013 my lab decided to tackle the topic in a comprehensive and large-scale manner. Villages in the developing world offer an especially appealing natural laboratory to evaluate the structure of negative ties because they are relatively closed social systems where people cannot easily avoid others whom they dislike. So, we mapped the social networks of 24,812 adults in 176 villages in a field site in western Honduras. Using our Trellis software, we asked respondents to identify "the people in this village with whom you do not get along well."[52] I had wanted to make the question even more pointed—perhaps "Whom do you dislike?" or "Who are your enemies?"—but I was advised against this by my project manager, a local expert, who pointed out that Honduras had the highest annual murder rate in the world (86.1 homicides per 100,000 people) and respondents would likely not want to risk answering such questions.[53]

We mapped each of the 176 villages separately and identified all the social connections, both positive and negative, within them, yielding the largest and most detailed study of antagonistic face-to-face ties ever conducted. The villages ranged in size from 42 to 512 adults. Using the three standard name generators, we found that people identified an average of 4.3 friends (which included kinship ties, such as a sibling or spouse), with a range from 0 to 29 (though most people had between 1 and 7 friends).[54] A total of 2.4 percent of the people reported having no friends by these metrics.

The good news is that animosity was much less common than friendship. On average, people identified 0.7 other people they did not like (which could also include people who were kin).[55] A total of 65 percent of the people reported having no one they disliked. This was the case for 71 percent of the men and 61 percent of the women, meaning that women were either more fractious (perhaps because they also generally report deeper relationships) or (as my sister, Katrina, insists) they simply had better memories about prior social interactions. One particularly vexatious person identified sixteen other people she did not get along with. This woman, who lived in a village of three hundred and twelve people, had four people who did not like her. But on the flip side, she nominated eleven people as friends and thirteen nominated her.

Furthermore, seen from the opposite point of view, on average, people were disliked by 0.6 other people, and most people (64 percent) had no one who disliked them.[56] The most disliked person was a woman who lived in a village of one hundred and forty-nine people—she had twenty-five people who identified her as an enemy.

She herself nominated four people as her friends and only two people as her enemies. Alas, just one person nominated her as a friend.

While the pattern of friendship ties was remarkably consistent across villages, animosity varied widely. Although the percentage of negative ties was 15.6 percent overall, this measure was only 1.1 percent in one village and as high as 40 percent at the other extreme. The environmental, social, and biological forces affecting the formation of positive ties generate greater conformity than those affecting negative ties. The current local environments and cultures of the various villages seemed to have shaped animosity much more than friendship. Maps of the resulting networks for four of our villages are shown in color plate 5.

Just as friends tended to reciprocate friendship, enemies tended to reciprocate animosity, but the rates of reciprocation were quite different: 34 percent and 5 percent, respectively. That is, if you named someone as a friend, it was likely that person had also named you as a friend, but if you named someone as an enemy, it was less likely that individual had also named you as an enemy. This difference highlights the fact that people have secret enemies more often than they have secret friends. People declare their friendships to each other but are less likely to state that they are enemies.

These detailed data gave us the chance to quantitatively explore certain old theories (and commonsense ideas) regarding social connections, namely, that these four principles should hold:

The friend of your friend is your friend.
The enemy of your friend is your enemy.
The friend of your enemy is your enemy.
The enemy of your enemy is your friend.

These social rules suggest another principle, first noted by sociologist Georg Simmel in the late nineteenth century and codified by psychologist Fritz Heider in 1946 and psychologist Dorwin Cartwright and mathematician Frank Harary in 1956.[57] The principle is known as balance theory. Certain kinds of triads are seen as balanced, or stable, others as unbalanced, or unstable (see figure 8.3). Over time, unbalanced triads will either break apart (one person will cut a tie) or one of the relationships will change (for instance, you will come to see a friend as an enemy if that former friend becomes friends with your enemy).[58] Friendship and animosity are in endless motion within our social networks.

For the first time using population-wide data at this scale, we were able to quantify that the friend of a friend was nearly four times more likely to be a friend than to be an enemy. Our data also confirmed the second and third rules listed above. However, we found no evidence for the fourth rule, which claims that the enemy of your enemy is your friend. This rule is logical, and it would seem to be true. But it is a bit of a delusion. The person who is one's enemy's enemy has a large chance of being a friend simply because friends are so much more numerous in any ordinary group. It's like trying to figure out whether black birds are more numerous than red birds in one location compared to another. Since black birds are generally more numerous than red birds, we must take into account this background prevalence before concluding whether black birds are more numerous in this particular spot. Once the greater frequency of friendship in general is taken into account, it turns out that one's enemy's enemy is *not* more likely to be a friend. In fact, that person is actually more likely to be an enemy.

This appears to relate to the likelihood that both enemies and friends cluster together within communities in networks. In our analysis, we were able to allow for a third possibility—namely, that people could simply be *strangers* to one another. And people were much more likely to be enemies of people they knew than of people they did not know. One must know someone in order to be either a friend *or* an enemy. This simple observation means not only that the enemy-making process requires contact and familiarity, but also that enemies are more likely to be found *within* one's own group than among people in other groups. We also found that the more friends a person had, the more enemies he or she was likely to have, with each ten extra friends being associated with one extra enemy. Overall, we

**Figure 8.3: Balance in Three-Person Social Systems**

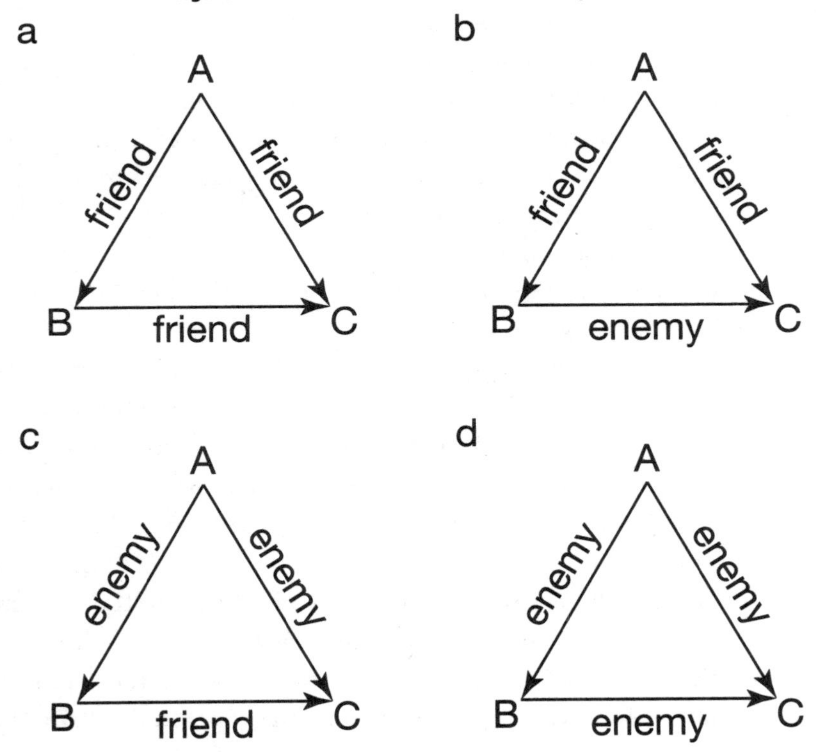


Four possible arrangements of three people (labeled A, B, and C in each triad) are shown. Each relationship between any two people can be positive or negative and can involve friendship or animosity (labeled *friend* and *enemy*). From the point of view of person A, for example, triangles in panels *a* and *c* are balanced, but those in *b* and *d* are not. For the unbalanced triangles, a tie must either break or switch valence (from friend to enemy or vice versa) for balance to be restored.

found that similar underlying social processes, arising from repeated interactions, resulted in both positive and negative ties.

### *Liking One's Own Group*

People choose their friends and then form groups and networks. And these groups interact with other groups in positive or negative ways. These interactions between the individuals of one group and the individuals of another group have equipped humans, over the course of evolution, with still other qualities, including the propensity for in-group bias that is a part of the social suite. But one could well ask why people feel differently about outsiders in the first place. Why do humans not like everyone? Can one not like one's own group without hating other groups? Did these twin tendencies—of ethnocentrism and xenophobia—coevolve as part of the same process and so must always appear together, or did they arise independently and for different reasons?

The preference for one's own in-group is a cultural universal. As we saw in chapter 1, children as young as two will prefer other children randomly assigned to the same T-shirt color. In-group bias is so widespread that, when it attracted the attention of experimental psychologists in the twentieth century, they began to wonder how easily this sentiment could be evoked and just how tiny the differences between "us" and "them" had to be. In a kind of race to the bottom to understand the limits of human degeneracy, a new line of experiments emerged involving what came to be known as the *minimal-group paradigm,* initiated by psychologist Henri Tajfel in the 1970s.[59]

With barely any coaxing at all, it turns out, human subjects treated others who were assigned to *artificially* and *arbitrarily* created groups differently. These groups could be based on the most trivial of features, such as whether the subjects preferred abstract paintings by Klee or Kandinsky.

The ability to elicit in-group bias with such meaningless affiliations was a startling finding. Why would humans treat some individuals more favorably than others over such a trivial social category? I suppose there are some legitimate differences between these two artists (Klee's "ironically refracted realism was alien to Kandinsky's idealism," according to one critic).[60] But most people (including the subjects in the psychology experiments) were unaware of the difference. And anyway, why should it have mattered? But apparently, it did matter. The whole point of the minimal-group experiments is that humans will discriminate against out-group members even though there is no logical reason to do so and even when the groups have no history together or even any communication between them!

Tajfel found that categorization, no matter how trivial, was enough to foster the emergence of ethnocentrism and xenophobia. Initially, he argued that this behavior could be explained by notions of social identity.[61] People were driven to make their own group positively distinct from others. The primary impulse was seen as the need to feel good about oneself, which—because one's sense of self-worth derives in part from one's sense of membership in a group—meant feeling good about one's own group. Treating others in one's own group more favorably helps that group seem better.

But these experiments also found something that depresses me even more than the existence of xenophobia in the first place. People in other experiments who were given the opportunity to assign rewards to in-group and out-group members preferred to maximize the *difference* in amounts between the two groups rather than to maximize the amount their in-group got, reflecting a zero-sum mentality that one group's gain is another group's loss.[62] What seems to be important to people is how much *more* members of one's own group get compared to members of other groups, not how much one's group has. Mere fealty to a collective identity is not able to explain this, because if all that mattered was how well off one's group was, its standing relative to other groups would not matter. But both absolute and relative standings are important to people. Not only must one's group have a lot, it must have more than other groups.

There is still another explanation for in-group favoritism. It can enhance self-interest in addition to self-esteem. When you assign rewards to your in-group, you expect to be more favorably treated by its members, and this is distinct from merely thinking your in-group is somehow superior.[63] Practicing in-group favoritism can lead to practical benefits for oneself. People think better of members of their own group, experiments suggest, and treat them better specifically because they expect the same from the others in that group. The group members feel they have a *shared fate.* It is this sense of togetherness and *expectation of mutual aid,* rather than a need to establish positive distinctiveness for one's own group, that leads to the higher evaluation of the in-group. So in-group favoritism might be strategic; you treat in-group members more favorably because you expect to be treated that way in kind. This is known as *mutual-fate control,* meaning that all the members in the group affect one another's fate even if they cannot force others to behave in a certain way or even communicate about it.[64] It's not just that in-group members' fates are bound together because outsiders affect them jointly; it's also that insiders affect one another's fates because they are part of the same group.

Group identity, like sustained friendships, provides a solution to the risks of unreciprocated cooperation. If you make your altruism contingent on the other person being a member of your group (which is distinct from them being your friend), you can increase the odds that he or she will be willing to pay you back.[65] So being a part of a group with the shared norm of mutual support helps facilitate cooperation, even with strangers—as long as the strangers are "one of us."[66]

### *The Robbers Cave Demonstration*

Another explanation for in-group bias—both evolutionarily speaking and in our daily experience—has to do with *conflict* between groups. Conflict and competition over limited resources (land, food, money, prestige) can give rise to intergroup hostility and to prejudice and discrimination toward out-groups. Feelings of resentment are especially likely to emerge when only one group can be the winner in a competition over resources. Experiments show that positive relations between conflicting groups can only be restored if "superordinate" goals, goals of interest to all the groups, are in place.

In the summer of 1954, psychologist Muzafer Sherif and his colleagues took a group of twenty-two boys who did not know one another and sent them to camp in Robbers Cave State Park in Oklahoma. The boys were deliberately isolated, unaware that they were living in an enjoyable and exciting but nevertheless constantly surveilled world in which their camp counselors were actually participant-observers engaged in a psychology experiment.

The chosen boys had just completed fifth grade; they all came from similar lower-middle-class Protestant backgrounds; and they all had similar intelligence, family structures, and academic skills. The investigators divided the boys into two groups carefully matched for size, athletic and other abilities, popularity, prior camping experience, and temperament.[67] The investigators realized, of course, that their subjects were not acultural; like stranded sailors or Antarctic scientists, they were already members of a culture. They all knew, for example, how to play baseball.

The demonstration proceeded in three careful stages over three weeks. In the first stage, the two groups of boys were set up in separate areas, each group unaware of the existence of the other. Both groups made up names for themselves—the Eagles and the Rattlers—and engaged in activities that fostered in-group solidarity and identification, including designing team T-shirts, choosing supplies for the camp canteen, and participating in poorly supervised outdoorsmanship that would strike fear in the hearts of today's parents (for instance, they portaged canoes through a rattlesnake-infested area). The boys settled on their own norms (such as morning reveille) and chose joint activities (such as finding hideouts), all of which built tremendous in-group solidarity. They even engaged in pinecone battles like the Buyukada boys of my childhood. Over the first week, group members increasingly referred to objects and places (such as flags and swimming holes) as "ours." Each group adopted its own symbols and preferred songs, and each developed an internal status hierarchy with leaders and followers.

Both the Rattlers and Eagles began to express the desire for competitive games, so in the next stage of the demonstration, the investigators put the two groups in conflict with each other. The counselors of each group revealed that there happened to be another group of boys nearby. Initially, one of the Eagles sweetly observed that "maybe we could make friends with those guys and then somebody would not get mad and have any grudges." Alas, his genial feeling did not last. When the two groups met for the first time the next day, derogatory name-calling between the groups began when this same boy referred to one of the Rattlers as "Dirty Shirt."[68]

The two groups competed in zero-sum games such as baseball, tug-of-war, football, tent-pitching competitions, and a climactic treasure hunt. Though the Eagles appeared to be the weaker of the two teams, the scientists engineered their victory. The rewards were pocketknives (much coveted by the boys) as well as a trophy and medals.

Over the course of the competition, the groups came to dislike each other more and more, hurling invectives and speaking of the out-group in increasingly derisive ways (as noted by the counselor-scientists surreptitiously listening to their conversations back in the cabins). The boys' attitudes were measured by having them assess the extent to which the in-group and out-group showed favorable qualities (being brave, tough, or friendly) or unfavorable qualities (being sneaky, "smart alecs," or "stinkers"). The members of each group wanted nothing to do with members of the other group and thought ill of them.

By the end of the second stage of the demonstration, the investigators had accomplished their objective: "There [were] now two distinct groups in an unmistakable state of friction with one another."[69] As Sherif observed:

> The derogatory attitudes toward one another are not the consequence of pre-existing feelings or attitudes which the subjects had when they came to the experimental site. They are not the consequence of ethnic, religious, educational or other background differentiation among the subjects. Nor are they the result of any extraordinary personal frustration in the particular life histories of the subjects, or of marked differentiation and physical, intellectual or other psychological abilities or characteristics of the subjects.[70]

And this intergroup hostility was noticeable despite the fact that there was also bickering and friction present *within* each group.

The final stage of the demonstration was designed to see if the now-hostile groups could be made to work together, effacing the negative attitudes toward the out-group. The scientists contrived to sabotage the huge water tank, located a mile uphill, that fed the entire area. First, they stopped up the faucet on the side of the tank, which the boys all used to drink from, and then they shut off the main valve and covered it with large rocks. The boys were divided into details and sent to search the plumbing lines from the camp up to the tank. When they got to the tank, they were very thirsty from the lack of water, as the experimenters had planned. The boys turned on the faucet to drink, but no water flowed.

And so they started to work together. They climbed onto the top of the tank with a ladder they found nearby (placed there by the psychologists-saboteurs) and ascertained that the tank was full. Then they began to collaborate to get the water flowing again. After they succeeded, the boys began to intermingle in ways they never had before, and mixed groups of Rattlers and Eagles caught lizards and made wooden whistles. After the water-tank exercise, the scientists obliged the boys to work together to choose a movie (*Treasure Island*) and to free a truck stuck in a muddy ditch (the truck held their dinner for the evening). One of the boys said to the others, "Twenty of us can pull it for sure," and so they did, a collective accomplishment that was followed, once again, by friendly conversation and backslapping. When the expedition ended a few days later, the majority of the boys wanted to ride home in one bus, not two, and the seating arrangement in the bus did not follow group lines. And when intergroup friendships were measured more formally, negative stereotypes of the out-group had declined and intergroup friendships had risen.

This demonstration showed that, when groups are united against a common enemy, negative attitudes abate—similar to the science fiction trope of an alien invasion or impending meteor strike suppressing ethnic hatreds on a global scale.[71] In fact, it is possible that the political polarization we see in the United States today, with factions demonizing one another and with increasingly strident and extreme language, may partly reflect the collapse of the Soviet Union and the end of the Cold War. When the country had a common enemy, it fostered more informal solidarity and more civility in domestic politics. This is a general principle. The boundaries of the group and the attendant in-group bias can be broadened by a shared agenda, which facilitates cooperation on a larger scale.

### *Conflict and Cooperation*

The Robbers Cave demonstration, as well as a broad set of similar experiments, suggests that group identity can be a cause of conflict.

But maybe it's the opposite. Maybe intergroup conflict is not a consequence of group identity but a cause of it.

As early as 1906, sociologist William Graham Sumner argued that the love and hate relationships within and between in-groups and out-groups were intertwined. As he put it:

> The exigencies of war with outsiders are what make peace inside lest internal discord should weaken the we-group for war.... Thus war and peace have reacted on each other and enveloped each other, one within the group, the other in the inter-group relations.... Loyalty to the group, sacrifice for it, hatred and contempt for outsiders, brotherhood within, warlikeness without—all grow together, common products of the same situation.[72]

A century later, evolutionary theorists Sam Bowles and Jung-Kyoo Choi used mathematical models to argue that conflict between groups for scarce resources was actually *required* for altruism to have emerged in our species' evolutionary past, even though such conflicts may no longer be a prerequisite for altruism to persist today. These distinct features—altruism and ethnocentrism, cooperation and in-group bias—have together become part of our species' evolved psychology.

The extreme sort of lethal intergroup conflict we see in humans—namely, outright war—is exceedingly uncommon in animals.[73] So it's a puzzle that humans can be so friendly and kind but also so hateful and violent. The only other species that comes close to this duality is the chimpanzee. Maybe, therefore, kindness and hatred are actually related. Mathematical analyses of models of human evolution suggest that, in the past, conditions were ripe for the emergence of both altruism and ethnocentrism, but—and here is the catch—only when *both* were present.[74] They needed each other.

Altruism involves helping in-group members at a cost to oneself, and ethnocentrism (or parochialism) is hostility toward out-group members. Periodic resource scarcity—for example, due to droughts or floods—is a key predictor of conflict in contemporary forager groups (and scarcity of resources, such as oil, is still a good predictor of war, even in modern times). We know that the climate during the Pleistocene (stretching up to ten thousand years ago) was volatile, so our ancestors lived in environments in which occasional competition for scarce resources favored groups that had brave and self-sacrificing members. It was useful to have in-group altruism in the service of out-group conflict. The modeling by Bowles and Choi shows that neither altruism nor ethnocentrism was likely to have evolved on its own, but they could arise together. In order to be kind to others, it seems, we must make distinctions between *us* and *them*.

Political scientists Ross Hammond and Robert Axelrod also showed (again with a simple mathematical model) that ethnocentrism facilitates cooperation between individuals, independent of a you-scratch-my-back-and-I'll-scratch-yours reciprocity.[75] They found that, even when the *only* thing that people could discern was group membership, not prior history of cooperation, the individuals who became most numerous in the population were those who selectively cooperated with their own groups and did not cooperate with other groups. In this analysis, in-group favoritism and selective cooperation can arise simply by equipping people with visible group markers.

There is thus plenty of evidence for the relationship between in-group bias, altruism, and competition. But, using mathematical models, my collaborators (including mathematical biologists Feng Fu and Martin Nowak) and I evaluated whether in-group bias and cooperation could arise even *without* competition between any groups.[76] The key for this to happen is the mere ability of individuals to alter their group membership. Fluid social dynamics can change yesterday's enemies into today's friends. Consider the case of Confederate general Lewis Armistead, who was mortally wounded at the Battle of

Gettysburg. As he lay dying, he gave a secret sign known only to Masons in the hope that it would be recognized by a fellow Mason, which it was—by a Union soldier named Henry Bingham. Bingham protected him and arranged for his transport to a Union field hospital.[77] It is possible to engender shifts in the importance assigned to specific groups and in the attention paid to their boundaries without the presence of a common enemy, simply because of the possibility of switching from one group to another.

Evolutionarily speaking, fluid group membership can provide a means for the preference for one's own group to emerge. Attention to group identity need not arise from intergroup conflict; it can arise from the capacity to adopt successful behaviors seen in other groups and to join that group. Paradoxically, all it takes for members of a group to like their own group and dislike others—and to care about group boundaries and group membership—is for people to be able simply to switch groups.

### *No Necessary Hate*

Prejudicial treatment of out-groups starts when people are very young and it does not seem to vary much with age, which suggests that the capacity for intergroup cognition is innate.[78] Additional evidence for this comes from brain-scan studies that show that there are particular regions of the human brain devoted to social categorization.[79] And as we have seen, the existence of even minimal groups can elicit in-group bias. While xenophobia can arise even without conflict, conflict between groups can surely exacerbate people's dislike of other groups.

But it's possible to have positive feelings about one's own group without necessarily having strong negative feelings about the out-group. While resources might be zero-sum, attitudes need not be. Some surveys of people's opinions, as well as lab experiments, show that in-group bias and out-group hatred are not necessarily correlated, regardless of the evolutionary path our species followed to get to where it is today.[80]

As Toshio Yamagishi and colleagues put it, "Feeling superior based on skin color [for instance] is far different from sending people of the other skin color to a gas chamber."[81] In his classic 1954 book *The Nature of Prejudice*, psychologist Gordon Allport made the same point about the range of attitudes toward out-groups: "At one extreme they may be viewed as a common enemy to be defeated in order to protect the ingroup and strengthen its inner loyalties. At the other extreme the outgroup may be appreciated, tolerated, even liked for its diversity."[82]

Still, there are specific ways in which in-group affinity can indeed facilitate out-group hostility. One is a sense of moral superiority. In some respects, people always feel like their own group is more friendly and trustworthy than any other group, so it is easy for in-group members to believe that they are just better in general. Groups can live in a state of mutual contempt and disgust, which might maintain peace by prompting avoidance and thus reducing direct conflict. But once contact or mixing is inevitable, if feelings of moral superiority are high, groups can be especially prone to slip into deep hatred, enslavement, colonialism, or wars of ethnic cleansing. Power-hungry leaders can cultivate this xenophobia, taking advantage of the fertile terrain provided by in-group identification to actively foster out-group hatred. It does not take much to tip the scales, and given an audience naturally and evolutionarily primed to in-group bias, malevolent leaders can point the finger at another group and lay blame on them for anything they wish.

Many of the vilest expressions of racism and prejudice are likely extreme forms of out-group hatred, not of in-group affection. Still, prejudice can be more subtle, involving not the presence of strong negative views regarding out-groups but the absence of positive views of them. Discrimination and bias can be sustained because humane and generous sensibilities—such as admiration, sympathy, trust, and friendship—are given solely to the members of the in-group and withheld from out-groups.[83]

These observations can lead to a paradox: societies that stress uniqueness and individuality and that provide a fertile terrain for friendship based on the personal and specific can actually be those where our *common* humanity is more easily recognized. It is not a coincidence that the philosophers of the Enlightenment emphasized the special worth of each individual and, in parallel, highlighted the notion of universal human dignity (even though, at least initially, this was not applied to all classes of people). In fact, findings from cross-cultural studies suggest that in-group bias and an emphasis on the distinction between *us* and *them* is higher in collectivist societies (including Communist societies), which stress the importance of group membership and subsume the individual within the group, than it is in individualist societies (where social interdependence is less salient), which stress autonomy.[84] Similarly, the more identities available for individuals to assume and the more cross-cutting they are (such that people who are discordant on religion might be members of the same political party, for instance), the more tolerant a society can be of outsiders and, hence, of everyone.

### *Universal Bias*

Humans often frame the natural world in terms of dualities (a tendency that itself may be innate!). Anthropologist Claude Lévi-Strauss observed that binary opposition (male/female, good/evil, hot/cold, conservative/liberal, human/animal, body/soul, nature/nurture, and so forth) is one of the simplest and most widespread ways that humans come to terms with complexities in the natural world.[85] Unsurprisingly, this tendency to categorize is also applied to social life, demarcating the difference between us and them, between friend and foe.

Friendship is a fundamental category, and philosopher Ralph Waldo Emerson was onto something when he said, "A friend may well be reckoned the masterpiece of Nature."[86] Yet scientists have tended to neglect the role friends have played in the life of our species. An extreme focus on kinship and marriage has obscured the more numerous relationships people have with unrelated friends. These friends are also, after all, the primary members of the social groups we form and live within.[87]

Friendship and in-group bias are indeed universal. Friendship is similar in nature and frequency around the world. Consequently, the social networks humans make are similar. Friendship has demonstrable genetic antecedents and a consistent developmental trajectory worldwide. And people everywhere form groups about which they feel more favorably, even if those groups contain some people they dislike. Such human universals arise primarily as a consequence of our species' shared evolutionary past. Just as ancient forces and environments have shaped the human mind to think and act in certain ways, these same forces have predisposed humans to *interact* in certain ways. Of course, what counts as a suitable choice of friend or as a proper in-group (and what is required to be a member of it) is constantly shaped by culture.

An evolutionary framework evoking between-group comparison or competition is required to understand the origins of this universal, as we have seen. Without that, ideas about in-group bias can border on tautological or be a kind of self-fulfilling prophecy—in-group favoritism makes people behave nicely to their group's members, and this in turn confirms the superiority of the in-group and makes it preferable as a source of friends. An evolutionary perspective allows us to discern ultimate, not just proximate, causes of universal sentiments. If members of other groups pose threats, then it would make sense to be able to detect who they are. This would result in evolution favoring cognitive tools for identifying group membership, and over time, this process would become instinctual. It would be to our advantage to evolve the capacity to tell *us* from *them* if there was any chance that *we* are better friends than *they*.

Living in groups presents different challenges than living in a solitary or even paired fashion. Humans adopted group living as a survival strategy. And to optimize success in this (social) environment, humans took on a host of adaptations (including physical traits and instinctual behaviors) and gave up adaptations suitable for solitary living. This trade-off made it possible for our species to have an enormous geographic range and to be ascendant on the Earth. Like snails carrying their physical environment with them on their backs, we carry our social environment of friends and groups with us wherever we go. And surrounded by this protective social shell, we can then survive in an incredibly broad range of circumstances. As a species, we have evolved to rely on friendship, cooperation, and social learning, even if those appealing qualities were born of the fire of competition and violence.

The true universality of these traits in our own species is also proven by their ubiquity in other social animals, from primates to elephants, as we saw in chapter 7. The ability to form and recognize alliances is essential to social animals, and the ability to categorize individuals as friends or foes, or as inside or outside one's group, is a crucial cognitive predicate for alliances. In this view, bias and prejudice in humans are evolved forms of this otherwise useful capacity. This observation once again reminds us not to fall for the fallacy of seeing whatever is natural as necessarily moral. Out-group hatred can be natural as well as wrong. Our species evolved cognitive systems for the fast detection and maintenance of alliances, but this system can be hijacked or deployed to form the basis of vile actions.

Yes, we generally prefer to be with people we resemble rather than those we do not; to like our friends and hate our enemies; and to value our own groups and revile other groups. But the bigger story here is that we are friendly and kind, and we have a psychology shaped by natural selection to be this way. These features of the social suite work together. They set the stage for us to cooperate with others and to teach and learn from others. We have not evolved simply to live in undifferentiated groups like herds of cattle; we have evolved to live in networks in which we have specific connections to other individuals whom we come to know, love, and like.

## CHAPTER 9

# One Way to Be Social

Until 1960, patients who had failing heart valves due to rheumatic fever or any other cause had no treatment options. They died of their condition. The human heart has four valves, and, when diseased, they can become either too tight or too leaky. Either way, back then, there was nothing a surgeon could do except make an incision into the patient's chest and blindly use a finger to widen the valve's opening if it was blocked, a procedure that was roughly as effective as banging on the TV to fix the cable connection.

It was therefore a huge step forward when, in 1960, surgeon Albert Starr and aerospace engineer Lowell Edwards invented a mechanical heart valve. Their first patient, a thirty-three-year-old woman, died within ten hours of receiving the artificial valve because a bubble of air entered her bloodstream. But their next patient, a fifty-two-year-old truck dispatcher, lived for over a decade before dying from a fall off a ladder while painting his house.[1] Within a few years, many thousands of patients' lives were made better or saved with prosthetic heart valves, and such valves (with various improvements) are still widely used today.

There was a problem with the Starr-Edwards valves, however. Introducing a foreign object into a patient's bloodstream prompted clotting, and those blood clots could then break free and migrate from the heart through the arteries to the patient's brain or kidneys, causing strokes or kidney failure. Patients had to take blood thinners for the rest of their lives to help prevent this. But that medication made them prone to serious internal bleeding, which could kill them just as efficiently as clotting or the underlying heart disease.

So in 1964, a young French surgeon, Alain Carpentier, decided to explore more natural alternatives. There were not enough valves from human cadavers available, so, after studying the anatomy of various animals, he settled on pig-heart valves because their size and morphology matched our species'. But transplanting animal tissue into humans presented a different problem—these foreign bodies would elicit a furious immune response. So the pig valves had to be rendered immunologically inert. Carpentier first attempted this with a process involving mercury, but he eventually succeeded by using the chemical glutaraldehyde to fix the pig tissue (a technique still in use today). Carpentier also devised a Teflon-coated metal frame over which the pig valve was slipped prior to implantation (the metal did not cause problems because it was out of the way of the bloodstream). He called this hybrid a bioprosthesis. Carpentier and his colleagues performed the first successful replacement of a human aortic valve with a pig valve in 1965, followed within a month by surgeries on four other patients.[2] Pig valves are now routinely implanted into tens of thousands of patients each year, precisely because human heart anatomy is so similar to that of our porcine donors.[3]

### *Continuity with the Animal World*

Most animals, including humans, have a bilateral body plan (that is, the right and left sides are symmetrical), which is something so basic we do not even consider it remarkable. Most vertebrates also have hearts, lungs, and other organs incredibly similar to our own. Sometimes these similarities are so direct that animals can be used as stand-ins for humans, as when people dissect frogs to learn about anatomy, test pharmaceuticals in mice, use cow insulin to treat diabetes in humans, or replace defective human heart valves with those taken from a pig. In 1985, when I was a medical student, we operated on dogs to learn about anatomy, anesthesia, and surgery. Over thirty years later, practicing on dogs has been, appropriately, phased out at medical schools. But back then, we did clumsy procedures, and I still remember—with great shame—the thud of the dog's body as I dropped it into the container for disposal after we had performed the assigned splenectomy, which we surely botched.

The similarities between humans and animals do not end with the anatomy and physiology of our bodies, however. Laboratories around the country are now (humanely) using dogs to study animal cognition and emotion in order to gain insight into our own. Slowly but surely, stark, dichotomous statements regarding the supposed differences between humans and animals have yielded to the recognition of countless nonhuman analogues of human behavior. Dogs and even rats have empathy. Crows, crocodiles, and wasps use tools. Gorillas use language. Chimpanzees and elephants form friendships.

These abilities are not precisely the same as ours, but they still amaze and trouble us as we sense our continuity with the animal kingdom in fuller ways. The recognition of this continuity makes it increasingly hard, morally, to ignore that animals' muscles, which we eat, are guided by their brains, which have thoughts and feelings. As we break down barriers between ourselves and the animal world, the human claims to superiority and dominion, not just distinctiveness, break down.

The idea that human social behavior resembles animals' is not new. For instance, similarities between human societies and ants' have been noted since ancient times. In Homer's *Iliad*, Achilles commanded an army of fierce, swarming Myrmidons, from the Greek word for "ant." Ant societies have continued to be a focus of scientific study and popular attention, especially since the 1960s with the groundbreaking work of biologist E. O. Wilson.[4] But humans are social in a materially different way than insects are.[5] The organization of social insects such as ants, bees, wasps, and termites differs from ours in its extreme division of labor and (most important) in the sterility of virtually all individuals within a colony. Plus, all the members of a colony are clones and therefore genetically identical.[6] As we saw, science fiction writers even treat insect societies as the most disturbing contrast to our own.

Human social behaviors have much more in common with early hominids and our primate cousins than with insects, of course. But we have seen that mammals other than primates—such as elephants and whales—have independently evolved similar ways of having friends, for example. If such widely separated species have converged on the same basic way of being social, it demonstrates that this pattern of traits—the social suite—is adaptive and coherent.

### *Convergent Evolution*

The last common ancestor of both humans and chimpanzees lived about six million years ago, while the last common ancestor of both humans and whales lived about seventy-five million years ago, and the last common ancestor of both humans and elephants lived about eighty-five million years ago (meaning that we are more closely related to whales than to elephants).[7] But since the common mammalian ancestor we shared with elephants or with whales did *not* live socially, all of these species must have *independently* arrived at these similar solutions to the challenge of living in groups. As noted earlier, this is known as *convergent evolution*. This process is the development of similar structures or behavioral strategies in very different species, and it can apply to general abilities, such as the capacity for flight—as seen in bats and birds—or to more focused anatomical and behavioral features, such as the protective scales and the tendency to roll into balls seen in armadillos and pangolins. Convergent evolution can also occur at very fine levels, such as the similar intracellular proteins used for vision in both fruit flies and humans.

Usually, convergent evolution applies to the evolution of similar body plans in animals or plants that occupy similar evolutionary niches. Anteaters and aardvarks have evolved exceptionally long tongues to reach into ant holes and termite mounds. Echolocation has independently evolved multiple times, in bats, dolphins, and even a kind of shrew.[8] Australian marsupials have evolved body shapes and behaviors to fill familiar ecological niches for burrowing, grazing, and predating—kangaroos fill in for deer, Tasmanian devils for foxes, wallabies for rabbits, wombats for woodchucks, and so on. New Zealand's iconic kiwi bird, with its fur-like feathers, heavy bones, and burrowing lifestyle, plays an ecological role occupied by rodents in other parts of the world.

Perhaps the most famous example of convergent evolution is the camera-type eye (see figure 9.1). The human eye is structurally similar to that of an octopus, yet our last common ancestor lived seven hundred and fifty million years ago, and that ancestor probably had only a very simple patch of tissue on the surface of its body that was able to do little more than detect the presence or absence of light. Such independently evolved structures are obviously not exactly alike. Octopuses do not have a blind spot, like humans do, because the octopus retina evolved in a more sensible way, with the light-detecting cells on the top layer of the retina, facing forward, rather than on the bottom layer, as in human eyes. The octopus eye also lacks a cornea, in part because vertebrate eyes evolved as an outcropping of the brain whereas cephalopod eyes were pushed in from the skin.

The independent evolution of eyes has occurred at least fifty times across different species—as if seeing the light is inevitable. Given these eerie convergences, some scientists speak of "the ghost of teleology looking over their shoulders," hinting at the question of whether there might be some purpose to evolution or maybe even a

**Figure 9.1: Vertebrate and Octopus Eye Anatomy**

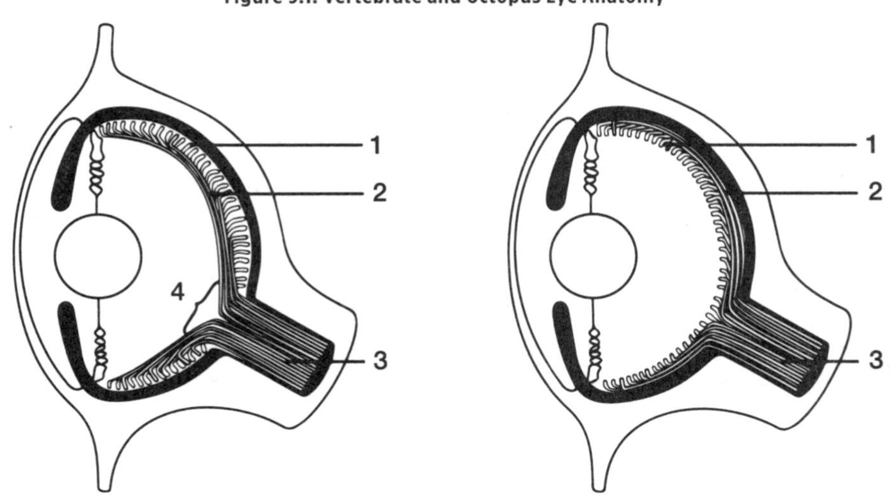


Vertebrate eyes (left) and octopus eyes (right) have many structural features in common, despite having evolved entirely independently. They also have some differences, such as the absence of an optic disk (#4 on the left panel) because of the different positions of the retinal layers (indicated by #1 and #2); #3 indicates the optic nerve. Octopuses also lack a cornea (not shown).

designer.[9] Scientifically speaking, evolution unfolds thoughtlessly, simply in response to chance mutations and environmental vagaries. But convergent evolution sheds light on other deep ontological questions, such as why animals have intelligence at all. Paleontologist Simon Conway Morris argues that, once life appears, it will inevitably culminate in intelligence, as it is a necessary solution to *any* environment.[10] As Morris notes, "Big brains may be, in at least some circumstances, adaptively useful, and are not just fickle blips of happenstance that in due course sink back into the chaotic welter of the evolutionary crucible."[11] Intelligence—perhaps even consciousness—*must* eventually arise.

The widespread evidence of convergent evolution suggests that if we could go back to the beginning of life on this planet and start all the commotion again, we would probably get many of the same features. In fact, Morris argues that we might find similar features in creatures on faraway planets orbiting other stars whose environments resemble Earth's. Alien planets might be filled with familiar-looking life-forms, beings with extremities, camera-type eyes, communication skills, intelligence, and social organization.

Not all scientists agree. Evolutionary biologist Stephen Jay Gould famously argued that if you could start evolutionary history again, the outcome would be totally different. His perspective highlighted the huge role of historical contingency and random events (such as the mass extinctions resulting from meteors that strike our planet every sixty million years or so).[12] The divergence of opinions on this matter depends in part on what level of convergence is being discussed, since we should not expect convergent evolution to result in absolutely *identical* structures. But the key point is that species of animals facing similar challenges—such as the challenge of living socially—will evolve in pertinently similar ways. This helps explain why the social lives of apes, elephants, and whales are so similar.

Of course, as we saw in chapter 4 in our discussion of seashell shapes, the feasible outcomes of evolution are limited. The trajectories of evolution are constrained by the environment itself. If there is light in many sorts of places, creatures will find a way to make eyes more than once. If there is only darkness, they will not.

The environment, obviously, plays a crucial role in natural selection. I'm reminded of this fact when I remember the iconic Greek fishing villages of my youth. As I grew older and traveled more widely, I was puzzled to discover what looked like Greek fishing villages in all kinds of unexpected locations: Nova Scotia, Hong Kong, Brazil, California. The layout of the brightly colored houses, the boats and nets and traps next to the living quarters, the little dockside tavernas, and the narrow, winding roads—it all felt so familiar to me. It seemed that everyone wanted to build a Greek fishing village. Then I realized that fishing villages converge on similar plans because they all serve the same functions: to allow people to launch boats into the water with relative ease, to return safely with their catch, to maintain materials in salty, windy, and wet conditions, and to cope with the hazards of living near the sea.

Like other aspects of living things, social behavior is also inevitable and similarly convergent. But the repeated evolution of similar forms of social living begs the question of how this similarity arose. The appearance of social life sets the stage for something else—the more social a species is, the more social it will become. To the extent that an animal's social interactions are an important feature of its environment (as they usually are), the evolutionary transition from nonsocial to social living can lead to a feedback loop: social animals are programmed to create a social milieu (with features of the social suite) and natural selection then favors cognitive and other traits optimized to cope with this particular social environment. This feedback loop can lead to a rapid convergence on the optimal solution. It's like saying that the most adaptively relevant environment for a snail is its shell, and since the snail also makes the shell, it has the seeds to its own future within it. The snail forms the shell, and over the course of evolution, the shell forms the snail.

In other words, once set on the path to being social, different animal species converged on a similar plan for social living. In a sense, the reason for this is that all social animals adapt to the *exact same environment.* That same environment is the presence of other members of their own species; the animals must live with and take account of others like themselves. The environment we humans have in common with elephants is not that humans are huge, vegetarian savanna dwellers, but rather that we live among and interact personally with other members of our species. Moreover, we take this all-important social environment with us wherever we go, so it is always affecting us and shaping our species' evolution. If the most relevant selection pressure is the *social* environment, that pressure will work similarly for primates, elephants, whales, and other mammals, and it will yield similar outcomes.

In chapters 7 and 8, we saw how the capacity to form friends leads to certain kinds of social-network structures with consistent mathematical properties across species and across societies. Let's now consider how other aspects of the social suite—individual identity, cooperation, hierarchy, and social learning—fit together and form a blueprint that is at the core of a good society.

### *Unique Faces*

You might take it for granted that you can differentiate yourself from others and tell others apart, but the ability to have an individual identity and recognize the identity of other individuals (especially beyond one's mate or offspring) is uncommon in the animal world. Why should animals be distinctive in appearance or even personality, and why should they be able to tell others apart?[13]

Facial recognition is a key part of human social and sexual interactions.[14] Humans who are unable to recognize faces, a condition called prosopagnosia, have a severe handicap.[15] Imagine running into an ex-boyfriend at a bar and asking him if you had ever met before. This kind of experience is routine for some people. They sometimes even have trouble recognizing their parents. One patient describes seeing a face as dreamlike—"incredibly vivid in the moment but it drifts apart seconds after I look away."[16]

But recognition of faces is only one side of the issue here. The other side is that faces must differ one from the other. Compared with other parts of human bodies, faces are unusually variable and distinctive. You would have a hard time recognizing your friends from pictures of their hands or knees but no trouble at all recognizing them from pictures of their faces (figure 9.2).[17]

The ability to express and recognize individuality evolves when that ability is beneficial. The features animals use to identify others can be categorized into two types: *cues* and *signals*. Identity cues are phenotypic traits that make it possible to tell one individual from another but that do not themselves confer a survival advantage. In humans, fingerprints are unique and can be used to identify individuals, but they did not evolve to signal that, and humans do not ordinarily identify people by noting fingerprint patterns.[18] Fingerprints, like the unique pattern of small blood vessels in everyone's eyes, are therefore merely a possible cue.

Identity signals, however, are phenotypic traits that facilitate individual recognition while also assisting an animal's survival. If you do not want others to mistakenly attack you, neglect to repay your kindness, forget that they had sex with you, or fail to recognize you as their offspring, then you need some way to indicate *This is me, not someone else.*[19] In order to achieve this, the relevant phenotypic trait should have a lot of variation so that it can be noticeably distinctive and memorable.

As we would therefore expect, facial traits show more variability than other parts of our bodies. And because every detail of one's face might be useful to convey one's identity, it is advantageous to have as many combinations of traits as possible so as to make it easier

**Figure 9.2: Distinctive Appearance and Individual Identity in Humans and Penguins**

Humans show striking variability in facial appearance, even in genetically homogeneous populations, such as in Finland. On the left (A) are portraits of six male Finnish soldiers, each of whom is clearly distinctive. By contrast, as an example, king penguins (B) have a much more uniform physical appearance. Nevertheless, although king penguins are not known to recognize individuals using visual signals, they do have highly distinctive vocalizations used for this purpose.

for every face to be unique. Each aspect of the face, from the distance between the eyes to the shape of the ears to the height of the hairline to the angle of the cheekbones and so on, should be combinable into as many patterns as possible so as to make individuals uniquely identifiable. This means that these variable facial features should not be related to one another.[20]

As an example, consider two facial traits: the distance between the eyes, and the width of the nose. If people with wide-set eyes always had wide noses, for example, the information conveyed by each of those two features would be redundant, leaving just *two* options: wide-set eyes and a wide nose, and close-set eyes and a thin nose. From the point of view of increasing the options to have distinctive faces, it would be better for these features not to be correlated. That way, there would be four possible types: wide-set eyes and a wide nose; wide-set eyes and a thin nose; close-set eyes and a wide nose; and close-set eyes and a thin nose. When there are many more traits under consideration than just these two, the details of human faces should be combinable into endless configurations (see figure 9.3).[21]

Variation in human faces is so important that rare phenotypes that help individuals appear special should be favored by natural selection. Consequently, there is indeed a lot of variation in the genes that code for facial features. By contrast, in genes that code for traits that should not vary much, such as pancreatic-enzyme function, there is less variation; all pancreases should, ideally, work the same, whereas all faces should, ideally, look different.[22] To be clear, individuality in an organism (whether in its face, personality, or any other attribute) can emerge for reasons other than genes.[23]

Human faces are distinctive in males and females and in the young and old, suggesting that facial distinctiveness evolved to serve multiple purposes. For instance, if it were crucial that mothers be able to recognize the faces of their own children when they are babies but not once they got older, humans could have evolved to have distinctive

**Figure 9.3: Correlated Dimensions in Facial Features (Such as the Nose) and Non-Facial Features (Such as the Hands)**

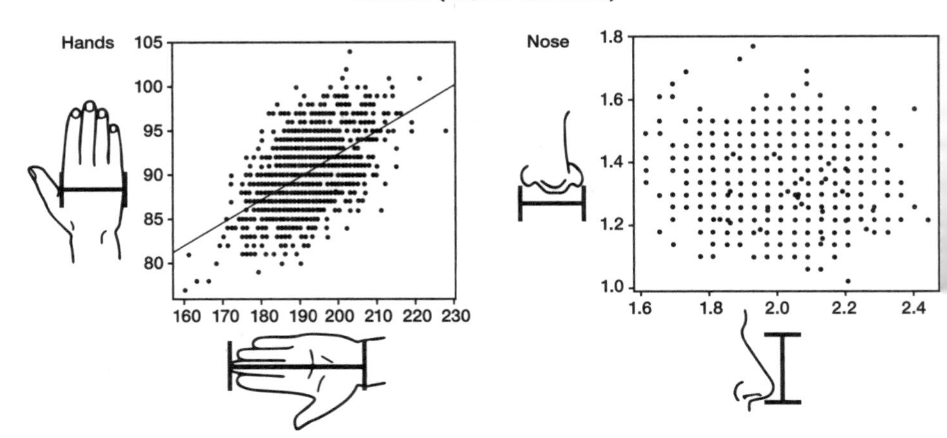


A comparison of the measured length and width of hands and noses in a sample of humans is shown. The scatterplots show the dimensions of the indicated aspects for American male service members. For most body parts, such as hands, larger individuals have larger dimensions, and, pertinently, the width and length of an individual's hand are correlated. The line in the left panel indicates that the relationship is significant and positive, such that wider hands are also longer. By contrast, in the right panel, the width and length of the nose are *not* correlated (and there is no line in the right panel).

faces early in development that later transitioned to a homogenous type of adult face. Or, similarly, facial distinctiveness might have evolved to appear only during individuals' reproductive years, when partner identification would be key. But that's not the case; individual recognition is important across our life spans and in many contexts.[24] Variation in facial appearance, the lack of correlation across facial features, and the observed variability in genes coding for facial features all point to the use of faces as identity signals because of the survival advantages conferred.

This also reinforces another important idea: how we interact with one another filters down to affect our genes. Our *social* interactions, which require and benefit from facial uniqueness, can affect what sort of genetic variants we have by maintaining diversity in regions of the genome that code for facial appearance. Across evolutionary time, human social interactions have shaped our species' bodies, not just our minds.

### *Self and Other*

Individual recognition is especially useful if a species can benefit from non-kin cooperation. If you could not be sure who was who, then how could you remember which people had been nice to you in the past or know whom to cooperate with in the future? How could you avoid the person who did not share food or who slunk away from the table when the check arrived? If you interacted only with kin, with whom you shared genes, this might not matter; you could just help everyone, even if people did not reciprocate, in accordance with the principles of kin selection we saw in chapter 7. But to facilitate cooperation among non-kin, identity is especially valuable as a way to reduce cheating.

As social complexity grows, the advantage of recognizing and remembering individual members of the species becomes increasingly paramount; friendships and alliances must be maintained, antagonisms must be recognized, and hierarchies must be heeded.[25] Jane Goodall repeatedly captured these recognition capabilities in her early work with the chimpanzees in Gombe National Park, noting that chimpanzees could recognize one another individually by their vocalizations.[26] Subsequent research in laboratory settings has confirmed that apes make use of a variety of signals to discriminate among individuals.[27]

In one series of clever experiments, five chimps and four rhesus monkeys were required to match identical portraits of unfamiliar peers. Both species were able to perform the task, with chimpanzees reaching an accuracy rate of 82 percent by their second exposure to the facial stimuli. A second experiment showed that the animals could identify another member of their species as they would in nature, in various visual orientations and configurations.[28] A study of squirrel monkeys showed different levels of electrical activity in the brain depending on the familiarity and social status of monkey (or human) faces.[29] Some primates can even match two unfamiliar kin to each other with moderate accuracy, as we might be able to detect that two men are brothers even though we have never seen them before.[30]

Elephants have identity too. They use low-frequency sounds to coordinate behaviors among family units and bond-group members across distances of several miles. Some of these sounds—announcing the reproductive readiness of both males and females—can be very loud (116 decibels) even though they are at frequencies humans cannot hear (12 to 35 hertz). Elephants can also uniquely identify one another using infrasonic "contact calls" (with harmonics extending into the audible range for humans). Some older matriarchs can uniquely identify over one hundred individuals with whom they have had previous interactions.[31]

We know less about how cetaceans identify one another, but one study found that every dolphin in a group of bottlenose dolphins in Scotland developed a signature whistle, like a name.[32] The animals responded to hearing their own signature whistles by calling back. These whistles function like an airplane pilot's call sign (*Delta 811 to Tower, come in*). Further evidence that these whistles serve a name-like function is that a great deal of each dolphin's repertoire of communications is its signature whistle, much like if you sent and received many short e-mails and your e-mail address was a part of all of them. The dolphins also copied and repeated the names of their social partners more often—again, just as you would be more likely to use the names of your friends when you communicate.[33]

### *The Mirror Test*

Social animals can recognize others' identities and can signal their own. But can they also recognize themselves and have a sense of their *own* identity? You may take for granted that the person you see in the mirror each morning is you. Humans can perform this magic trick by around eighteen to twenty-four months, but it's very rare in all but the most social animals.[34]

Mirror self-recognition has so far been consistently demonstrated in humans and apes, elephants, and dolphins.[35] Animals that recognize themselves when seeing their image in a mirror generally go through several stages reflecting increasing curiosity and awareness. But some species can even pass the final mark test, in which an animal preferentially touches or explores a mark placed on its own body that can be seen only in the mirror, thus indicating that it recognizes itself (this typically involves anesthetizing the animal and applying the mark).

In 1970, primatologist Gordon Gallup Jr. argued that if chimpanzees were able to recognize themselves in a mirror, they would exhibit fewer *social* responses to the reflected image (such as vocalizing, threatening, and bobbing) and more *self-directed* responses (such as self-exploration). And indeed, Gallup observed these self-directed behaviors in front of a mirror:

> grooming parts of the body which would otherwise be visually inaccessible without the mirror, picking bits of food from between the teeth while watching the mirror image, visually guided manipulation of anal-genital areas by means of the mirror, picking extraneous material from the nose by inspecting the reflected image, making faces at the mirror, blowing bubbles, and manipulating food wads with the lips by watching the reflection.[36]

Subsequent scholarship corroborated the presence of self-recognition in chimpanzees and extended these findings to other great apes such as orangutans, bonobos, and gorillas (virtually all monkeys fail the mirror self-recognition test even though, as we saw above, some monkey species can individually recognize *other* members of their species).[37] Mirror self-recognition appears to depend, paradoxically, on early social exposure to others. Chimpanzees born in the laboratory and raised in isolation generally do not show signs of self-recognition.[38] A sense of self and of being distinct from others may actually require the presence of others from an early age.[39]

A gorilla named Koko was famously taught to use sign language, and Francine Patterson and Ronald Cohn took advantage of her communicative abilities to investigate self-recognition. They placed black pigment on Koko's upper right gum, a target area that she touched just twice without the mirror but fourteen times with the mirror. Perhaps even more telling, when Patterson gestured to Koko's reflection in the mirror and asked, "Who is that?" Koko replied, "ME THERE KOKO GOOD TEETH GOOD."[40] Even more astounding, when Koko's gorilla playmate Michael saw a pink mark that had been placed on his forehead during a mirror test, he apparently signaled that he thought he had a wound on his face and then reacted with symptoms of stress; he used sign language to describe the killing of his mother by poachers, which he had witnessed. Self-recognition indeed.[41]

However, we cannot dismiss the possibility that an animal might recognize its reflection as itself without having a true sense of self; for instance, via a phenomenon known as kinesthetic matching. Perhaps when an animal looks in a mirror, it is simply noting that the mirror image moves when it moves its body, whereas the other objects in the mirror do not move.[42] Proof of this idea comes from a very simple humanoid robot named Nico; it can watch itself in a mirror and use simple mathematical programming to learn the difference between self and other without any social understanding at all. When it moved, it "learned" that only its own reflection and nothing else in the mirror moved. Still, this robot could not pass the mark test. And if it was as easy as matching visual stimuli to motor action, why don't cats and dogs pass the mirror test?[43]

In 2006, Frans de Waal, a pioneer in understanding animal social interactions, and his colleagues Joshua Plotnik and Diana Reiss were the first to explore mirror self-recognition in elephants, working with three animals at the Bronx Zoo: Happy, Maxine, and Patty.[44] A huge mirror was placed in their enclosure. When the mirror was covered, they ignored it. But when it was uncovered, their interest was piqued. For example, Patty spent nearly half her time in front of the mirror when it was uncovered and only 3 percent of her time there when it was covered. The elephants also explored its surface and searched behind it, putting their trunks over and under the mirror; they even tried to climb over the wall and look behind it. The zookeepers had never previously seen the elephants try to look over or under enclosure walls, so the mirror was clearly something different. After four days of exposure, the elephants gradually appeared to learn that the image in the mirror was not a manifestation of another animal.

The elephants moved side to side and up and down, watching their images do the same thing, and they played peekaboo with their images, moving their heads into and out of the line of reflection, like children. As the chimpanzees did, the elephants used the mirror to explore parts of their bodies they could not otherwise see. Maxine explored the inside of her mouth and used her trunk to pull her ear slowly forward while watching her reflection. Only Happy passed the mark test, however. She was Happy and she knew it. While it might seem disappointing that only one of these elephants passed this stage, it's important to note that scientists are studying the outer limits on the capacities of a species, not the inevitability of the behavior in every animal. It's similar to the idea that, while not every human can make a beautiful abstract painting, do laser physics, or speak many languages, it's still noteworthy that some can.[45]

Self-recognition in two dolphins was documented at the New York Aquarium in Brooklyn in 2001. Because dolphins cannot touch or point to their bodies, the scientists observed the dolphins carefully as they twisted and turned in front of the mirror and repeatedly approached it to examine the parts that were marked.[46] In subsequent tests, the dolphins played in front of the mirror, opening their mouths, blowing bubbles, and watching themselves swim upside down.

Finding such strong parallels among apes, elephants, and dolphins provides compelling evidence for convergent evolution in the cognitive capacities to recognize one's self. The neuroanatomy of the dolphin brain is very different from a human's, yet dolphins still evolved this ability.[47] Individual identity is an essential part of the social suite.

### *Identity and Grief*

Another way we can understand identity and individuality is through the expression of grief. Nonhuman primates and elephants grieve for dead animals they were close to but not for those to whom they had no special attachment.

Grief is familiar to me both because of my work as a hospice doctor and because of my own loss, at age twenty-five, of my forty-seven-year-old mother, who died after a nineteen-year struggle with Hodgkin's lymphoma. Grief is a rare and special emotion because it is tied to the loss of particular, identifiable people. A person can be angry at a stranger, but does not ordinarily grieve a stranger's death. Intense, very painful, and long-lasting, grief has been described by many people as a physical sensation with crushing feelings in their chests, aching shoulders, and tears so copious that their faces hurt.[48]

There is much evidence that grief is physiologically harmful and can even increase one's subsequent risk of death, so ancestors of ours who did not feel grief should have outsurvived those who did. Therefore, how could this sensibility have evolved? Some theorize that grief motivates people to connect with others in order to reduce the pain. In this account, grief is adaptive because it keeps humans, a social species, from being alone—just like the pain you feel when you touch a hot pan is adaptive because you jerk your hand away from the aversive stimulus. Less plausibly, some hypothesize that the anticipation of grief might compel people to work harder to keep loved ones alive, which would be advantageous. Still another theory views grief as a signal from those who are suffering, a sort of plea for help. I think the most likely idea is that grief is a by-product of a psychological system that evolved to make humans feel bad when separated from their living kin, since *staying together* was presumably adaptive. In this view, grief relates to social cohesion, and it is the price we pay for social intimacy.

Archaeological evidence confirms that personalized responses to death have been a powerful and long-standing experience, stretching back to the famous Egyptian funerary procedures more than six thousand years ago.[49] One man, known to archaeologists as M9, who died at age thirty in Vietnam four thousand years ago, suffered from a congenital condition known as Klippel-Feil syndrome that causes paralysis. In the opinion of the archaeologists who excavated his skeleton millennia later, his disability was "so severe as to be inconsistent with life without the long-term interventions of dedicated caregiver(s)."[50] His extremities were atrophied, and his jaw did not work properly. Since there were no draft animals among his people, he was surely carried by others and was incapable of obtaining food and water on his own. His remains showed no evidence of abuse or neglect and there was no indication of pressure sores or related infections in the bones, which suggests that he was kept clean by others. Such a person could not have survived without the assistance of a *community* of others to care for him, people to protect him when others went foraging, people to carry him when others tired of doing so. We must conclude, I believe, that he was loved. And it seems likely that he was mourned as a valued individual within the group when he died and was carefully buried.

There are further examples of disabled people who clearly had sustained help in the distant past: A forty-year-old Neanderthal with an amputated right arm and other injuries who died forty-seven thousand years ago in Iraq.[51] A seventeen-year-old boy, known to archaeologists as Romito 2, with severe dwarfism who died eleven thousand years ago in Italy.[52] A fifteen-year-old boy who was paralyzed due to a neurological condition called spina bifida (now easily prevented by pregnant women simply eating breakfast cereal with folic acid) who died seventy-five hundred years ago in Florida.[53] The loving care required to maintain them all their lives probably made its mark after their death too—and many of these individuals were buried in special ways, suggesting an unusual attachment to them. I should note, of course, the many counterexamples. There are countless ancient burial mounds, seemingly devoid of special sentiment, filled with the discarded bodies of infants or children, especially females.

Grief, too, has parallels in other social species. Some of the most famous descriptions come from the work of Jane Goodall:

> Flint [a young chimp] . . . stopped and stood motionless, staring down at an empty nest. . . . The nest was one which he and Flo [his mother] had shared a short while before Flo died. . . . He had travelled for a while with Figan and, in the presence of his big brother, had seemed to shake off a little of his depression. But then he suddenly left the group and raced back to the place where Flo had died and there sank into ever deeper depression. By the time Fifi showed up, Flint was already sick, and though she groomed him and waited for him to travel with her, he lacked both the strength and the will to follow. Flint became increasingly lethargic, refused most food and, with his immune system thus weakened, fell sick. The last time I saw him alive, he was hollow-eyed, gaunt and utterly depressed, huddled in the vegetation close to where Flo had died. . . . The last short journey he made, pausing to rest every few feet, was to the very place where Flo's

> body had lain. There he stayed for several hours.... He struggled on a little further, then curled up—and never moved again.[54]

Flint died three days after his mother.

Goodall, of course, was not alone in witnessing and recording deep emotional and physical reactions to loss in primates.[55] One study, using fecal samples collected from a troop of more than eight baboons in Botswana, found that glucocorticoid levels (a measure of stress in both baboons and humans) were higher in female baboons most closely related to a deceased animal.[56] Even more interesting, these females appeared to reduce their stress by increasing the frequency of grooming and the number of grooming partners. It seems like the baboon equivalent of a human wake, and it's not lost on me that, in many cultures, women engage in hair-modification rituals (such as cutting their hair or pulling it out) during bereavement. The grieving female baboons who expanded their grooming networks had substantially reduced glucocorticoid levels compared to grieving females who did not.

Cetaceans, primates, and elephants, like humans, appear to treat bodies of deceased loved ones gently, not as mere inanimate objects. One orca whale mother, assisted by her pod, was observed buoying up her dead baby for three days in a manner one expert described as follows: "They know the calf is dead. I think this is a grieving or a ceremonial thing done by the mother.... She doesn't want to let go."[57] Primatologists have seen adult female chimpanzees, gorillas, and snub-nosed monkeys carrying the corpses of infants (both related and unrelated) long after the corpses have started to decompose.[58] Chimpanzees have even been filmed cleaning a corpse's teeth in a mortuary-ritual-like behavior.[59] The parallels to human grief rituals, such as the Jewish tradition of sitting shivah and the Muslim ritual practice of washing the body in preparation for burial (also seen in many other religions), are striking.

Zoologist Cynthia Moss has explored elephants' response to death. In one vivid account, she describes a herd of elephants visiting the bones of its deceased matriarch, Emily:

> They stepped closer and very gently began to touch the remains with the tips of their trunks, first light taps, smelling and feeling, then strokes around and along the larger bones. Eudora and Elspeth, Emily's daughter and granddaughter, pushed through and began to examine the bones.... All elephants were quiet now and there was a palpable tension among them. Eudora concentrated on Emily's skull caressing the smooth cranium and slipping her trunk into the hollows in the skull.[60]

In her memoir *Coming of Age with Elephants*, Joyce Poole describes a similar vigil by an elephant mother over her dead newborn: "Every part of her spelled grief."[61] Poole has "no doubt that elephants have conscious thoughts and a sense of self,"[62] and she described elephant grief, on another occasion, as follows:

> The family approached [Jezebel's] remains and then suddenly stopped and became silent...and then spent the next hour turning the skull, the jaw, and the long bones over and over. The elephants, who appeared to be in a sort of trance, neither interacted nor vocalized and seemed to focus only on the dead elephant. Jolene, Jezebel's daughter, appeared to be the most absorbed of the group.... Why would an elephant stand in silence, over the bones of its relative for an hour if it were not having some thoughts, *conscious* thoughts, and perhaps memories?[63]

Like the chimpanzees Flint and Flo, some elephants have reportedly died of broken hearts too. One captive matriarch stopped eating and starved herself to death after her protégée died in childbirth.[64]

Elephants may even have a kind of collective mourning, like humans do. Large groups of elephants have shown symptoms associated with grief, including abnormal startle responses, depression, and extreme aggression.[65] These population-level psychopathologies appear related to the decimation of elephants throughout Africa due to widespread poaching and habitat loss; between 1900 and 2005, the population plummeted from ten million to just half a million individuals. In human populations, mental-health professionals refer to *historical* and *intergenerational* trauma to describe the chronic collective burden imposed on later generations by widespread death from warfare, enslavement, or famine. By this measure, the sheer scale and magnitude of loss within elephant society and the grief it has caused is not unlike the exceptionally high rates of PTSD found among human survivors of areas devastated by violence.[66]

### *Animal Cooperation*

Virtually all observational accounts of social behavior in apes abound with evidence of cooperation. Among the most influential experiments in this regard is primatologist Meredith Pullen Crawford's classic 1937 study in which two chimpanzees in a cage were positioned at some distance from a crate full of food.[67] Each chimpanzee was given access to a rope connected to the crate; however, the crate was designed to be too heavy for one chimpanzee to move and required coordinated effort from both of them. To successfully reel in the food, the chimpanzees had to work in a synchronized manner, which they did without any outside guidance.

Crawford's experimental model—creating situations in which cooperation yields a mutually desired outcome—has been reproduced many times. One experiment placed pairs of capuchin monkeys in adjacent sections of a single test chamber, separated only by a mesh partition. In front of each of the monkeys was a cup holding food and a bar that, when pulled, would move the two cups closer.[68] Once again, given the weight of the apparatus, both capuchin monkeys needed to pull the bars simultaneously, and high levels of cooperation were observed. To ensure that this cooperation was not merely a chance occurrence, the researchers showed that the animals performed much better when they could see each other through the partition, suggesting that each monkey understood the other's role in the task and that they maintained their cooperation through forms of visual communication.

Frans de Waal and his colleagues adapted the classic Crawford experimental setup for pairs of elephants who had to walk in two lanes up to a table and cooperate to jointly pull a rope for a food reward in two bowls (figure 9.4).[69] The elephants quickly learned that in order to be successful, they both had to pull the rope at the same time.

But how could the scientists be sure that the elephants actually understood that cooperation was needed? Perhaps they were merely learning "See the rope, pull the rope, get the food" and were acting this way in parallel, without any recognition that joint effort was required. A second experiment addressed this by varying the time at which the elephants were released into their lanes. If the elephants had learned only to pull on the rope and not that the other elephant had to be present and pull on the rope at the same time, then the first elephant would start pulling as soon as it got to the table. But the researchers found that the first elephant did indeed wait for its companion to arrive, pulling on the rope only when cooperation was possible.

But what if the elephants had simply learned to pull the rope only when standing next to another elephant, not realizing the role a partner's cooperative efforts played in their success? To address this, the investigators came up with a third experiment in which only one elephant had access to the rope. The end of the rope for the elephant's partner was coiled up and out of reach, though it was visible to both elephants. This made obtaining the food impossible. If the elephants understood that cooperative effort was absolutely required to achieve the objective, they would not bother to pull on the rope. And sure enough, when a partner did not have access to the rope, the first elephant did not pull on its own end.

Some of the elephants even developed their own novel strategies to get the reward. One young female elephant reached 97 percent success not by pulling on her end of the rope but by approaching it and simply stepping on it firmly. This prevented the rope from being

**Figure 9.4: Experimental Apparatus to Test Elephants' Ability to Cooperate**

Three views of an apparatus to evaluate elephant cooperation are shown. View 1 shows a ground-level perspective from beyond a table upon which food rewards are placed in bowls attached to the top of the table. View 2 shows a bird's-eye perspective. As part of the procedure, the two elephants started from the release point located ten meters behind the table and walked down two separate, roped-off lanes (shown in views 1 and 2 as a dashed line). View 3 shows a side perspective from the base of the barrier. The table apparatus consisted of a sliding table that could be moved by grasping ropes. By design, only coordinated pulling of *both* rope ends by the two elephants at the same time would move the table toward them so that they could retrieve the food. Pulling one rope only resulted in the rope becoming unthreaded through the pulleys.

pulled away, but it also forced her partner to do all the work of pulling! In humans, we call this exploitative strategy *defection,* since it involves taking advantage of a cooperative partner. Indeed, the diversity of strategies adopted by the elephants suggests they understood the nature of the task and were not merely engaged in rote learning. The elephants were super-cooperative in another way: they were unbothered by unfair distributions of food (that is, if there was more food in one bowl than in the other). They continued to cooperate with each other regardless. This is rather shaming to those primates (including us) who would be quite annoyed by this situation.

### *Human Cooperation*

Our species started out in dispersed groups of a few hundred people with smaller bands cooperating on diverse tasks, but today there are billions of us in an interconnected web. From forager societies to nation-states, cooperation is a central organizing principle of human life. Nowadays, we vote and form governments, pay taxes and care for the needy, and participate in large-scale religious observances, all because we are willing to cooperate with complete strangers.

Cooperation and altruism have long puzzled scientists because there is no simple explanation for how humans evolved to cooperate, given that natural selection generally favors self-interested acts.[70] All the members of a group may be better off if everyone contributes, but each person individually may be better off if he or she does not contribute. As a result, we might expect groups to fail because everyone has an incentive to free-ride on the efforts of others. Yet both modern and ancient human societies rely on cooperation at a level unseen anywhere in the animal world.

*Cooperation* is formally defined as a contribution to an outcome that benefits all members of a group (even a pair of people), regardless of whether the other people themselves contributed to the outcome. Those who contribute (cooperators) pay a cost for doing so, and those who do not contribute (the defectors or free-riders) pay nothing, like the sneaky elephant described above. Since defectors gain more than cooperators, they have an evolutionary advantage if the gains can be used to increase survival and reproduction. So it is puzzling to see so many acts of cooperation all over the place, in humans as well as in other species.

Decisions about cooperation must have had a critical effect on survival and reproduction in our ancestors. *Should I join a perilous hunting party to bring back food? Share what I have foraged and have less for myself? Defend the camp when it's attacked at some risk to my life?* The answers to these questions have had evolutionary implications for hundreds of thousands of years. You might wonder how this logic works in a modernized society, when human reproductive capacity is no longer closely related to material gains. But it is important to remember that for most of our history as a species, up until about two centuries ago, humans all lived on the edge of death. We are still marked by this history, evolutionarily speaking. So we are left with the question: Why didn't selfish defectors take over the population and drive out the cooperators? That is, why aren't we all selfish today?

One theory has to do with family. A mother who dives into a frigid river to save her child clearly pays a personal cost (possibly even her own life) to provide a benefit to her child, and with every heroic mother's death, that behavior—and the genes that contribute to it—would have tended to become more and more rare. However, even if the mother dies, her genes live on in her child. This process, as we saw in chapter 7, is called kin selection.

However, most of our interactions are with *unrelated* individuals, not kin. A favorite question among economists is why a solitary trucker at a remote truck stop tips a waitress whom he is unlikely ever to see again. Yet he almost always does.[71] Strangers in anonymous modern cities are mostly cooperative and decent to one another, and they even band together to an extraordinary degree during natural disasters and other crises. Sometimes, these impulses take on dramatic forms, such as the firefighters from all parts of the United States who rushed to New York City after September 11, 2001, and the ordinary citizens who left their homes and drove to Louisiana and Texas after Hurricane Katrina in 2005 just to help strangers.[72]

So the only way to explain the human tendency to cooperate as a result of kin selection is to assume that our species evolved its cooperative behavior very long ago when humans lived mainly in small, familial groups and that our genes still happen to carry those cooperative traces. But that explanation fails to account for the social networks of foragers like the Hadza who researchers believe live close to the way of our species' ancestors. Their life is full of interactions and friendships among genetically unrelated individuals, as we saw. Family members are important, but the Hadza spend only a fraction of their time with family every day.[73] Some other mechanism is needed beyond kin selection.

Another theory about the evolution of cooperation, one for which there is much evidence, is called direct reciprocity, and it has to do with *repeated* interactions across time. The basic idea is that the promise of cooperation tomorrow can incentivize cooperation today. When you get to know someone and build a relationship of mutual cooperation, you both gain. This reinforces trust and expectations that the relationship will continue to be beneficial—regardless of whether that person is kin. In fact, the best strategy is to start out cooperating with someone and thereafter copy whatever that person responds with.[74] If he or she cooperates, you reward that person with a future round of cooperation. If he or she defects, you punish that person with a future round of defection. This tit-for-tat strategy is both widespread and effective. Moreover, the rounds of cooperation can, in fact, be widely spaced in time, and the goods exchanged can be extremely varied.

But the problem with direct reciprocity as a theory for large-scale human cooperation is that we sometimes interact with people only once (especially in modern life). Direct reciprocity cannot explain the truck driver's tip, or why a person moves aside to let a stranger pass on the sidewalk, or why someone derives satisfaction (a "warm glow") from being generous to a homeless person. Possibly, such decency is just a spillover from the instinct to act nicely to people you interact with repeatedly, an instinct humans evolved when all of their interactions were indeed repeated and ongoing.

But there is another explanation for this kind of cooperation that is based on people living in groups. *Indirect reciprocity* assumes that interactions are observed by some people in the group who then inform other members of the group. In other words, people gossip. When group members share good or bad stories, the conversations they have can affect other people's reputations. One person is kind to another because it might lead someone *else* to be nice to the first person later. Mathematicians who create abstract models of such processes have shown that natural selection favors the use of reputation as a tool to decide whether or not to cooperate with others.[75]

Both direct and indirect reciprocity work because they keep people from taking advantage of one another. You can stop interacting with a person who defects or you can simply never start. In both cases, a society based solely on two-person encounters could evolve to support very high levels of cooperation. But these mechanisms become less effective as populations increase in size. Direct reciprocity breaks down because cheaters can cheat and then move on to new victims; there are many marks in large populations. Indirect reciprocity breaks down because it is harder to transmit information about and keep track of everyone in the society; cheaters can more easily hide in larger groups.

Moreover, many of the most important actions of humans require whole groups. Ten people might be needed to bring down a wildebeest or protect the camp from invasion. And these acts of cooperation produce what are called public goods, which are enjoyed by everyone. A large animal may yield enough meat for everyone in the camp. Fighting off rivals may protect many people, even if some of those people did nothing to help with the defense. So the benefits accrue for the entire community, but the costs are still borne by individuals who decide to contribute to the public good by risking their lives.

Furthermore, individual and group incentives may be at odds. This kind of problem goes by many names; it's a public-goods or common-resource or collective-action problem. How do people collaborate to protect shared meadows where their livestock roam? Classical models that assume all people are purely self-interested predict that lands will be overgrazed, seas overfished, and air polluted because individual incentives are contrary to what is best for the group. Ecologist Garrett Hardin famously called this the "tragedy of the commons."[76] Individuals act selfishly because the benefits of the acts accrue to them but the costs are divided across the whole group. These types of group-wide interactions and collective efforts cannot easily be addressed with the notions of reciprocity and cooperation we have been considering so far.

### *The Importance of Punishment*

It is difficult to maintain cooperation as group size increases.[77] Each person becomes individually less important for cooperation. If it takes two people to build a shed, a decision by one of them not to participate could have a big effect on whether or not it gets built, and the defector cannot avoid detection. But if it takes one hundred people to build a dam, a single person's decision to defect is unlikely to matter that much; the free-rider can remain anonymous and can probably enjoy the benefits of the dam being built without paying the costs. As a result, evolution tends to favor free-riders more in larger groups.

One way to overcome these difficulties is to arrange people in a social network where they do not, in fact, interact with *everyone* within the larger population they are a part of. This is known as adding structure to a population. To encourage cooperation, it is very helpful to foster a smaller sense of scale, for people to have *particular* friends drawn from among the larger group. When there are strong pairwise ties within broader social groups, cooperation is less anonymous and less diffuse, and it can flourish.[78] Here again we see the appeal of *Gemeinschaft*, or community.

A second way to overcome the problem is to allow *punishment* of any free-riders. But who does the punishing? In modern societies, cooperation is enforced through third parties and complex institutions. We have laws that dictate contributions (such as taxes), police to enforce the laws, and courts to levy penalties. In contrast, in the ancestral human environment, absent those formal institutions, people had social norms and peer pressure.

Small groups in our evolutionary past probably did occasionally have leaders who made decisions on behalf of the group and meted out punishments, but this was probably not usually the case. Humans bristle at overly authoritarian arrangements, and we have evolved an ability to reject dominant individuals with exclusive power to punish. This is what allowed natural human societies to be much less hierarchical than those of our primate cousins (as we saw, for example, in the case of the very egalitarian Hadza).[79]

The responsibility to punish free-riders was, and is, more equally distributed in human groups. And our desire for retribution is quite standard. Think about how you feel when you see someone cutting in line to board a flight. Even if you are in a different line, you probably feel annoyed; you think the cutter should get punished, and you might even be willing to do something about it. At the least, you could make a snippy remark to the interloper or commiserate with another nearby stranger about how rude it is to cut in line. Or you could intervene more directly. But why would you take such a risk? From an evolutionary point of view, it does not make sense to get involved if you are not immediately affected by that person's behavior. This willingness to pay a personal cost to punish someone else who is hurting a third person is known as *altruistic punishment*.[80]

Cooperative and punishing behaviors can be quantified by researchers with ingenious, simple games administered to people in both laboratory and real-world settings. For example, people can play the so-called ultimatum game (invented by economist Werner Güth and colleagues in 1982) in anonymous pairs.[81] Here is how the game works. A windfall endowment (say ten dollars) is given to player 1 (the proposer), who then gets to decide how much money to keep and how much to give to an anonymous partner, player 2 (the receiver or the responder). Player 2 can decide whether or not to accept the offer. Player 2 is told that if he or she accepts, the money will be split as player 1 proposed. But if player 2 rejects the offer, *neither* player gets anything. Player 1's behavior—namely, the proposal regarding how to split the money—can be used to measure altruism or cooperation. And player 2's behavior can be used to measure how willing the person is to engage in altruistic punishment, since a rejection means player 2 has paid a personal cost (getting nothing) to punish a stranger who did not cooperate to make an offer perceived as fair.[82] The game is played with real stakes, but it is anonymous, so there are no reputational effects that would cause players to worry about future interactions, and it is usually played as a one-shot game, so there is no prospect of reciprocity either (unsurprisingly, when players do interact repeatedly, cooperation increases).

The invention of this simple game spawned a wide array of similar ones. The so-called dictator game does away with the ability of the receiver to act at all. The first person simply decides what portion of a windfall to give to a complete stranger. That's it. That's the whole game. Here, too, people are rarely selfish, and most players share some part of their money. In contrast to the ultimatum game, however, the dictator game allows researchers to tell whether proposers are making offers out of a sense of fairness or a fear of rejection, since the possibility of rejection has been removed.

Classic economic theory cannot explain any of the behaviors observed in these games. If people are rational and selfish, then the proposer in the ultimatum or dictator game should offer the least amount possible, and the decider in the ultimatum game should accept *any* nonzero amount. Some money—even a penny—must be better than none. If self-interest were the only factor in how humans chose to play the dictator game, then the dictator would choose to donate none of the money. But everyone knows that this is not what actually happens. Moreover, in the ultimatum game, the receivers will forgo a reward if they are upset about being ill-treated. The proposers know this, and so they offer more than the minimum. People care about fairness. In American samples, typically more than 95 percent of dictators give some money to the other player, and most share it fifty-fifty, with the average gift being about 40 percent of the endowment.[83]

But how universal are these behaviors? For years, fifty-fifty splits were observed to be typical, at least among college students in industrialized countries. But then, in the mid-1990s, anthropologist Joe Henrich decided to try the ultimatum game in a field site in the tropical forest of southeastern Peru with the Machiguenga, a nonliterate group that did not engage in a market economy. Surprisingly, they behaved much more in keeping with economic theory, acting *very* selfishly. The most common offer was 15 percent of the pot, and despite many low offers, the offers were rarely, if ever, rejected.[84]

This unexpected finding—referred to as the Machiguenga outlier—sent social scientists scurrying. It prompted a large-scale, multiyear collaboration by anthropologists at fifteen sites on four continents to measure the consistency and variation in how the dictator, ultimatum, and other games were played. In this worldwide analysis, just 5 percent of the offers in dictator games were zero; 56 percent fell somewhere between zero and a fifty-fifty split; 30 percent were fifty-fifty splits; and 9 percent were a bit *more* than a fifty-fifty split. The Tsimané of Bolivia, like the Machiguenga, made low offers, with an average of 26 percent of the pot offered. Americans in rural Missouri were more generous, offering about 50 percent.[85]

The scientists also evaluated the cultural and ecological factors that might explain this cross-cultural variation. The overarching finding was that, while people were not purely selfish in any society, there was meaningful variation across societies, and a lot of that variation had to do with how market-oriented the society was and how important cooperation with non-kin was to subsistence (both of which fostered generosity).

Results were similar in the ultimatum game. Whereas offers in the ultimatum game in college-student samples in the industrialized world typically ranged from 42 to 48 percent of the pot, in the cross-cultural sample the range was from 25 to 57 percent, on average. The fraction of rejected offers also varied. For instance, among Kazakh pastoralists, there were no rejections once the amount offered surpassed 10 percent of the windfall, but among the Aché horticulturalists of Paraguay, offers had to be at least 51 percent of the windfall before none were rejected. The hunter-gatherers of Lamalera on the island of Lembata in Indonesia were among the most generous people studied. They live in seaside settlements of roughly a thousand people and hunt whales in the open ocean with small boats and harpoons. This dangerous and athletic work requires intensive coordination and cooperation in large groups, and, perhaps consequently, the Lamalera were the most cooperative people measured. On average, in the ultimatum game, they offered roughly 57 percent of their windfall.[86] Generally speaking, cultures that made higher offers had lower rejection rates and vice versa. Proposers therefore quite naturally made offers accordingly. People around the world know what sort of people they are dealing with.

Now, these two basic games we have been considering can be modified to add a third party (an observer or punisher) who is able to mete out rewards or punishments. As usual, the dictator or proposer is given an endowment (ten dollars, for example) and can choose to share any portion of it with the recipient. In this version of the game, however, the observer is also given an endowment (five dollars, say) and can choose to punish the proposer by spending some of this money—for example, paying one dollar of the observer's allotment in order to remove two dollars of the proposer's money if indeed the proposer has been judged to have acted selfishly toward the receiver.

How does the ultimatum game change when a punisher is added to the mix?[87] Punishment is present everywhere in the world, and there is a clear pattern across societies where an increasing fraction of people in each society is willing to punish the proposers as the proposers' offers get smaller. Overall, roughly 66 percent of the punishers were willing to give up 20 percent of their own windfalls (which corresponded to half a day's wages in each society) to punish the proposer if the proposer allocated zero to the receiver. Keep in mind that the punisher receives *no benefit* for doing this and must actually pay a cost. Nevertheless, the fraction of people willing to punish proposers who offered zero varied from 28 percent among the Tsimané horticulturalists of Bolivia to over 90 percent of the Maragoli farmers of Kenya.[88] And cultures showing more punishment also showed more altruism.

Across the cultures, what was it that motivated people to punish—compensating the wronged party or punishing the miscreant? It turns out that people show a stronger desire to restore justice and compensate the wronged parties than to rebuke those who behaved badly.[89] Furthermore, while altruistic punishment provides short-term justice to identifiable victims, more important, it can actually promote the emergence of cooperation at the collective level more generally. In an experiment in which some people were randomly assigned to be able to punish cheaters, cooperation levels started and stayed higher in groups that simply had such people among them, even if the punishers did not have to administer any punishments.[90] Punishment works as an institution—changing behavior by its very existence.

Still, we have the same evolutionary problem with altruistic punishment that we do with cooperation: Why doesn't natural selection weed out all these self-appointed sheriffs (who are paying a personal cost in a way that seems to harm their interests)? Evolutionary biologists Robert Boyd and Peter Richerson have shown mathematically that punishing behavior can indeed evolve because punishers can share the cost of punishment among themselves.[91] If everyone takes a turn imposing costs on cheaters, then the costs per punisher are much lower, and the benefits in terms of increased cooperation in a group may be enough to offset these lower costs.

But where did the punishers come from in the first place? To understand how punishment could emerge at all, the original analysis must be made more realistic. Instead of just giving people the choice of whether to cooperate or defect (that is, whether to be kind or mean), it is very instructive to also allow people to choose whether to interact at all.[92] Adding a so-called loner strategy to the Boyd and Richerson model created evolutionary cycles of altruistic types. Free-riders (defectors) do really well when there are a lot of cooperators because there are more people for them to take advantage of, so the number of free-riders tends to go up as evolution progresses. But eventually, there are too many free-riders and no cooperators. With no one to support them, free-riders generally do worse than loners, so loners then increase. But as loners replace most of the free-riders across time, it gets easier for cooperators to survive, so cooperation goes up again. And then the cycle repeats.[93] Each of the types—cooperators, defectors, and loners—can survive, but they do so only because they are coevolving with another type that beats its nemesis. *The types need one another to exist.* It's just like rock-paper-scissors: Rock loses to paper. Paper loses to scissors. Scissors loses to rock. This three-way cycle ensures that no type will ever completely disappear from or dominate the population.

An intriguing implication of such a process is its ability to maintain diversity. No type dies out because each has a specific coevolutionary environment that will allow it to come back from the brink of annihilation. A type that is unfit today may be fit tomorrow, and vice versa. Now consider what happens when we add a fourth type to this mix. To the cooperators, defectors, and loners, we add the punishers. The presence of loners in a population ensures there will sometimes be very few defectors (as noted above). In such a situation, this would reduce the cost of punishment to the point that punishers can do better than both loners and defectors, and this then gives them a chance to thrive—until the cycle repeats itself.[94] In sum, in a fundamental way, allowing people the option to opt out of social interactions altogether is what makes it possible for punishment to be a viable behavior, which in turn supports people having interactions and cooperating with one another to begin with! The possibility of being alone reinforces the ability of a group to be together.

All this work regarding the conditions for the evolutionary emergence of cooperation, its rational basis, and its cultural universality speaks to the role of cooperation in the blueprint for social life. Growing evidence also suggests that cooperative behavior in humans is stable and heritable, partly explained by our genes.[95] Cross-cultural variation also occurs, of course, mostly as a result of ecological constraints or historically specific factors, just as in the case of cross-cultural variation in marriage patterns. The actual rates of cooperation and punishment vary around the world, but the fact of cooperation—along with our species' evolved sense of justice and tendency to form connections with others—does not.

## *Teaching and Learning*

Identity, friendship, and cooperative interactions all serve a further purpose: they support the capacity for social teaching and learning and, thus, for culture, a capacity that reaches an apogee in our own species.

One distinctive reason that animals form social groups is the enhanced learning that can take place.[96] Social learning is likely to be more effective than solitary learning when the costs of acquiring information are high and when peers are sources of reliable information. If it is difficult for you to learn to make stone tools on your own, it's much better to just copy someone else. If I put my hand in the fire and you observe that it's painful to me, you can learn to avoid doing the same thing; you have acquired almost as much knowledge as I have but paid none of the cost. Social learning is very efficient.

The situation can be made even better. Teaching is a distinct behavior that can make learning even more efficient. One person can learn from another more easily if the second person is affirmatively trying to teach the first. *Teaching* can formally be defined as a behavior that (1) is primarily or exclusively performed in the presence of a naive individual; (2) costs the teacher or provides no immediate benefit; and (3) facilitates the ability of the learner to acquire information or skills more efficiently than would otherwise be the case.[97] While formal schools are rare (or nonexistent) in forager societies, teaching (and so-called natural pedagogy) from a very young age is widely observed.[98]

Teaching is actually a kind of cooperative behavior, and it is uncommon in the animal kingdom, since it is costly. Still, it has evolved independently in ants (which teach others the location of food by a kind of tandem running), meerkats (which teach others how to handle dangerous prey), pied babblers (which teach chicks to associate a particular call with food), and other animals, such as primates and elephants.[99] The evolution of this behavior can be driven by kin selection, like other altruistic acts. And it seems likely that animals that have already evolved the capacity to learn socially from others' inadvertent acts would be set up for the further evolution of explicit teaching.

To be clear, even animals that do not manifest all aspects of the social suite can learn socially. For instance, dogs learn hunting techniques from one another. This point was powerfully driven home to me not long ago by our miniature dachshund, Rudy. He had lived with us for seven years and had always stuck low to the ground; he had never had much interest in things up high and out of reach. One day, after a friend had a meal with us at our house, our friend's elderly beagle, Leila, waddled into the dining room, jumped onto a chair, hopped onto the table, and began to help herself to the left-over food on our plates. Rudy watched with keen concentration, then hopped up onto the table in exactly the same manner and proceeded to wolf down a plate of half-eaten pasta. This was a permanent lesson for him. Since then, we have had to clear all plates promptly.

Of course, social teaching and learning is much more advanced in primates.[100] Here is a description of a young chimpanzee learning to open nuts:

> Nina tried to open nuts with the only available hammer, which was of an irregular shape.... Eventually, after 8 minutes of this struggle, Ricci [her mother] joined her and Nina immediately gave her the hammer. Then, with Nina sitting in front of her, Ricci, in a very deliberate manner, slowly rotated the hammer into the best position with which to pound the nut effectively. As if to emphasize the meaning of this movement, it took her a full minute to perform this simple rotation. With Nina watching her, she then proceeded to use the hammer to crack 10 nuts (of which Nina received six entire kernels and a portion of the other four). Then Ricci left and Nina resumed cracking. Now, by adopting the same hammer grip as her mother, she succeeded in opening four nuts in 15 minutes.[101]

Another study of nine chimpanzees showed how tool use can be socially acquired.[102] Separately, each chimpanzee was given a straw and a container of juice. Two different techniques for juice extraction emerged—four of the chimpanzees sucked the juice up through the straw, as humans do, while the other five dipped the straw into the container and then licked the juice off the tip. The straw-sucking technique is more than fifty times more efficient than the dipping technique, but those who initially used the dipping method failed to discover the more effective method after five days on their own. However, four of the five dipping chimpanzees switched to the superior method after being paired with one of the straw-sucking chimpanzees. Moreover, the speed and ease of learning was closely correlated with the amount of attention the dipping chimpanzee pupil gave to the straw-sucking chimpanzee teacher.

Chimpanzees in various communities in natural field sites in Africa (such as Gombe and Tai) have been observed learning from one another to use particular techniques, tools, and positions in order to catch and eat ants by plunging sticks into anthills.[103] There is a clustering of behaviors within the communities whereby techniques of tool use are uniform *within* given communities but vary *between* communities; this also supports the conclusion that behaviors spread socially. And this is another way of describing *culture*—knowledge that is transmitted between individuals and across time, that can be taught and learned, and that is distinctive to groups.[104]

Another compelling example of social learning comes from a group of long-tailed macaque monkeys living near a temple in Indonesia who show extraordinarily entrepreneurial (I'd even say criminal) acumen. They have learned from one another how to run a racket with visiting tourists. They will swoop down and steal hats, glasses, cameras, and so on and return the items only if bribed with food.[105] It's highly distinctive, if not unique, and it's clearly a socially transmitted behavior. Scientists studying the site noted that members of a new group of immigrant macaques began to see that they, too, could exchange stolen goods for food. In fact, one primatologist at the temple studying this larceny had to buy back her own research notes.[106]

Crop-raiding behavior in elephants is another example. Villagers will attack and spear elephants they spot raiding, and, in Amboseli Park in Kenya, 65 percent of adult elephant mortality is caused by humans as a result of such conflict. Since crop raiding can be deadly for elephants, it would be adaptive for them to learn how to raid, not through trial and error, but by copying other elephants who know how to avoid getting caught. Indeed, it seems the elephants have learned to raid in the middle of the night, especially on moonless nights, and to form bigger-than-usual groups for raiding parties. Elephants also manifest other qualities we associate with efficient learning. They seem to put more credence in the behavior of peers who are deemed more reliable, such as older or more experienced elephants. And they even put more credence in the raiding strategies that they observe to have been adopted by multiple contacts.[107]

How might the structure of interactions among animals affect the likelihood of social learning? Imagine a null network, one in which no one is connected to anyone else. Without connections, social learning is obviously not possible. One step up from that, a network with only isolated pairs of individuals would not be very effective either. At the other extreme, imagine a fully saturated network in which everyone is connected to everyone else. This is also suboptimal, as it exposes each individual to an overwhelming cacophony of inputs and requires everyone to maintain many social ties. Something in between these extremes would be better. Maybe just a few ties for each animal arranged in a particular way would do the trick.

In fact, *network density* (the fraction of actual ties out of all possible ties), *community structure* (the existence of subgroups within the network with more ties within than between the subgroups), and *homophily* can all affect the likelihood of ideas or behaviors spreading and, therefore, the emergence of culture.[108] Someone is much more likely to adopt a practice if many of his or her peers like it, and the foregoing structural features all affect the probability of having such peers. The role of homophily is complicated, however. On the one hand, homophily can facilitate learning, since peers may face similar challenges and learn relevant solutions that they can share with one another. On the other hand, too much similarity inhibits innovation. A balance of similar and dissimilar contacts is therefore ideal.

Let's return to the crop-raiding example. In Amboseli, a network of fifty-eight male elephants was mapped using the association index.[109] An elephant was more likely to be a raider when its closest associate was a raider and also when its second-closest associate was a raider. We can see in figure 9.5 that the raiders appear more likely to be connected to one another, as are the non-raiders. Raiding behavior differed across the subgroups, and one of the six community clusters had many fewer raiders than expected, as if this particular cluster practiced a non-raiding norm.

**Figure 9.5: A Network of Male Elephants Showing a Strong Community Structure and Crop-Raiding Behavior**

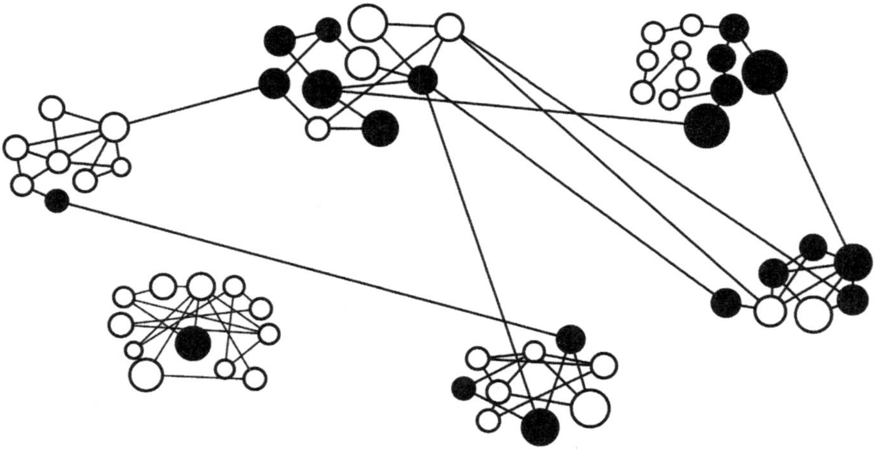

The circles (nodes) represent fifty-eight individual male elephants, and the size of the node is proportional to the age of an individual male. Lines indicate social relationships based on time spent together. Black circles indicate crop raiders, and the white circles indicate non-raiders. The elephants are naturally grouped into six clusters. Crop-raiding behavior varies across groups.

Intelligent animals living close together can benefit from the memories and knowledge of older group members or others who have made discoveries. The transmission of memories regarding the location of water sources can mean the difference between life and death for elephants facing drought conditions.[110] Similarly, during the widely spaced years in which El Niño strikes the Pacific Ocean, food becomes scarce near the equator, and groups of sperm whales will migrate very quickly to good feeding grounds thousands of miles away, places that older female members appear to remember from prior El Niños.[111]

Social species have evolved a set of interrelated attributes, including friendship, cooperation, intelligence, and transmission of knowledge via social learning.[112] Intelligence in such species may relate not only to the necessity of being able to track the identity of other group members and live socially, but also to the ability to retrieve specific memories and teach and learn from others.

### *Animal Culture*

The ability to network, cooperate, and learn socially all make possible yet another element of the social suite: the capacity to acquire and maintain culture. For instance, six long-term field studies of chimpanzees in Africa (at Bossou in Guinea, Tai Forest in Ivory Coast, Kibale in Uganda, Budongo in Uganda, Gombe in Tanzania, and Mahale in Tanzania), lasting from eight to thirty-eight years, have demonstrated that these separate populations have their own distinctive cultures.[113] At least thirty-nine out of the sixty-five behaviors that were studied, ranging from tool use to grooming practices, were found to be customary in some of the six populations but absent in others. These behaviors were learned and transmitted among the animals. Plus, there were hints that *combinations* of behaviors were distinctive, the way that, in the United States, gun ownership and religiosity are often found together, or veganism and support of marijuana decriminalization.

Cultural variation in animals can be explained in several ways. First, it's possible the variation has a genetic origin. For example, a local population of a bird species might sing a different song or use tools in a distinctive way because of genetic differences that affect the larynx, beak shape, or instincts. We know that New Caledonian crows reared in isolation still show stereotypical (and genetically encoded) tool-use behavior, for instance.[114] And since some subpopulations of chimpanzees have been isolated from others for hundreds of thousands of years, they would have had plenty of time to evolve genetically based behavioral repertoires. Alternatively, ecology (not genes) might explain distinctive local behavioral traditions. Perhaps chimpanzees in one location make a practice of sleeping on the ground, yet those in another location do not simply because of the presence of leopards in the area.

But we also have the possibility of true culture to explain the variation. To review, culture is a set of beliefs, behaviors, and artifacts that can be arbitrary or adaptive, that are shared by members of a group and are typical of it, and that are socially transmitted. For patterns of behavior to be seen as culture in animals, there must be evidence of transmission via imitative learning or teaching. Some have argued that language and symbolic thinking are also necessary, but this definition strikes me as too anthropocentric, like the overly strict criteria for friendship we reviewed before.

Animal culture research is controversial. With limited ability to do experiments, researchers find it hard to definitively distinguish among the foregoing interrelated explanations.[115] We expect correlation between ecology and cultural behaviors because cultural traits are often adaptive, as when chimpanzees invent and transmit ways to use sticks to fish out termites in areas where termite mounds are found but not in areas that are termite-free. But variation across ecological circumstances could arise for noncultural reasons too. Maybe chimpanzees eat biting insects in certain ways depending on the biting proclivities of the local ants, and the chimps might learn to do so by trial and error, *individually*, not because they are taught or copy the practices of other chimps.

Genes and culture might also be correlated. Birds could have the innate, genetically based capacity to sing certain songs but still need to be primed by learning from others. Experiments with rhesus macaques show that they learn to fear snakes by observing other monkeys, but they cannot easily be taught to fear other objects.[116]

Complicating matters, over time certain behaviors that previously were learned may actually *become* genetic. This is known as the Baldwin effect, and it works as follows: In any given generation of a species, some animals might be born, by chance, with gene variants that make it easier for them to learn a particular behavior. For instance, some birds might learn a song more easily than others because they have brains more capable of producing the typical introductory note of the song. If this behavior is adaptive (for example, if the song helps the birds attract mates), then it confers a fitness advantage on them, and the genes that make it easy to learn the behavior get reinforced and expand in frequency as the population reproduces. The same thing continues to happen. In each generation, those animals fare best who are most easily able to learn the target behavior, because they innately manifest more and more of it. Having acquired the innate capacity to produce the initial note, some birds might now be born with other gene variants that make it easier for them to produce the motifs of the song. These too could become innate, across generations, in response to selection. Eventually, over time, the entire behavior—and not just its predicates—comes to be encoded in the genes.[117] What was previously a nongenetic action has become a genetically encoded one.

In short, either/or explanations for shared behavior in animal groups (that the behavior is caused by ecology *or* genetics *or* culture) are likely to be as unproductive as thinking in a sharply categorical way about the nature-versus-nurture debate. Some animals clearly have culture, even if the origins of that culture can be multifarious.

Ethologists distinguish between traditions and full-scale culture.[118] Culture is broader than tradition, encompassing many behaviors. It's not uncommon for birds or mammals to have one learned tradition of some sort (like migrating to a standard hunting ground or singing a conventional song), but it's rare for groups to have whole constellations of distinctive practices. Yet a group of chimpanzees might have one particular, learned tradition about how to prepare twigs for use in hunting ants and another tradition regarding how to use the twigs, just as different cultural groups of humans might have different, related traditions about how to prepare dumplings and what implements to use to eat them.

This in turn raises the question of whether animals might even have *cumulative culture*. This is the sine qua non of human culture—our capacity to build on prior innovations. There are some hints of this in other primates. For instance, it seems that chimpanzees first invented the use of stones as anvils on which to crack nuts and then embellished the practice by using smaller stones to prop up the anvil. Or Japanese monkeys may have progressively developed a procedure to process sweet potatoes by washing them in ever more complex and effective ways.[119] But humans have gone from melting metal to fabricating fuselages.

Culture also requires a particular cognitive apparatus. That is, a cultural animal, as opposed to an animal simply manifesting a tradition, should have the ability to recognize that it can teach and learn behaviors. Chimpanzee mothers clearly understand this; they will explicitly teach their young to use anvils to crack nuts, thoughtfully placing relevant materials within reach of their young, like Montessori teachers.[120] And they have eager pupils. This awareness of the value of social learning is as fundamental to culture as an awareness of the existence of individual identity is to self-recognition.

With all this background, we can see that the six chimpanzee populations mentioned earlier indeed have distinctive cultures (color plate 6). Some behaviors, like dragging a large branch as a threat display, were present at all sites and so a genetic explanation could not easily be excluded. Perhaps this behavior is innate, like a gorilla pounding its chest. Or maybe it is a product of a Baldwin effect that began long ago when some chimp happened to drag a branch on one occasion, somehow intimidating a rival and enhancing the branch-dragger's survival. Other behaviors, however, varied from site to site, including using a stone hammer and anvil to crack nuts; using leaf parts to fish for termites; using a probe to extract liquids; using a stick to enlarge the entrance to a termite mound; using large leaves as a seat; using leaves as towels to clean the body; throwing objects at targets; and bending back saplings and releasing them suddenly as a way of making noise to warn others. These behaviors all appeared in distinctive combinations in these six widely separated populations. Importantly, chimpanzees are not the only primate with the capacity for culture, as studies of orangutans and capuchin monkeys have shown.[121]

Sometimes, by sheer luck, researchers observe new innovations being created and adopted in chimpanzee groups. The chimpanzees of the Sonso community in the Budongo Forest in Uganda ordinarily fold wads of leaves with their mouths to make a sort of sponge for dipping into tree holes to get water. In 2011, researchers observed the dominant male of the group, Nick, make a sponge of moss and use it to drink from a small waterhole on the ground, a behavior not seen in twenty years of observation. From Nick, the behavior spread to Nambi, the dominant female, and then (very quickly over the next six days) to six other chimps out of the twenty-seven in this group, preferentially along preexisting social connections.[122]

Social transmission of an innovation and the emergence of a new practice have been noted in cetaceans too. Among a group of humpback whales, a new fishing technique, known as lobtail feeding (in which the whales bang the surface of the water prior to making their usual bubble net below the fish), emerged in 1981 and then diffused from whale to whale along preexisting social connections.[123]

Another powerful example is a unique hunting technique developed by a thirty-member orca pod off Patagonia who, astonishingly, capture sea lions by beaching their five-ton bodies onto the shore; they are the only pod of orcas in the world known to hunt this way, but they have passed this technique on to subsequent generations.[124] Additional evidence of culture among whales, mostly in feeding and acoustic behaviors, has been documented by ethologist Hal Whitehead and his colleagues.[125]

Though most of the cultural traditions we see in animals are practical ways to enhance survival—new methods to gather food, for example—some new cultural traditions seem *arbitrary*. A group of chimps at Chimfunshi Wildlife Orphanage Trust, a sanctuary in Zambia where the chimpanzees live in forested enclosures, developed a useless practice of putting a long blade of grass into one ear, like an earring (see figure 9.6). This appeared to serve no practical purpose and was thus akin to a fad in humans. Starting from one inventor, Julie, in 2007, it spread to seven other chimpanzees by

**Figure 9.6: Julie: The Chimp Who Invented Grass Earrings**

2012.[126] Eventually, eight of the twelve adults in the group adopted this innovation. Notably, this spread was observed in only one of the four groups at Chimfunshi, which were all isolated from one another.

This case reminds me of an argument I have had with a friend for a number of years about my lab's research trying to change public-health practices in the developing world.[127] In our work in Honduras, where we have mapped the social networks of 24,812 people in 176 highland villages, my colleagues and I are using techniques to identify influential individuals (based on their network position) whom we can target for behavior change. Our hope is that, if we can get these structurally influential individuals to change their behavior, then the whole village in which they reside will copy them. We are trying to change local cultures so as to increase breastfeeding rates, improve management of childhood diarrhea, and promote paternal involvement during pregnancy, all of which are known to reduce infant mortality.

But this friend, Adam Glick, has pointed out that the real test of our methods would be to get people to do something arbitrary and useless. "If people can be persuaded to wear propeller beanies," he said, "I would believe that this method for inducing social change actually works." What he means is that the claim that something is learned socially is especially strong when the practice that is learned has no purpose. It is unlikely that many people would *simultaneously* and *independently* adopt a useless fad by virtue of individual learning, but it is *not* unlikely that a useless practice could spread by social learning.[128]

In modern societies, where cultural traditions are generally not aimed at immediate survival, many of our practices may be arbitrary. The circumstances of modernity coupled with communication technology may have created runaway conditions for our cultural abilities. The rapid swings in clothing fashion and bodily adornment that we see in modern societies—at rates that far outstrip any analogous changes in attire and self-decoration in traditional societies—provide a good example of this.

### *Humans and Animals*

As a thinking species, humans are fascinated by what is and is not humanlike in other animals. At times, the focus is on how humans are unlike other animals. At other times, even given the same observations, the focus is on how similar humans are to other animals. To me, both the differences and similarities are illuminating.

Humans share with certain social animals several aspects of the social suite beyond the pair-bonding and friendship we saw in previous chapters. We have the capacity to express and recognize individual identity in ourselves and in others too. This is reflected in our capacity for grief, which is connected to the deaths of individuals whom we see as unique and special to ourselves. We cooperate with our friends and within our groups. We have evolved a variety of behaviors—including sometimes opting out of social interactions altogether and punishing those who do not cooperate—that facilitate this. Our connection and cooperation with others allows us to learn from them, which sets the stage for us to evolve an interest and willingness to teach them too. Since teaching is costly to teachers and does not necessarily benefit them, it is actually a kind of altruism. All this affords one final miracle: the capacity for culture, which can be very complex, and cumulative, in humans. We are all beneficiaries of the collective knowledge that our species created over eons and transmitted from person to person across time and place and that now resides among us.

All these universal features in human groups work together, reinforcing one another, making it possible for us to be a successful species. These features also seem so manifestly good too. In fact, these elements of the social suite, which support our capacity for transmissible culture, set the foundation for a second inheritance system, parallel to (and sometimes intersecting with) genetic inheritance. We pass on *both* genes and culture to the next generation. And, as it turns out, each can affect the other, as we shall see in chapter 11.

Our similarity to animal societies helps pinpoint features that are essential to the structure and function of our own societies. If our societies have discernible traits that are even seen in other social animals, this supports the idea that those traits must be fundamental in our species. More important, however, this brings to the fore features that are universal among all human societies, highlighting our common humanity. Paradoxically, when we resemble other animals with respect to the social suite, it binds us all together. The more like these animals we are, the more alike we humans must be to one another.

## CHAPTER 10

# Remote Control

The male bowerbirds of western New Guinea are extraordinary creatures with a "passion for interior decoration," according to BBC *Planet Earth* host David Attenborough. Bowerbirds make elaborate structures known as bowers. These architectural marvels are built around a sort of maypole on the forest floor and have a large conical shape stretching as much as six feet across, with supportive pillars and a thatched roof of orchid stems. Inside, the bird will carefully arrange piles of beetle wings, tropical acorns, black fruits, glowing orange flowers, and even a "lawn" of carefully planted moss. The pièce de résistance is a bunch of lush pink flowers.

The bird itself is nothing special; its feathers are a mix of drab neutral tones. But, as an admiring Attenborough explains, the bowers "are better than feathers. Individual birds of paradise have no option over the shape and color of their plumes. They have to display with what their genes have given them. Bowerbirds, however, can choose. If a male decides he has a better chance of seducing a female with pink rather than blue, well, then, he can decorate his bower that way."[1]

The approximately twenty species of bowerbirds found in Australia, New Guinea, and Indonesia are genetically pre-wired to make bowers of distinctive overall types, among them the maypole style, described above, and an avenue style with thatched sides. Female choosiness is the main driver of the evolution of this astonishing behavior, but there is latitude for individual male choice. Even if their genetic programming compels them to find green objects rather than red ones, they still can choose within a genetically prescribed range from materials available in their environment. Luckily, the variety of materials in the New Guinea forest is vast and includes fruits, berries, fungi, butterflies, beetle wings, pebbles, shells, seedpods, caterpillar feces, and even human debris, like bottle caps from nearby campgrounds. Ornithologist Richard Prum described an individual bowerbird in one species who preferred to use bleached white fossil shells from a nearby sedimentary cliff rather than equally accessible white pebbles.[2]

The latitude for choice has been explored with ingenious experiments by ornithologist and geographer Jared Diamond. The birds must make species-specific types of bowers (such as tall towers of sticks or low towers of woven grass), but by placing poker chips of different colors near various bowerbird groups, Diamond showed that individual birds have preferences about what color chips to use, where to place them within the bower, and even how to mix the colors. This variation could have arisen in several ways—for instance, genetic variation among bird populations, cultural transmission between birds, or simply the vagaries of individual taste. One bird, Diamond observed, "made a separate pile of beetle heads, which other males discarded."[3] Others specialized in orange objects (fruit, flowers, seeds, or fungus), in yellow or purple flowers, or in butterfly wings.

Some species of bowerbird build bowers in which the objects get larger the farther away one gets from the entrance, a technique that fosters a visual illusion that flattens the space when seen from the entrance.[4] This technique appears to affect their mating success, and the females can detect differences in how well they do this. Our own species did not create paintings with perspective until the fourteenth or fifteenth century.

Prum and other ornithologists argue that male bower-building behavior and female choosiness coevolved to prevent sexual coercion.[5] For instance, avenue-type bowers allow the females to come close to males to inspect their thoughtful collages, colorful décor, and elaborate dances without risking a forced copulation, because the female can enter the tunnel from one end and observe the male through the front opening, but the male cannot approach her unless she lets him. If he were to enter the back of the bower, she could simply fly out the front. Male birds might have evolved to construct their bowers this way because females preferred to avoid sexual coercion, physical harassment, and forced copulation. Female choosiness redirected male behavior to create beautiful structures that suited the *females* better. These evolutionary events resemble the food-provisioning strategy we saw in chapters 5 and 6, in which the males of our species, instead of becoming bigger and more aggressive, might have evolved to use nonviolent acts of love and kindness to attract females.

But in my opinion, the most amazing thing about bowers is that these workings of sexual selection are manifested *outside* the birds' bodies (unlike colorful feathers in peacocks or size in silverback gorillas). Genes can have effects at a distance, beyond the bodies of animals in which they are found. These microscopic bits of code can shape the macroscopic world far above them.

### *Genes Use Bodies to Change the World*

The effects of genes can be understood at many different levels, often reflecting scientists' arbitrary interests. Biochemists might call it a day after studying how genes affect cells and observing phenotypes at the very first stage, in which genes are translated into their corresponding proteins. But why stop there? Medical geneticists may ignore the effect of genes on proteins and instead study how different genes affect muscle function or brain structure or disease symptoms. Zoologists interested in whole animals may do breeding manipulations to study a phenotype they are interested in, such as the coat color of foxes or the monogamy behavior of voles. Behavior geneticists may ignore these intermediate levels and look even farther downstream, at complex traits like risk aversion or novelty seeking.

But if genes express themselves at multiple levels—from protein to anatomy to physiology to behavior—why not go one (admittedly large) step further and look at the effects of genes outside the organism's body? In 2005, political scientist James Fowler and I termed the effect that genes have outside bodies, specifically on the structure and function of *social groups*, an exophenotype.[6]

This type of idea was first advanced by evolutionary biologist Richard Dawkins in his profound 1982 book *The Extended Phenotype.* Dawkins made a case that it should theoretically be possible "to free the selfish gene from the individual organism which has been its conceptual prison."[7] From this perspective, a beaver may be wired to make a useful dam just as it is wired to have a functional pancreas. Dawkins speaks of extended phenotypes, but I prefer to use the term *exophenotype*, by which I mean the nonincidental, genetically guided changes that an organism makes to its surroundings in order to improve its prospects for reproduction and survival.[8] Dawkins cautioned that, "since it is not a factual position I am advocating, but a way of seeing facts, I wanted to warn the reader not to expect 'evidence' in the normal sense of the word."[9] Yet such scientific evidence for these ideas has started to appear.

In a sense, in our species the social worlds we make are an exophenotype similar to bowers. Understanding the power of genes to shape not just the body and mind of an organism but also the world around it opens up new ways of seeing human social systems. When we look at our friends, at the organization and function of our groups, at the society we are embedded in, and at other societies around the world, it is possible to recognize the impact of our genes.

The precise mechanism by which genes affect traits such as height or weight is often obscure. The mechanism remains hidden in part because organismal development involves multiple genes interacting with one another and with the environment in a gloriously complex manner over the course of an organism's life. This is true even for traits that are unaffected by external factors, traits such as eye color. Genes work in convoluted ways. For instance, people with blue eyes have a mutation in the *HERC2* gene, but this gene does not straightforwardly code for the production of a protein with a blue color that fills the iris, nor does it code for melanin (the substance that gives a brown color to skin, hair, and eyes). Instead, the mutation in this gene sets off a cascade of downstream effects by *reducing* the expression of another gene, *OCA2*, the gene that produces the P protein, which is in turn involved in the production of melanin. After many steps, the *HERC2* gene mutation indirectly causes a reduction in brown color in eyes, rendering them blue.[10]

Like a Rube Goldberg machine with its linked contraptions, even simple genetic changes set off a chain reaction of downstream effects. And people have understandably tended to focus on what happens within the body in these genetic cascades. But the effect of *HERC2* does not stop at the edge of a person's skin. If you have blue eyes, it is due to the foregoing genetic mutation, and it appears that blue-eyed Europeans primarily descended from a common ancestor who lived between six thousand and ten thousand years ago.[11] But for the gene to have become as prevalent as it is today, it probably offered some sort of advantages to offset the known *disadvantages* of blue eyes, which include greater light sensitivity and an increased risk of macular degeneration and eye cancer.[12]

The advantage of blue eyes is still unknown, but it seems to have something to do with what blue eyes do outside the body of the owner. One idea is that blue eyes were unusual and therefore attention-getting, so they conferred greater reproductive success over evolutionary time.[13] Also, people prefer the type of eye that has a pronounced limbal ring (the dark line between the iris and the white of the eye which is a marker of both health and youth) regardless of eye color, and limbal rings are generally more apparent in light-colored eyes.[14] Or perhaps blue eyes make it easier for people to read one's facial expressions, thus potentially enhancing communication accuracy.[15] But regardless of how people react to blue eyes, if the genes for blue eyes systematically change others' reactions *at all,* then these genes have a downstream effect on other people, and it might be this indirect effect outside the body that is making the phenotype advantageous.

A skunk warding off predators or a flower attracting bees is changing the behavior of other organisms in order to improve its own fitness. Every time human beings use speech to persuade other humans to do their bidding—as when a baby babbles in an adorable way to attract adult attention—it reflects the evolution of a similar power ensconced in our genes. Your mouth, by an indirect path, can control the muscles of another person to affect your survival.

It's important not to extend this idea too far.[16] A crucial consideration when speaking of downstream effects outside the body is that the genes should encode the *specific* exophenotypes and that these encoded exophenotypes should affect reproduction or survival. Dawkins notes that the buildings people construct are not extended phenotypes, because humans do not, in fact, have genes that affect the propensity to build dwellings or genes that encode for whether those dwellings will be round huts made of ice or square huts made of branches (at least, no genes that we know of). The houses humans fabricate are not like the shells snails secrete or the bowers birds build.[17]

But while genes do not code for particular ways humans modify the physical world, they *do* affect the social world we make for ourselves. People create more or less hospitable environments for those around them to occupy. How they go about doing this affects their prospects in life. Moreover, people are shaped by other people's genes. We live in a sea of genes, and other people's genes may be even more critical to our destiny than our own.

To further build the case for the role of evolution in the social order of humans, I want to describe the many ways that living things exhibit exophenotypes. First, let's consider that genes may influence how organisms manipulate the *inanimate material world* around them. Birds build nests. Beavers build dams. Spiders build webs. These are all examples of animal artifacts made using natural materials that are fabricated or found and that are placed outside the body.[18] Second, genes may influence how organisms manipulate not just the inanimate world but also the *living world* around them. For example, parasites sometimes induce modifications in the body of the host they occupy, such as when particular pathogens that infect snails oblige them to make thicker shells. In these cases, one organism can act on the body (and, as we shall see, even on the behavior) of another organism that it is in contact with. Third, and perhaps most remarkable, genes can influence organisms at a distance. The genotype of one living thing can affect the phenotype (including the behavior) of another member of its own species or of a different species without ever coming into physical contact with it.

If human genes systematically affect other humans' actions, then evolution has a lot to do with the kind of social lives we construct for ourselves. It means that key features of our societies are genetically encoded. Our ancestors created social niches for themselves, and these niches became part of our species' evolutionary environment, feeding back to modify the sorts of genes humans carry today. The advantages of being friendly are greater in an environment that an individual has filled with other (nice) people. This whole process has contributed to the social suite being expressed universally, lying at the core of our societies the world over.

### *Animal Artifacts*

Animal artifacts are material objects created by animals as a deliberate result of their actions; bird bowers are an example. Like any phenotype whose variation is influenced by a genotype, artifacts can help or hurt the chances that the organism will survive and reproduce and that the relevant gene variants will propagate. If a spider evolved to have bigger mouth parts at the front of its body to more easily catch prey, it would be seen as a phenotypic trait arising from genotypic change that affects its survival. But the web a spider places outside its body is really no different; by using the spinnerets at the back of its body, it effectively expands the reach of its mouth to catch food. Its web is like a big mouth, and the spinning of webs and their structural forms are under the control of natural selection.

It is quite clear that such artifacts reflect the genetic makeup of the species. The order Araneae, the group to which all spiders belong, exhibits extreme diversity across one hundred and five families and over thirty-four thousand species.[19] As shown in color plate 7, differences between orb-weaving species are expressed in multiple ways, most importantly in the manner the spider draws its web and in the primary sequences of proteins constituting its silk. This variation is a powerful example of an adaptive radiation, the process whereby organisms with a common ancestor diversify into multiple new forms and species in order to exploit different environmental niches (for example, spiders adapting to catch different sorts of prey).[20]

Animals can create artifacts with more than just inanimate materials that they produce themselves, like spider silk. As we saw with the case of bowerbirds, over the course of evolution a gene might enhance its own propagation by leading to, say, the addition of a sexually attractive blue feather on a bird's body, a bird painting his bower with blue pigment derived from berries, or even a bird assembling blue-tinged pebbles in a pattern. A gene has several ways of propagating (when seen from the standpoint of different species), some of which work inside the body and some of which work outside it.[21]

Research linking identifiable genetic variation to animal artifacts is still very rare. We do not usually know which genes play a role in the shape of webs or style of bowers. However, there are a few exceptions. One excellent example is found in the oldfield mouse.[22] Oldfield mice and deer mice, two closely related species, both make burrows. Deer mice make a simple burrow with a short entrance passage leading to the nest. Oldfield mice make a burrow with a longer entrance passage and an escape passage in the rear (as shown in figure 10.1), allowing egress should a predator come in the front. Evolutionary biologist Hopi Hoekstra and her team showed that burrow style was *not* learned socially. They raised oldfield mice and deer mice free of parental influences, and when the test mice reached adulthood, they were given building materials. Each species made its own characteristic type of burrow.

Since the mice are closely related enough that they could mate and reproduce, Hoekstra was able to crossbreed them, and she consequently showed that the making of such burrows is encoded genetically. Hoekstra found that a small set of genes, localized in just three regions of the genome, accounted for entrance-tunnel length, and that just one area of the genome, perhaps just a single gene, some-

**Figure 10.1: Cross Section of a Typical Oldfield Mouse Burrow**

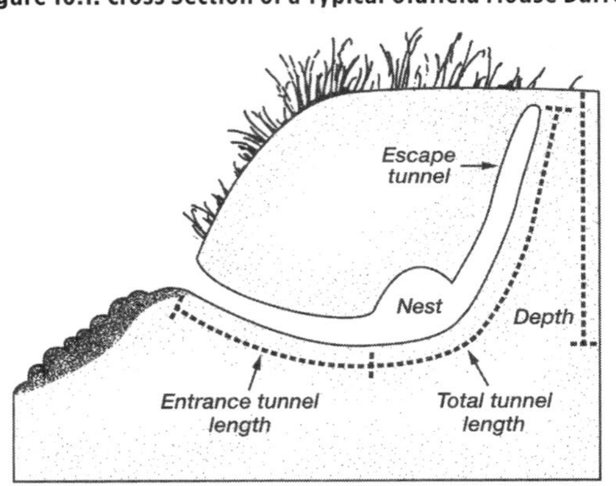


Key physical features (including the escape tunnel, not present in other species) of oldfield mouse burrows are indicated.

how coded for the presence of an escape tunnel! Leaving aside the mystery of how genes wind up coding for such complex behaviors resulting in artifacts at all, this experiment shows that genes can actually control the nature of the artifacts that animals make and that variations in specific genes correspond to variations in specific types of artifacts.

### *Parasite and Host*

Once we realize that genes express themselves both inside and outside the body and can work on the material world, we can understand phenomena like parasitism and symbiosis in a different light. Instead of rearranging their *physical* environment to suit their interests in some way, organisms can rearrange their *biological* environment. They can act on living tissue surrounding them rather than on inanimate matter.

Snails infested with a certain kind of parasitic fluke have thicker shells. The shell phenotype is clearly influenced by the snail's own physiology and genes. But the change in the host's morphology may also be seen, at least in part, as an adaptation *on behalf of the fluke.* From the point of view of the fluke, it wants the snail to shift resources away from reproducing (which does not benefit the fluke) and toward survival (which does benefit the fluke), and thicker shells make it more likely that both the snail and the fluke will survive. One might think that the host is doing things differently to deal with the threat within its body, activating cells that produce the shell in an effort to protect itself somehow, analogous to an oyster shellacking a grain of sand to make a pearl. But this may have it all wrong; it may be that the parasite, not the host, is driving the changes, and for its own benefit! Fluke genes manipulate the living cells that produce the shell of a snail in a way that is analogous to bowerbird genes manipulating seedpods on a forest floor.

When a parasite evolves to induce anatomical changes in its host in ways that enhance the parasite's prospects for survival, we can say that the phenotype of the host is under the control of the genotype of the parasite. From the conventional point of view of the *host*, the modifications in its own structure or function can be ascribed to *its environment*—which now contains these pesky parasites to which it is responding or adapting. From the point of view of the parasite, however, these changes in the host's body can be ascribed to its own genes.

Snail flukes do more than affect snail shell thickness, however. They also exert control over the snail's *behavior.* They somehow cause the snails to move toward light rather than, as they normally do, away from it. The flukes then migrate to the eyestalks of the snail and hang out there, pulsating and (it is thought) attracting the attention of hungry birds. All of this is bad for the snail because it means birds are more likely to see the snail and nibble the snail's eyes. But it is good for the flukes, because it means they can enter the bird's digestive system, which is essential to the next phase of the flukes' own life cycle.

This behavioral control of the snails seems like science fiction, and, in fact, the evolution of parasites to modify their hosts' behaviors (rather than just their bodies) is thought to be exceedingly rare, found in less than 0.5 percent of parasite species.[23] But still, scientists are finding more and more examples. Mice infested with the parasite toxoplasma lose their innate aversion to cat urine, which sets them up to be eaten by cats, which become the next hosts of the parasite.[24] Hairworms can cause their cricket hosts to jump into water and drown, which furthers the hairworms' life cycle.[25] Some parasites appear to achieve their objectives by evolving the capacity to release totally alien (to them) chemicals that imitate the hormones of their hosts, thus changing their behavior.[26]

One of the most baroque examples of parasitic behavioral control involves so-called zombie ants. Certain species of ants have the misfortune of being prone to infection by a particular species of fungus (*Ophiocordyceps unilateralis*); the fungus prompts them to climb a plant to a certain height and then bite into a vein on the underside of a leaf in a death grip.[27] The fungus then kills the ant and grows a large mushroom-type stalk from the ant's head from which its spores can rain down on future ants (as shown in color plate 8). This phenomenon was first observed by Alfred Russel Wallace, a naturalist who advanced a theory of natural selection at the same time as Darwin but who received much less recognition for his efforts. Here, we have an example of a species without a nervous system (the fungus) evolving to control the behavior of the species with a nervous system (the ant), converting the latter into a spore-delivery platform. We know from fossilized leaves showing telltale marks of ant bites on their veins that this fungal phenotype is tens of millions of years old.[28]

Is it possible that some *human* traits and behaviors may actually be the genetic by-product of other organisms' genes?[29] Is it possible that humans sneeze not to expel pathogens from the upper airway for our own benefit, as I was taught in medical school, but for the benefit of the pathogens that might thereby spread through the air? Maybe our behavior is being manipulated by those very pathogens. Humans think that sneezing is something that *we do to the germs* to clear those pesky invaders from our bodies and promote our own well-being. But maybe it is something the *germs do to us* to help facilitate their own spread and increase their fitness. Some intestinal worms may decrease human fecundity in the furtherance of their own interests.[30] Maybe, when people are infected with certain pathogens, they behave in ways that prompt their loved ones to come close and take care of them (manipulating the human tendency to love, befriend, and cooperate), which in turn facilitates the further spread of the pathogen that made them sick. It might not be a coincidence that people who are sick act like babies, eliciting help from caregivers.

Some scientists have even proposed the highly speculative hypothesis that microorganisms could increase their evolutionary fitness by encouraging certain aspects of human religious behavior that favor their transmission, acts such as mortification of the flesh, rolling on the ground, or the kissing of an icon or holy relic by large numbers of people.[31] Microorganisms might even enhance our desire to form aggregations, thus facilitating their spread while also shaping our social lives.

### *Action at a Distance*

The parasites discussed so far provide examples of exophenotypes that are relevant to the fitness of two animals (the host and the parasite) when the two animals come into direct contact with each other.[32] But it's also possible for one animal to affect the behavior of another animal of the same or different species in a reproducible manner that benefits itself without the two actually coming into contact at all—for instance, by releasing a scent. Animals might affect the appearance and behavior of other animals without ever touching them, and many aspects of animal communication by means beyond odors are subsumed here.[33] For instance, an elegant experiment with stickleback fish has shown that placing a fish infected with a parasite that modifies the fish's predator-avoidance behavior (making the fish more likely to be eaten by a bird, which is the next obligatory host for the parasite) affects the predator-avoidance behavior of *other* fish in the school, even if they are not infected.[34] By infecting one individual, parasites can control the actions of many.

Seen through this lens, the entire living world could be viewed as an interacting, if not coherent, network. This perspective evokes the idea of a butterfly flapping its wings on one side of the Earth and setting off a hurricane on the other side. Perhaps it is possible that each of us is floating in a vast and complex sea that includes not just our own genes but the genes of every living thing on the planet. Of course, it's not possible to take every single living organism and every one of its features and actions into account. Instead, we need to focus on relevant exophenotypes, which we can define as those that have a direct effect on an organism's fitness and that involve specific interacting species, such as the flukes and the snails. Otherwise we would have a *reductio ad absurdum* in which anything an organism did to act upon the world might be seen as an extension of itself that affected every other organism. This would apply even to banal acts like depositing feces on the ground (as one of my children observed, although in less clinical language).[35]

How can we focus our scientific gaze? A crucial consideration is this: for a phenotype to be relevant, it must result in the differential propagation of the underlying genes. Variants of genes can increase or decrease in frequency in a population as a consequence of their phenotypic effects, but some phenotypic effects may be incidental and not have any impact on the frequency of alleles in the next generation. To illustrate this point, consider an example of footprints suggested by Dawkins. A genetic mutation that results in a change in the shape of a seabird's foot will likely affect the bird's survival and hence how the gene is propagated; for instance, it might affect how easily the bird can move in wet sand. This mutation will also unavoidably affect the shape of the footprints the bird leaves behind. If the shape of the footprint has no influence on the fitness of the animal, it is not relevant to its survival and can therefore be excluded from our collection of candidate exophenotypes.[36] However, if the newly modified footprint *does* influence the survival of the bird—say, by making it easier for predators to track it—we would then consider the footprint a relevant exophenotype, affecting the bird's survival. Predators, too, could evolve, say by developing better vision to see smaller footprints. The important question is whether the exophenotype influences fitness enough that it causes the underlying gene to become more or less prevalent in the population.

Such exophenotypic effects can also occur within species so that one individual affects other individuals, and these effects can create new selection pressures redirecting the course of evolution. By damming a creek, a beaver creates a pond, which gives the animal a bigger shoreline on which to forage for food and a safer route to swim from place to place. This pond can thus be seen as an exophenotype, different than bodily phenotypes like the size or shape of the beaver's teeth. By modifying the size and depth of the pond, the beaver can create selection pressures on other animals, such as the fish and insects in the pond. But similar effects can occur within its own species. The beavers' damming behavior may result in new environments that create selection pressures to change body parts of future beavers; perhaps they develop bigger lungs that allow them to swim underwater farther in this larger habitat. In turn, bigger lungs might be advantageous only in ponds with bigger dams, which would privilege dam-building behavior. As a result, the capacity to build dams (an exophenotype) and the possession of bodily features that allow a beaver to swim far (a conventional phenotype) may coevolve.

### *Regulating One Another's Genes*

Something similar to beaver dams, I think, applies to the social structures humans create. If gene variants predispose people to possess personality traits like extroversion, and if some long-ago human ancestors were extroverts who insisted that their friends become more social and brokered introductions among their friends, then they might have affected their friends' fitness. This is a downstream effect of their genes. And if their being more social enhanced their own and their friends' survival in some way (perhaps they recruited all of their friends for a large and effective hunting party), then the tendency to form groups of a particular type is an exophenotype. For instance, by mapping networks of the Nyangatom pastoralists of Sudan (as shown in color plate 4), my colleagues and I have demonstrated that the organization of risky, dangerous, but possibly profitable raiding parties against the Nyangatom's neighbors depends on the social-network architecture of the group. Men join raiding parties because their friends have joined, and the composition of parties is associated with their success.[37] Any gene that influences interactive or communicative behaviors is by definition having an effect on an organism other than the person within whom that gene resides and can provisionally be seen as a gene that shapes social systems.

An example from chickens can help us understand indirect genetic effects and hence such social, action-at-a-distance exophenotypes. The feather condition of a chicken is clearly a phenotype, with genes playing a role in a bird's plumage. Environmental factors, including the chicken's diet and light exposure, are relevant too, of course. But chicken feathers are also affected by something else in their environment: the behavior of other chickens. Other hens can aggressively peck at a bird's feathers, damaging them (some chickens even go as far as cannibalism). In the past, farmers have resorted to the unfortunate (and now largely illegal) practice of trimming the beaks of hens to prevent this.

An experiment that varied chickens' cage mates explored this idea. It showed that feather quality is a product not only of a bird's own genes but also of the genes of the birds nearby.[38] Some genes might make the hen produce deformed feathers, and others might make her act submissively, eliciting bites that could ruin feathers that were otherwise of good quality. That is, the indirect effect arose from the propensity of *other* hens to express pecking behavior. In fact, in this example, those indirect effects were more powerful; the bird's feather condition depended more on her neighbors' genes than on her own!

The effect of genes *within* an individual organism often depends on the actions of other genes, what geneticists call epistasis. As we discussed earlier, the genes that make eyes brown will not work if one has a certain variant of the *HERC2* gene. Several genes must work together to determine one's eye color. There are countless examples of interactions like these within a single individual (like the example of the genes for going gray not being expressed if one also has the genes for going bald). If one gene changes, it may have an effect on the action of other genes, and hence on the expression of the attendant phenotypes.

As we have just seen with the chickens, however, such effects can also occur *between* individuals. I call this *social epistasis,* since the impact of genes in one individual may be affected by the genes of another individual. One chicken may peck and another may molt. One beaver may build a dam and another may then have the opportunity to exploit its ability to hold its breath for longer. In pursuing this line of thought, we can actually abandon the conventional "within-body genetics" that has dominated thinking in evolutionary biology for so long, and we can come to see groups of interacting individuals as being affected by one another's genes.[39] The genes of other individuals now become yet another feature of the environment that every gene must face in the course of its own evolution.

In fact, genes may even *require* the presence of particular genes in other organisms in order to be useful and in order for natural selection to favor them. Imagine that you were the first human on the planet to have a mutation in a gene that conferred the capacity to speak (by modifying the anatomy of your larynx, say, or by changing a neuronal pathway in your brain). While the capacity to speak is surely polygenic and the result of a very lengthy process of evolution that started before our species split with the Neanderthals over three hundred thousand years ago, some specific genes may indeed play a salient role.[40] If you were the first person on the planet to have a genetic mutation that gave you the capacity to speak, it would be nearly useless to you unless you had someone to speak to, someone who could understand and, perhaps, communicate back. The prospect of a *conversation* would, all of a sudden, make this mutation much more advantageous and facilitate its spread in the population across time. And the more it spread, the more useful it would be to individuals in succeeding generations.

The capacity for speech is what is known as a network good, meaning that its value rises in relation to how many others have it too. An example of a network good is an e-mail account. If you are the first person on the planet to have one, it is useless to you. But as soon as one other person has one, its value becomes nonzero. And as more and more people get e-mail accounts, yours becomes more and more useful. It can be the same with genes that affect our social lives. They may require the presence of the same or other genes in other people to have their beneficial effects.

There are other possibilities for social epistasis beyond communication-related genes. Some genes work better in an individual if his or her friends have those particular genes themselves. Genes that regulate your immune system may perform better (or worse) depending on the immune systems of others around you. For instance, if you have a mutation that makes you susceptible to a particular pathogen but all of your friends are resistant to that pathogen and so cannot pass it on to you, your mutation is irrelevant. And any genes related to altruism and reciprocation behaviors may be like this, benefiting people who have them and are also interacting.

### *Creating Social Worlds*

The examples of interindividual, indirect genetic effects discussed above require something else before we can consider them exophenotypes. They must also affect the survival or reproduction of the individual expressing them. In the chicken experiment, if the Darwinian fitness of the pecking hen depended on the feather quality of the hen she threatened, then we would say it was an exophenotype. In the case of our species, if your genes predispose you to introduce your friends to one another, and if this social web you weave affects your survival, then it is a social exophenotype.

Humans make social networks as reproducibly as bowerbirds make bowers. To properly analyze these social exophenotypes, we would ideally need a set of different hominid species—as in the examples of the oldfield and deer mice or the different species of spiders—and compare *across* species to show how social structure varied. But an illustration *within* our own species can still shed light on this topic by demonstrating the role of genes in social exophenotypes.

It is a straightforward observation that some people have more friends than others. People vary in their taste for friendship and in their skill at it. In one study, my lab examined eleven hundred adolescent twins from one hundred and forty-two schools around the United States to explore how genes help explain the kind of real, face-to-face social networks that people make.[41] As we saw in chapter 8, about half of the variation in how many friends people have was attributable to their genes. This variation can be shown visually. Figure 10.2 maps the social network of a group of a hundred and five people (the circles) and a few hundred friendship ties between them (the lines connecting the circles). On average, each student is connected to six other close friends. But some people, like person D, have more friends than others, like person B.

It's not just the number of friends that two people have that distinguishes them, however. As noted in chapter 3, the mathematical algorithm used to draw networks in two dimensions on a page places those who are more connected in the center of the figure and those who are less connected at the periphery. When your friends become better connected, it increases your own connection to the whole network. Scientists say it makes you more central because having better-connected friends moves you away from the edges and toward the center in a social network. Your centrality can be quantified by counting not just the number of your friends, but also the number of your friends' friends and the number of their friends and so on. In our study, we found that genes could affect this; about a third of the variation in how central people are in the network depends on their genes. That's the difference between being persons C and D in figure 10.2. Both have six friends, but one is more central and one is more peripheral.

One way this difference might arise is if C and D differed in their innate taste for popular friends. D might want friends who have few other friends—perhaps so that D can get more of their attention. In contrast, C might want friends who themselves have many friends. While C might get less of their attention, those friends might be helpful in other ways; for example, they might be better conduits of useful information, given their own numerous connections. Depending on what sort of friends you pick, your centrality in the network will be affected.

**Figure 10.2: Friendship Network of College Students Indicating Variation in Position**

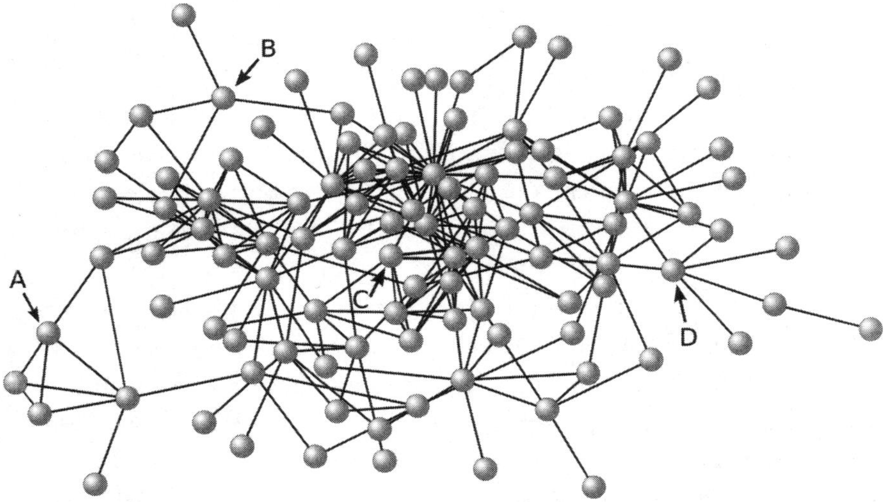

In this natural network of close friendships among 105 college students living in the same dorm, each circle (node) represents a student and each line (tie) a mutual friendship. Even though persons A and B each have four friends, A's four friends know one another (there are ties between them) whereas none of B's friends do. That is, A has greater transitivity than B. Also, even though persons C and D both have six friends, they have very different locations in the social network. That is, C has greater centrality than D; this is reflected in the fact that C's friends have many friends themselves, whereas D's friends tend to have few or no other friends.

Finally, we found something else about how variation in genes affects positions in social networks. As we saw in chapter 8, people may vary in their tendency to deliberately introduce their friends to one another, though social networks naturally perform this function whether people deliberately do this or not (for instance, since your friends spend time with you, they are simply more likely to meet). Recall that this is transitivity, a measure of whether your friends are friends with one another. You can appreciate this by comparing A and B in figure 10.2. A's friends are generally friends with one another, but B's friends are not connected among themselves and are not friends. As we saw, about half the variation in this feature of the social environment can also be explained by our genes.

The transitivity finding suggests that your genes could affect the social lives of *other people.* When we think of niche construction, we usually think of animals engaged in acts to modify their physical environment, such as beavers building dams of sticks to enhance foraging opportunities or ants warming their nests to support fungi. But people might analogously engage in niche construction by manipulating the *social* world around them just like other animals manipulate the physical world. Returning to our spider example, some webs are better for catching flies than others, and the spiders with genes that programmed them to make those webs are more likely to survive and pass on those genes. Similarly, some ancestral humans may have been better able to shape the social webs around them and engage in collective activities like hunting large game or warding off competing groups. Humans with genes that helped them do that would have been more likely to survive as well. Eventually, genes and alleles predisposing humans to make beneficial sorts of social arrangements would arise and expand. This behavior could affect individual survival, not just compared with other individuals but also compared with related species that do not form such networks. Social niche construction would be adaptive.

### *Domestic Bliss*

In the wild, the silver fox largely conforms to the cultural stereotype of foxes—elusive, cunning, even mean. In captivity, the creature is averse to human contact, and when it does cross paths with a human, it typically results in the human getting a firm bite. But the silver foxes at a farm outside of the Institute of Cytology and Genetics in Siberia yearn for human contact; they lick people's faces, wag their tails, and whine to be picked up and held. This behavioral transformation is the product of a breeding experiment begun by the Soviet biologist Dmitry Belyayev in 1958. Using silver foxes, Belyayev—later joined by Lyudmila Trut (who is now the project's supervisor)—reproduced and condensed millennia of canine domestication into a matter of decades.

Domesticated animals are mainly distinguished from their wilder cousins by the ability to tolerate the presence of humans. This transformation from wild to domesticated usually occurs via a selection process that favors nonaggressive members of a population. This leads to a number of changes in the animal's anatomy, physiology, and behavior. Neuroendocrine changes related to aggression arise in diverse body systems, especially those involved in social behavior and reproduction. These are followed by anatomical changes, like smaller teeth, floppy ears, and changes in pigmentation (like the splotchy black, white, and brown piebald fur found in so many domesticated cats, dogs, cows, and horses). The animals also show increased social tolerance, playfulness, and cooperation with humans and other animals.

This constellation of traits is known as the domestication syndrome. Many of these changes are not directly targeted for selection but are the by-products of a primary selection against aggression. For instance, it's not entirely clear why piebald fur or a curly tail would go with docility (perhaps a smaller, curly tail is less useful as a counterbalance, which might have been helpful in a hunt or a fight).

But this is one reason that scientists suppose that all these traits may collectively be related to certain regulatory genes expressed during development that affect multiple processes (remember: one gene can influence many traits). Moreover, when humans intentionally select animals that are more placid, it turns out that they select animals that display traits and behaviors, such as play, that are ordinarily seen in more juvenile members of the species. Just compare dogs to wolves, cats to cougars, cows to buffalo, or pigs to boars, and it's easy to see the contrast with the domestic animal's more aggressive and less playful counterpart.

The domestication of wolves into dogs was probably set into motion by initial chance mutations in certain wolves that made them less fearful of approaching human camps and less aggressive—the sort of preadaptation we have seen before. Such animals gained a selective advantage through access to human garbage and thus exploited a novel ecological niche provided by humans. As a second stage, humans may have chosen the less aggressive among these animals and domesticated them further by breeding them, bringing along the other traits that distinguish dogs from wolves, including floppy ears, wagging tails, adult playfulness, and so on.

Belyayev and Trut simply bypassed these thousands of years of evolution. First, they visited commercial fur farms throughout Estonia and carefully selected thirty male and one hundred female foxes, all of them friendlier than average. With the founding generation of foxes in place, Belyayev and Trut bred successive generations of silver foxes for tameness, following strict selection guidelines. Every month, experimenters tested the docility of the fox pups by offering them food while trying to handle them individually. They also tested their sociability by observing whether they preferred humans or other foxes when they were allowed to roam freely in a large enclosure.

Upon reaching sexual maturity, the foxes were assigned a tameness score and sorted into one of three classes according to temperament: those that fled or bit when touched, those that tolerated being handled but showed no friendly or positive emotions in response, and those that enjoyed contact with the experimenters and wagged their tails and whined (labeled class 1).[42] The researchers restricted breeding to only the tamest subset of each generation, allowing no more than 5 percent of the male and 20 percent of the female offspring to breed further. The physical and behavioral changes brought about by selective breeding were so drastic and rapid that, after just six generations, the researchers established an even higher tameness category, class 1E, which was made up of the "domesticated elite"—namely, foxes "eager to establish human contact, whimpering to attract attention and sniffing and licking experimenters like dogs."[43] After ten generations, over the course of a few years, 18 percent of the foxes belonged to this domesticated elite, a statistic that rose to 35 percent by generation twenty and crested at about 80 percent by generation thirty, a few decades later.

As with most domestic animals, morphological changes accompanied the behavioral changes; the foxes acquired floppy ears, rolled tails, short legs, feminized cranial features, and novel coat colorations. Likewise, the domesticated cohort reached sexual maturity earlier, followed different mating patterns, and showed alterations in hormonal and neurochemical processes. The development of such variation, despite selection *only* for behavioral temperament, lends support to Belyayev's hypothesis that the same underlying genetic mechanisms that regulate behavior can also influence physical appearance. In fact, the changes in the foxes seem to have occurred in similar regions of the genome as with the domestication of dogs thousands of years ago.

Interestingly, Belyayev's ideas were not popular at the time he began his work, and he had to describe his experiment as being about physiology, not genetic inheritance.[44] Soviet doctrine at the time was heavily influenced by Trofim Lysenko, a Russian scientist who rejected the genetic and evolutionary theories of Mendel and Darwin in favor of a theory of "acquired characteristics" similar to that advanced by Jean-Baptiste Lamarck. This theory posited, for example, that if a parent learned to run faster to catch its prey, then its children would inherit this newfound ability to run fast. Natural selection had nothing to do with it.

Lysenko wielded tremendous political power during the Stalinist era in the USSR, denouncing his scientific opponents, some of whom were executed. In 1948, the Soviet government outlawed opposition to his theories. Genetics was officially declared "a bourgeois pseudo-science" and virtually banned until the mid-1960s. Given the patently political ways in which his theories were advanced, the term *Lysenkoism* has come to be used to describe the manipulation of science to reach predetermined, ideological conclusions. I believe that the rejection of the role of genetics in contemporary human social life is a type of Lysenkoism, a topic I explore in chapter 12.

### *Good Chimp, Bad Chimp*

Belyayev's experiment confirms that humans can direct and speed up evolution via intentional domestication to make the species more peaceful, friendly, and cooperative. However, some species may have undergone similar changes *without* an exogenous plan. This is a kind of natural *self-domestication* that can reduce aggression within a species and favor the selection of other juvenilizing traits. Some researchers have argued that bonobos are a self-domesticated version of chimpanzees and, more particularly, that humans have also followed this path.[45]

Bonobos and chimpanzees last shared a common ancestor about one million years ago, and they exhibit many striking similarities in their appearance.[46] But bonobos have developed very different behaviors related to aggression, and they have other features suggestive of domestication syndrome, including reduced tooth size and white tail tufts. Adult bonobos have extended reproductive seasons and generally have a lot more sex, including non-conceptive heterosexual sex and occasionally homosexual sex, than chimpanzees. They are also more playful and engage in more prosocial behaviors, such as voluntarily sharing food.[47] Interestingly, bonobo brains have more gray matter in regions implicated in perceiving distress than in others—suggesting that they may experience feelings of empathy that are stronger than those in chimpanzees.[48]

Chimpanzees, by contrast, are very aggressive. They use complex displays to intimidate rivals that can often escalate to outright violence, leading to the wounding or killing of their own group members. Male chimpanzees patrol their territory in groups and will fight with rival groups, killing both young and adults. In comparison, bonobos are strikingly peaceful. Their display behavior is more limited and involves nonviolent acts such as running and dragging branches. Unlike chimpanzees, they do not interfere with one another's mating. Intergroup interactions are also less aggressive, and while such interactions often include attempts to intimidate rivals with displays, they do not generally escalate into outright violence.

Some researchers hypothesize that different environmental pressures pushed the two species in different directions. Although both faced similar predation risks, bonobos did not have to compete with gorillas in their terrain (bonobos are confined to a geographic region in Democratic Republic of Congo, whereas chimpanzees range through large parts of the middle of the African continent), and bonobos had a more plentiful food supply than chimpanzees. This may have reduced the bonobos' impetus to take risks to acquire food and to display a variety of aggressive behaviors.

Other researchers argue that sets of bonobos might have joined forces against, and subsequently killed, more aggressive members of their troops, reducing the number of aggressive individuals in the gene pool. Or females might have developed a preference for mating with less aggressive (more cooperative) males, as we saw earlier in the case of humans and other species. While the process of self-domestication can be driven by natural selection in response to environmental pressures, the actions of the animals themselves can play an extremely important role in the *speed* of selection, just as in Belyayev's foxes. Instead of waiting around for natural selection to weed out the wild ones, individuals within a species can do the weeding for themselves directly, if not wholly intentionally.

Such a process might have occurred in our own species as well, via both female choosiness and the collective opposition to (and, in prehistoric times, the killing of) overly aggressive individuals. As a species, we are built to tolerate only mild hierarchy. Anthropologist Richard Wrangham has argued that such a self-domestication process in humans has altered our species' behavior and biology. In fact, the particular pattern of genetic changes seen in the domestication of many animal species is also seen in humans, with many similar genes being affected, further supporting the hypothesis that humans self-domesticated.[49] Domestication may have fostered neurological changes that made humans not only more docile, but also more attentive to others, thus making our species more responsive to training and more prepared for social learning. Humans have also become more juvenilized compared to ancestral hominids. And, even in the past few thousand years, the tendency toward less aggression has sped up, dramatically reducing violent interpersonal conflict to its lowest level in recorded history.[50] While as much as a third of humans during the Paleolithic period died from intentional violence, only about one out of a thousand people in even our most violent communities die from it today.[51]

Animals—whether spiders spinning webs, birds making bowers, fungi controlling ants, or humans arranging their social networks—are genetically programmed to act on the world and reshape it in ways that make it more convivial and that enhance their survival. The social environment humans create is partially under the control of our genes. This environment in turn feeds back and affects us, making some ways of being social more optimal than others and selecting for the genetic variants that support that. As humans, we have changed ourselves. Across long and short stretches of evolutionary time, our own genes—and our friends' genes—seem to be working to build a safer and calmer world.

## CHAPTER 11

# Genes and Culture

If you were a farmer with a hoe and an acre of land, you could spend a day planting a small amount of seed. If you had a mule in a harness and a plow, you would be more productive with a day's labor. And if you had a tractor and fuel, you would be still more so. Your increased productivity relates, of course, to the increased capital (hoe, mule, tractor, and the knowledge to care for and use them) in each of these scenarios.

The beneficial effects of knowledge and technology are usually obvious to people who witness such innovations. One journalist writing in 1931 described the impact of a tractor upon her own extended family:

> For a month each year, a "harvest gang" of fourteen men was added to [my mother-in-law's] own big family....And how those harvesters were fed! Rows and rows of vegetable dishes, huge platters of chicken with dumplings or dressing, pie and pudding at noon, pie and cake at night, with plenty of preserves and pickles all the time....
>
> The clatter of wagons...meant to the woman, who was sweltering in her inconvenient kitchen, the preparation of food for twenty to twenty-five men with the keenest appetites....

> All that belongs to the past. On the same farm (except that it is now considerably larger), her son manages with a tractor hand or two. The harvest weeks are still a strain, but not because of labor problems.... A nephew of high-school age runs the tractor and does a very competent job of it, too. A college boy stands guard over the power-driven elevator at the granaries. An eleven-year-old swells with the importance of his position on the truck that brings the wheat from the fields. A man, two slight youngsters, and a child are doing what fourteen men and a herd of army mules did fifteen years ago....
>
> Her husband's and my husband's parents gave their bodies to the making of that big farm. She and her husband do not have to do that. When they are forty, their steps will not drag and their shoulders round into a weary droop. The machines are saving them from that.[1]

Prior to 1850, power on American farms came mostly from humans, including enslaved people. From 1850 until the invention of the tractor in 1892, draft animals took over as the main source of power. Surprisingly, it still took until 1945 before tractor power eclipsed equine power as more important to American farming.[2]

Farmers initially resisted tractors for several reasons, including their expense or novelty. Many felt horses were superior, noting that "you couldn't grow gasoline, but you could grow feed for horses. And you could raise your replacement horses." Horses could be easier to use too: "When the horses come to the end [of a row], they'd turn around themselves and go back the other way."[3]

Still, in the end, the tractor rapidly won out. And the advantages of that transition did not just accrue to the beleaguered women in the kitchen or the worn-out men in the fields. The real price of food went down for *everyone*. If you were born in 1962 instead of 1862, you would pay less for groceries and could use your money for anything else you wanted, like buying a radio and listening to jazz, both of which had also been invented in the same interval.[4] Now, fifty years later, the situation is even better: Americans spend 9.5 percent of their income on food, down from 16 percent in 1962 and far down from 42 percent in 1900.[5] If this progress had not been made, the median-income family in the United States today would be spending two workdays per week just to earn money for food.

There are countless human inventions beyond the tractor that work this way, creating benefits (and costs) that redound on everyone. Imagine you were born twenty thousand years ago, and you spent your whole life doing what was necessary to survive and to enjoy yourself. You could accomplish only so much. Mostly, you would spend the day foraging for food, socializing with your group, making up stories about the workings of the world, crafting hunting and digging tools from stone, bone, and wood, and perhaps playing a drum or a bone flute.

Now imagine that you were born five hundred years ago. You could accomplish much more with your time, having inherited the cultural and technical achievements of the preceding eons: the agriculture and all its attendant technology (the plows, granaries, threshers, and so on); the species of animals and plants that had already been domesticated; the cities, stonework, roads, reservoirs, and metallurgy; and even navigation techniques, geometry, and astronomy. All this knowledge and stuff would be your birthright just because you were born when and where you were.[6] You would have a mule instead of just a hoe.

Now imagine you were born very recently. Simply by the felicitous timing and location of your birth, you have been handed a figurative tractor with which to plow your field. You do not even have to pay for it. It's just there. All of the science and art made by all prior humans and recorded in diverse ways (in oral tradition, in books, in online repositories) is yours to learn. You have available to you (in most parts of the world) a deep understanding of the cosmos; electricity and modern medicine; highways and maps; and plastics and nano-fabrication alongside bronze, iron, and steel. If you learned calculus in high school, you know so much math that if you went back in time five hundred years, you would be the most knowledgeable mathematician on the planet.

Sometimes, knowledge has accumulated for so long and in such complex and obscure ways that it seems like magic. Among the people living in the village of Ilahita, in the interior lowlands of New Guinea, there was a secret men's cult known as the Tambaran whose members periodically built "spirit houses." Anthropologist Donald Tuzin witnessed such a construction in the 1970s and marveled at their use of a secret and ingenious technique to dig very deep and narrow postholes for the building. He blurted out, "Who ever thought up that idea?" One of the men looked at his fellows as if this were a truly stupid question and said that no one had thought up the technique, and the others agreed, as Tuzin explained:

> "Come now," I said, warming to the topic, "*someone* must have thought of it. It didn't just come from nowhere." My interlocutor explained that such ideas are received from tradition; they come from olden times, not from ordinary men. "I understand," I persisted, "but someone at some time must have invented it!" A couple more exchanges of this type followed, and then: "Let me ask you this," said one of the men, with friendly condescension. "Could *you* have invented this technique?" "No," I admitted, unable to avoid the trap. "That is true of us, too," the man said, resting his case.[7]

This is cumulative culture. People endlessly contribute to the accumulated wealth of knowledge that belongs to humanity, and each generation is generally born into greater such wealth. (Of course, there are periodic reversals too, when information disappears, as happened after the collapse of the Roman Empire; for example, Europeans lost the knowledge of how to make concrete and for seven hundred years lived in Roman dwellings that they had no idea how to construct.) Many social scientists believe that this is the deepest origin of economic growth: people have ever-increasing amounts of cultural and intellectual capital at their disposal with the passage of time, and so they can plow ever more land in a day.[8]

Sometimes, humans bequeath to their descendants knowledge or products that are valuable in and of themselves, without any further modification needed. One of my favorite examples of this is the transmission of warnings, via myths or inscriptions, regarding major natural disasters that occur much less frequently than a single life span. The northeast coast of Japan is dotted with so-called tsunami stones, which are large flat rocks, some as tall as three meters, with inscribed warnings about where to build villages so as to avoid tsunamis that can kill tens of thousands of people or where to flee to when they strike. In Aneyoshi, a stone was erected a century ago with the warning DO NOT BUILD YOUR HOMES BELOW THIS POINT! When a tsunami struck in 2011, killing twenty-nine thousand people, the water stopped just one hundred meters below the stone, having totally destroyed everything in its path. The people in all eleven households built above the marker survived.

"They knew the horror of tsunamis, so they erected that stone to warn us," said Tomishige Kimura, speaking of former residents of his village. Sometimes, the ancient wisdom is transmitted by words, not rock, as with the village named Namiwake, or "Wave's Edge," located on a spot nearly four miles from the ocean, marking the devastating reach of a tsunami in 1611.[9] Similarly, the Aboriginal tribes of the Andaman and Nicobar Islands in the Indian Ocean have passed down an oral tradition for millennia that advises people, when the earth shakes and the sea recedes, to run to locations on higher ground in the forest. These tribes all survived the great tsunami of 2005 that killed thousands of residents of more technologically advanced communities.[10]

As discussed in chapter 9, a few animal species have limited forms of culture. But our species' type of cumulative culture, across generations and in its elaborate form, is very rare, if not unique.[11] Even more remarkable than the impact of human culture on the course of our own lives, however, is the fact that it has redirected the evolution of our entire species. The cultural environments that humans construct for themselves and have been constructing for millennia have become a force of natural selection, modifying our genetic heritage.

This idea, which we will now explore, is different than the notion of exophenotypes. In the case of exophenotypes, genes code for something specific, like an artifact or a social behavior, that then affects the evolution of the organism. But when it comes to culture, our genes equip our species with the capacity to *flexibly* produce things. While beavers are genetically programmed to build dams, humans are not genetically programmed to domesticate cows. But when we do so, the existence of domesticated cows affects our evolution nonetheless.

### *Culture Matters*

Our species' ability to survive in diverse habitats, from the Arctic tundra, where humans hunt seals, to the African deserts, where they build wells, has depended only a little on physiological adaptations such as a higher adiposity and shorter stature to conserve heat among humans living in the far north. Rather, our species' survival across the world hinges on its capacity for culture, a capacity that is ingrained and that has led to astonishing inventions like kayaks and parkas. No other species depends quite so much on creating and preserving cultural traditions.

Ecologist Peter Richerson and anthropologist Robert Boyd define *culture* as "information capable of affecting individuals' behavior that they acquire from other members of their species through teaching, imitation, and other forms of social transmission."[12] A key part of this definition is its interpersonal quality—culture is a property of groups, not individuals. Other scientists put more emphasis on material artifacts such as tools or art, but of course, cultural knowledge is the antecedent of the creation of artifacts.

In our effort to explore the blueprint for human society, we have, so far, been looking beneath the veneer of culture. In chapter 1, I started by suggesting that culture might, by analogy, explain why two hills are three hundred and nine hundred feet tall, respectively, but not why they both sit on a ten-thousand-foot plateau. In this perspective, culture is an overlay on a more fundamental set of processes. But the ability to *fashion* diverse cultures—our pre-wired inclination to produce culture—is itself a crucial attribute of our species. This capacity for culture, rather than its products, reflects our evolved proclivities for social interaction, cooperation, and learning.

But it turns out that this cultural veneer penetrates deeper than simply shaping our behavior over our own lifetimes. Culture can actually affect the genes we carry as a species. It is as if the hills could move the mountains on which they sit. This interaction between genes and culture is known as the theory of gene-culture coevolution. It is also sometimes called dual-inheritance theory, a reference to our capacity to inherit both genetic and cultural information from our ancestors. This joint perspective on genes and culture has three key elements.

First, as noted above, our capacity to create culture is itself an adaptation shaped by natural selection. Few species have culture, and those that do are the ones that have the genes that make it possible. We have evolved cognitive and psychological features, along with their underlying neural substrates, to make us capable of being cultural animals. Because we are a long-lived, group-living organism that has generational overlap (that is, multiple generations living together at the same time and place) and extended parental care (with a relatively long period of development), we can benefit from, and therefore likely evolved, the ability to manifest culture.

The second element is that culture itself can evolve over time in a process that follows a logic similar to that of natural selection. Just like genetic mutations can lead to improved disease resistance, chance discoveries can lead to better tools. And then superior ideas can outperform weaker ones—and be selected for. As a further process (also analogous to genetic evolution), "unusual" cultures, just like unusual species, can arise from the random *drift* of practices and ideas. Despite these similarities, however, cultural evolution has the special feature of allowing lateral transmission (of ideas or practices) between individuals within the same generation, not just to individuals in subsequent generations, as in the case of genetic evolution. Also unlike genetic evolution, cultural evolution can be explicitly directed (for example, by a powerful ruler, like Constantine the Great, who prompted his subjects to convert to Christianity).

The third element of gene-culture coevolution theory is that this cultural system of inheritance itself becomes a feature of our species' evolutionary landscape, exerting a force of natural selection on us.[13] We have evolved genetically in response to the culture that we are genetically prepared to create. Genes affect culture and culture affects genes.

Let's consider these three elements.

### *Humans Are Cultural Animals*

Some scientists believe that humans evolved the capacity for culture in order to cope with the variability of their environment. Animals living in a more consistent environment need not have the ability to learn from one another; natural selection could just endow them with useful instincts that worked reliably generation after generation. But our species must cope with variability and needs a brain that does so.

Humans have a constellation of psychological traits built for culture, including the tendency to conform to what others are doing, to develop and obey local norms, to privilege high-status or older people who could be potential teachers, and to pay attention to the people whom others are paying attention to. We also evolved to preferentially attend to those who show confidence in their behavior (whether that confidence is warranted or not).[14] We have a predilection for what is known as *over-imitation*. From a very young age, children will routinely copy even arbitrary, trivial, or unnecessary actions taken by adults they are closely observing. Although seemingly misdirected and foolish, over-imitation is an evolutionary adaptation that plays a fundamental role in the development and transmission of culture in our species.[15] Some of these psychobehavioral attributes are also present in other animal species who have culture, as we saw in the case of elephants who learn from others.[16]

The presence of social learning and culture has changed the nature of status in our species.[17] *Status* can be defined as the relative ability to gain or control valuable or contested resources within a group. In most animal species, status is usually equivalent to *dominance* and is measured by physical power and the *costs*, or potential harm, that an animal can inflict on others (think of a deer with enormous antlers). But status can also be measured by the *benefits* an animal can offer others, which we call *prestige* (think of an elephant matriarch directing her group to a supply of water). The existence of cumulative culture in our species makes prestige an especially appealing way to acquire status because human societies are so rich in commodities and information. In dominance hierarchies, subordinates are afraid of their superiors and try to avoid them. But subordinates in prestige hierarchies are attracted to their superiors and try to get near them—to befriend, observe, copy, and otherwise benefit from them (think of eager students going out of their way to hear a speaker). Acquiring more valuable knowledge can also make animals (including humans) more popular, so the relationship between learning and network position can go both ways.[18]

A different evolved psychology is required for the prestige type of status. Our capacity for culture has refashioned the way our species' brains work over the course of evolution. If we evolved to value prestige, we should see traces of this from a very young age. Cleverly designed experiments show just that.[19] For instance, preschoolers are twice as likely to try to learn from a popular than an unpopular adult. This preference holds for copying the manipulation of toys and tastes for food. Insecure politicians exploit these types of cues, making ostentatious displays of fawning audiences in order to enhance their prestige in the eyes of others.

Prestige also works across domains. It's commonplace that people perceived as competent in one domain will be presumed to be good models in other domains ("When you're rich, they think you really know," as Tevye observed in a song in *Fiddler on the Roof*). When I practiced medicine as a hospice doctor, I was commonly at the bedside of patients when they died. Possibly because I was good at providing end-of-life care, many families would ask me theological questions and put great stock in whether I thought there was an afterlife, which struck me as odd. In a study of three villages in Fiji, people were asked to whom they would go for advice regarding fishing, yam cultivation, and medicinal plants. Some people in the villages were seen as the optimal models for each practice, respectively, as you would expect if expertise was both concentrated and recognized. But prestige also transferred across domains. An individual seen as knowledgeable about yams was more than twice as likely to be regarded as someone from whom to learn to fish.[20]

In most species, the trade-off between dominance and prestige is usually presented as the shift from bodily to cognitive resources. In other words, species may emphasize one or the other type of status (although few have a prestige system). Among humans, this trade-off can vary to some extent between and within cultures, depending on the environment, and for both men and women. In a violent environment with high mortality, physical formidability might be worth much more to oneself and one's potential partners. The variability of circumstances that our species has faced across evolution (and that we continue to face, given our geographic range) may also have contributed to our manifesting a mild hierarchy, perhaps reflecting a balance of prestige and dominance, rather than the more extreme form of hierarchy seen in other primates who have lived in more constant environments.

Still, both types of status can lead to greater reproductive fitness in our species.[21] The sex appeal of prestige in both men and women is important because evidence that prestige is associated with reproductive success supports the claim that humans evolved to value those who can teach things.[22] In one study, undergraduate women in the United States preferred prestigious men to dominant men as sexual partners, especially when they were looking for longer-term partners.[23] The evolutionary relevance of prestige is also illustrated by reproductive success among the Tsimané, our familiar forager-horticulturalists of Bolivia.[24] Adult Tsimané men in two villages (eighty-eight men in all) were each shown photos of all the other men and asked who would win a physical confrontation, how many allies the other men would have in a conflict, who had the most influence in community decision-making, who commanded the most respect, and miscellaneous other questions about hunting skills, generosity, and their wives' attractiveness. Both prestige and dominance were associated with reproductive success. Men in the top quartile of dominance had 2.1 more surviving children than men in the bottom quartile. Men in the top quartile of prestige had 2.6 more surviving children than men in the bottom quartile. And wives of both dominant and prestigious men were rated as more attractive. Our brains—capable of transmitting and learning knowledge, which is culture at its core—are sexy.

### *Culture Evolves*

A person's ability to adapt to diverse and sometimes deadly environments is not due simply to brainpower. Even the smartest among us could not devise everything required to produce something as mundane as a cup of coffee—the cultivation and processing of coffee beans, the transportation system to bring the beans to a store and the electricity to the coffeepot, and the complex procedures required to fabricate the plastic cup.

A Hadza adolescent can survive in surroundings that would kill me in a matter of days. Anthropologist Joe Henrich describes a predictable pattern of what is known as the Lost European Explorer Files: accounts of European explorers who got separated from their teams and were isolated in a demanding environment.[25] Their supplies run out, their clothes disintegrate, and their tools, if they had any, become unusable or are lost. They are unable to make proper shelters and they cannot find food and water. They do not know any of the elaborate preparation and cooking procedures sometimes necessary to detoxify local plants and animals, and so they fall ill. Sometimes, they resort to cannibalism. Very often, they die. Perhaps their bleached bones, frayed journals, or collapsed dwellings are found years later.

However, if they are lucky, these forsaken people are offered the benefits of local knowledge from generous natives (as we saw with the party seeking help after the wreck of the *Sydney Cove*) from whom they get food, shelter, clothing, medicine, or information about how to acquire or fabricate these necessities. As Henrich puts it, "These indigenous populations have typically been surviving, and often thriving, in such 'hostile' environments for centuries or millennia."[26] It's culture, not genetic evolution, that is the difference between life and death in such situations. Genetic changes sometimes do emerge to make some humans better adapted to particular environments (for example, the adaptations in oxygen metabolism seen among the peoples of the high mountains of the Tibetan plateau), but these are so much slower than cultural changes.

Many human cultural products are so complex that they can seem as remarkable as the genetic evolution of the eye. Consider the preparation of poison arrows and blowgun darts among Amazonian tribes such as the Yagua. In most tribes, dart and arrow poisons are formulated from complex mixtures of ingredients that include not only the primary active ingredient from plants (which is generally curare), but also other plant substances and sometimes snake, frog, or ant venom. Dozens of components might be required, and the formula varies from tribe to tribe. The preparation is complicated, involving heating and cooling, admixture of other substances, and specific steps of physical processing (like pounding or pulverizing). Even the time of day at which the items are to be collected might be specified and might be important for some reason—for instance, to increase the concentration of some chemical. The ingredients and methods might have practical, ritualistic, or symbolic relevance, but the truth is that ethnobotanists understand little of this. Some of the additives might increase the toxicity; others might facilitate quicker penetration into the victim's bloodstream; and others might help the mixture become more glutinous so as to adhere better to the darts in the first place.[27] Some ingredients may simply serve as markers for proper preparation—for instance, by providing a color change when a substance has been properly cooked. And all this complexity relates just to the poison, not to the equally intricate fabrication of the darts and blowguns, the necessary equipment to safely transport them, or the knowledge to use them![28]

The practitioners themselves often cannot account for what each step does or how they came to have this knowledge. They just *know* that they must perform the steps in the proper sequence, with the proper ingredients, in the right way, and at the right time in order for the whole process to work. It's possible these practitioners have fallen victim to the over-imitation strategy and there is no pragmatic basis for their methods.[29]

Given the extraordinarily intricate level of detail involved in an enterprise like poison-arrow hunting, it's hard to miss the conclusion that cultural patterns are honed and pruned over many generations (*adapted*, if you will) in a way that reminds one of evolutionary change. In fact, the idea that human culture might "evolve" in a manner similar to species goes back at least to Darwin in *The Descent of Man*.[30]

Culture can evolve by a number of mechanisms. Individuals and groups may stumble on new ideas and technologies by chance, analogous to mutation. Even ideas conferring no particular advantage may become solidified (as is thought to be the case for many styles of music across cultures), analogous to genetic drift. Successful ideas (such as monogamous marriage systems, as we saw in chapter 5) that enhance the well-being of their hosts (or at least are not harmful) may be even more likely to endure, especially if the groups espousing them are in competition with groups espousing less advantageous ideas.

Culture can also manifest convergent evolution, just like genetically based phenotypes in animals, as we saw earlier. Consider the form of fishhooks, which have been independently invented by humans more than once. There are several ways to catch fish, but only one basic way to design a hook. Similarities that emerged simultaneously in two or more locations can typically be ascribed to environmental factors unrelated to direct knowledge transfer.[31] However, scientists can use the observed similarities in fishhook design to reconstruct settlement histories and interactions between populations that may have transmitted manufacturing knowledge.[32]

To understand how this works, we have to consider two basic features of human technology. Technologies involve *functional* traits, which carry an adaptive value related to what they do, and *stylistic* traits, related to how they look (and which are adaptively neutral).[33] The former traits tend to arise through convergence because they indicate people are addressing similar problems, while commonalities among stylistic traits tend to reflect shared cultural descent.[34]

Fishhook technology has been most widely studied in the Asian Pacific region where populations have been pushed to rely extensively on the sea.[35] The earliest fishhook specimen in Oceania (a broken-shell fishhook from East Timor) is between sixteen thousand and twenty-three thousand years old and appeared along with the mastery of advanced seafaring. These U-shaped hooks are intended to catch a fish by the rigid parts of its mouth. The same site provides evidence for systematic fishing as far back as forty-two thousand years, so it is possible that fishhooks were used even earlier.[36] At a site west of Berlin, archaeologists have also uncovered ancient fishhooks fabricated from mammoth ivory that date to about twelve thousand years ago.[37] This discovery, along with the discovery from Timor, points to widespread barbless-hook usage at the transition period from Paleolithic to Mesolithic, roughly ten to fifteen thousand years ago.[38]

Early archaeologists thought that the similarities in shell fishhooks found in the Pacific rim, including in Chile, California, and Polynesia, indicated that the technology probably was invented only once and spread like a central light source radiating outward. Some even proposed that the diffusion could have been accomplished by the victims themselves; tuna could have snared hooks in Oceania, broken free, gotten caught in some other way on the western shores of the Americas, and revealed their painful secret to other humans opening them up there.[39] This is an intriguing idea, but, because these fishhook artifacts surfaced in areas separated by vast distances and during significantly different periods, it seems much more likely that these common designs were actually the products of independent invention and convergence.[40]

Cultural evolution resembles genetic evolution in still another way. Population size matters, just as it does in the evolution of species. Small populations are very vulnerable to the loss of knowledge because the death of one person with the key information means the knowledge is permanently lost to everyone. If the only person who knows how to navigate by the stars drowns, the remaining members of a small seafaring group will perish too.[41] Amadeo García García, a member of an isolated tribe in the Peruvian Amazon, became the last living speaker of the Taushiro language in 1999, when his brother

Juan died. "I am Taushiro," he told a reporter almost two decades later. "I have something that no one else in the world has." Beset by malaria, enslavement, and other calamities, the Taushiro in the end were simply not numerous enough to endure; nothing is left of their language except the recordings Amadeo and a few others left behind with linguists.[42] Bigger populations help insulate against this risk. One study of twenty Polynesian languages confirmed this mathematically, showing that larger populations gain new words at higher rates and smaller populations lose words at higher rates.[43]

Moreover, in bigger populations, if an individual stumbles on a better technique for building a fire, finding water, or tracking animals, there is usually someone else around to observe, copy, and remember the discovery. Consequently, bigger populations are better suited to social learning and to maximizing opportunities for valuable innovation. Given the unavoidable error in social transmission, populations have to keep transmitting just to avoid the loss of information with the passage of time. Teachers and students must work to keep a complex tradition alive. Larger populations mean more pupils to learn from the best members of a society, and they also provide for the occasional pupil who can surpass the teacher.

The potential role of size in cultural innovation and complexity can be tested by returning to the natural experiment of islands in Oceania. The traditional cultures of ten widely separated islands, originating from the same ancestral culture and with similar marine ecologies, were studied. The number of marine-foraging tools and their complexity were assessed (as summarized in table 11.1). These tools ranged from a simple stick used to pry shellfish off rocks to a sixteen-component untended crab trap involving a baited lever. As predicted, larger population size was associated with both the number of tools and their complexity.[44] Other factors were important too. Islanders who hunted mobile prey or who had to cope with a more variable environment required more tools.[45] Generally speaking, given the greater variability in weather and terrain, it is harder and more dangerous to find food the farther north or south one gets from the equator, and thus latitudes farther away from the equator have greater tool complexity, other analyses show.[46] Culture is useful in coping with environmental variability.

Experiments have confirmed such observations. For instance, in one inventive experiment, 366 men played a computer game in which they had to manufacture a virtual arrowhead or fishing net; the former involved fifteen steps and the latter thirty-nine steps (performed in a precise sequence). The players could choose whom to observe and learn from in their group. And the group size was experimentally varied so as to involve two, four, eight, or sixteen players. As group size increased, cultural knowledge was more often preserved, improvements more often made, and complexity in tool packages more often maintained.[47] When it comes to cultural innovation and preservation, size matters.

**TABLE 11.1: POPULATION SIZE, TOOLS FOR MARINE FORAGING, AND TOOL COMPLEXITY FOR TEN CULTURES IN OCEANIA**

| CULTURE | POPULATION | TOTAL NUMBER OF TOOLS | AVERAGE TOOL COMPLEXITY |
|---|---|---|---|
| Malekula | 1,100 | 13 | 3.2 |
| Tikopia | 1,500 | 22 | 4.7 |
| Santa Cruz | 3,600 | 24 | 4.0 |
| Yap | 4,791 | 43 | 5.0 |
| Lau Fiji | 7,400 | 33 | 5.0 |
| Trobriand | 8,000 | 19 | 4.0 |
| Chuuk | 9,200 | 40 | 3.8 |
| Manus | 13,000 | 28 | 6.6 |
| Tonga | 17,500 | 55 | 5.4 |
| Hawaii | 275,000 | 71 | 6.6 |

Tool complexity ranges from 1 to 16 in this sample of fishing and marine-foraging tool kits, depending on the number of independent components of the tool, such as 1 for a simple stick to pry shellfish from a rock to 16 for a multipart crab trap with a baited lever. Both generally increase with population size.

Finally, the maintenance and evolution of culture depends on the number and structure of the social-network ties among individuals and how easily and freely information is exchanged. This means that cultural transmission can hinge on how cooperative and friendly a population is.[48] Consequently, elements of the social suite lie at the foundation for our capacity to develop culture. This leads to a final, ironic conclusion about culture. These stable, universal features of humanity—cooperation, friendship, and social learning—are precisely what make the amazing variety of cultures possible. Our species' capacity for culture, based on teaching and learning, is a key part of the social suite even if the specific components of culture—so variable, as we saw in chapter 1—are not. Sustaining complex cultural knowledge requires a large and interconnected set of thinkers and innovators. Our blueprint is the foundation of cultural evolution. Human society is like a Rubik's Cube that is bound together and obeys a few particular principles but that is nevertheless configurable into 43,252,003,274,489,856,000 combinations.

### *Genes and Culture Coevolve*

Having established that humans are, genetically speaking, uncommonly capable of culture compared with other animals and that culture itself can vary across time and place, in part due to processes akin to evolution, let's now consider how genetic and cultural inheritance might interact.

Some have argued that culture did not start playing a role in our species' evolution until the agricultural revolution, but the effects probably appeared much earlier in the ancestral line leading to modern humans, perhaps even one million years ago.[49] There are ways to discern the impact of culture on human bodies, even without historical or archaeological evidence, by thoughtful examination of fossilized human bones.[50] We know that, beginning about nine hundred thousand years ago and until roughly five hundred thousand years ago, the climate started fluctuating substantially. Over time, this fluctuation would have created selection pressure for an animal that was nimble enough to cope with diverse environments, not optimized for just one. Social learning is especially adaptive when a species faces an environment that is changing so frequently that genetic evolution—in which individual mutations lead to small differences in form and function—cannot do its job quickly enough.[51]

The well-worked-out examples of the impact of culture on evolution are riveting. Once humans learned to control and then create fire, perhaps as early as 1.8 million years ago (the date is contested), they started to cook, and the caloric content of what they ate rose substantially because heating meat and vegetation makes food more nutritious. The kind of teeth, mouths, and stomachs hominids had at that time, suitable for chewing raw meat off bones and masticating twigs, could, and did, evolve in new directions once cooking arose. Teeth got daintier; masseter muscles (still the strongest muscle in the human body) got weaker (and jaw shape consequently changed); and the stomach got smaller (and the configuration of human ribs was consequently modified). Cooking freed up energy to power the demanding human brain, which got larger.[52]

Just as researchers can use teeth and other bony fossils to estimate when cooking started, they can use foot anatomy to estimate the onset of running practices related to tracking animals over long distances. Humans are unusual among mammals in having the capacity to be marathon runners. Though we cannot outrun even our household pets in a short sprint, we have all sorts of adaptations that equip us to run for a long time (these include slow-twitch muscle fibers, which are useful for endurance, and the ability to regulate body temperature during long periods of exertion).[53]

Foragers in sub-Saharan Africa can catch antelopes simply by chasing them over long distances in what is known as a persistence hunt. The prey sprint away, out of sight of the hunters. But if the hunting party can track the animal and keep at it for, say, eight hours while the animal repeatedly sprints ahead, eventually the animal will collapse from exhaustion and the hunters can simply spear or strangle it. I've seen film footage of exactly this, shown to me by physical anthropologist Dan Lieberman: the hunter simply walked up to a large kudu (a usually dangerous animal) and skewered it as if it were a marshmallow. This is an entirely different strategy than lions or leopards would use when pursuing the same prey.

Though endurance running is important to that style of hunting, more than anatomical evolution is required. The humans must be able to track a particular, individual animal when it is out of sight, not, say, pointlessly chase first one kudu, then another, then another, exhausting themselves in a series of sprints against fresh opponents. This is where culture comes in. The ability to track animals (based on footprints, scat, debris, and knowledge of their behavior) is acquired painstakingly over generations, and this is carefully taught and transmitted (which is why it takes many years to become a proficient hunter). This cultural invention then makes physical changes for long-distance running in our bodies adaptive and useful. Modern scientists can use changes in foot anatomy, preserved in the fossil record, to make inferences about when an otherwise useless ability to engage in slow long-distance running appeared and when, perhaps, the cultural practice of tracking emerged.[54]

Another very good example of gene-culture coevolution is lactose tolerance in adults.[55] Lactose is the key sugar in milk. Using the enzyme lactase, babies can digest the lactose in breast milk, but they usually lose this ability as they get older because they do not need it once they are weaned. Worldwide, the ability to digest milk in adulthood is uncommon and is generally absent in East Asia and Africa. Overall, perhaps 65 percent of the world's population (including a significant fraction of North Americans and Europeans) lose their lactase enzymes by adulthood.[56] The ability of *adults* to digest lactose confers evolutionary advantages *only when a stable supply of milk is available.* And this would not have been the case until after milk-producing animals (sheep, cattle, goats, camels) had been domesticated and a host of related cultural features, such as knowledge about how to care for such animals and collect their milk, had arisen. The advantages of being able to drink milk are numerous, ranging from a source of valuable calories to a source of necessary hydration during times of water shortage or spoilage. Amazingly, just over the past three thousand to nine thousand years, several adaptive mutations have occurred in the *LCT* gene (coding for lactase) independently in Africa, Europe, and Central Asia, all conferring this ability.[57] And these mutations are principally seen in herders, not in nearby populations that have retained a forager lifestyle.

Here is another example where we can discern the genetic changes that have arisen recently in response to cultural practices that abet those changes. The Sama-Bajau are a group of people who live in scattered communities in Indonesia, Malaysia, and the Philippines. Many of them spend their whole lives on the water, from birth to death, living on houseboats with their nuclear families, traveling in flotillas, and subsisting on food from the sea.[58] These people are sometimes known as sea nomads, and many of them are exceptional free divers, often spending five hours a day underwater.[59] It is unclear how long they have had this lifestyle, but it has been at least one thousand years. And in this time, because of the cultural innovation of living at sea and diving for food that has been made possible by boat technology, navigation, and foraging techniques, these people have evolved genetic mutations that appear to equip them with the ability to cope with oxygen deprivation. They can dive to depths of seventy meters and forage on the ocean floor with just a set of weights and some wooden goggles.

In a brutal reality, those divers who were better able to get food this way appear to have left more offspring. One of the adaptations that made this possible was a bigger spleen, a trait that seems to be partially under the control of a gene known as *PEE10A*.[60] This mutation, and adaptation, is seen just in the Bajau, not in nearby, similar populations who live a terrestrial life. A big spleen helps because it can function as a reservoir of oxygenated red blood cells, and it can pump them out during the hypoxia that accompanies long dives, enhancing diving ability and survival. Once again, we have an example of cultural innovations (both behavioral and technological) affecting human genes, likely over quite recent historical time.

Ancestral Polynesians who were driven to explore the open ocean with their sophisticated navigation and large canoes (as we saw in chapter 2) may have created selection pressures on genes that better equipped their bodies to cope with periods of starvation or cold while at sea. Their journeys likely imposed selection for more efficient handling of the energy needs of the body via what are sometimes termed *thrifty genes*. Plus, living on islands—despite people's romantic fantasies—is extremely harsh because adverse weather (like hurricanes) can wipe out all of the available food for substantial periods, which imposes a set of constraints similar to long sea voyages. Adaptive though they were for the long journeys and isolated living centuries ago, these genetic changes are a prescription for diabetes and obesity today, now that the descendants of those Polynesians have built settlements on land and obtained sources of food that are more stable.[61]

Other examples of gene-culture coevolution abound. There is some speculation that people who speak tonal languages (like Mandarin Chinese) face a different adaptive environment than those who do not and that variants of two particular genes affecting brain structure in ways that enhance fluency in such languages may be selected for.[62] Innovations in farming or material technology can also have effects. It's possible that the invention of agriculture may have made the ability to be patient (and wait for crops to grow) more adaptive and that it affected the utility of genes undergirding this disposition.[63] Moreover, the domestication of crops typically increases the amount of starch in the diet and therefore affects the adaptive landscape for variants of genes that code for certain enzymes, such as amylase, which makes it easier to digest starchy foods.[64] The effect of crops can be even more convoluted. When yams were domesticated, West African forests were converted to farmland for cultivation. This had the unintended effect of creating more puddles, which offered better breeding environments for malaria-bearing mosquitoes, which in turn intensified selection pressure for the otherwise maladaptive sickle-cell hemoglobin that helps people to resist malaria. As in the cattle and lactase-persistence example, nearby populations of humans that did not engage in farming did not show a parallel increase in genetic variants for sickle-cell disease.[65]

The forms of social order that our species has invented in the past seven thousand or so years—involving cities and markets and modern telecommunications—are surely also a force in our natural selection. People's experiences of interacting with strangers thanks to those inventions are, in my view, likely changing our genes. For instance, as a species, we may be getting smarter because we live in cities, where urban culture is getting ever more complex, stimulating, and demanding. We almost certainly are evolving to have different immune systems (and related physiological traits) as members of our species live in denser and denser aggregations, and as people move over ever longer distances between cities, which gives rise to new sorts of epidemics. For instance, mutations appear to have recently arisen in our species that confer resistance to epidemics of typhoid newly made possible by urbanization and far-flung trade in Europe in the past few thousand years.[66]

Perhaps surprisingly, religion may also provide an example of gene-culture coevolution. The genetic predicates for having brains capable of understanding the concepts of religion, on the one hand, and religious beliefs and practices, on the other, may coevolve in tandem. Henrich has argued that the religions of most large-scale civilizations made possible by the agricultural revolution seem to be concerned with precisely the sorts of moral principles (like reciprocity and respect for authority) needed to organize larger societies, which involve more frequent interactions with strangers. Stated another way, as humans evolved the capacity to live socially and then developed the cultural innovation of forming larger urban collectives, certain kinds of religions became more useful, and this in turn put pressure on the kinds of genes and brains prone to such religions. The religions of small-scale societies, however, generally involve less moralistic and more capricious gods who are often seen as part of the natural world.

The Hadza are one of the only societies described as having no religious structures, leaders, or ceremonies.[67] According to anthropologist Coren Apicella, among those Hadza who report believing in a god, about half of them state that they are either unsure or do not believe that the god has supernatural powers, such as the ability to know what people are feeling or the ability to control what happens to people after death. The Hadza lack any belief in the afterlife. They often gave earthly descriptions when queried. When Professor Apicella asked them, "What happens to a person after they die?" they responded, "People cry." When she asked, "But then what happens?" they said, "The person goes in the ground." And when she persisted and asked, "And then what happens?" they said, "*Mwisho*"—which means "the end" in Swahili.[68]

Having religious beliefs (including the promise of heavenly rewards for the disadvantaged) and rules, traditions, and institutions that allow a large number of non-kin to work together makes it easier for a group to defeat other groups with more divisive or inconsistent ideologies. Religions are a cultural feature that secondarily allow unrelated individuals to expand the circle of cooperation, exchange goods, and maintain a division of labor. And success in competing with other groups (and reducing conflict within one's own group) may ultimately come to be reflected in our genes, insofar as certain genetic variants fare better in environments made more cooperative in the first place.

Cultural norms regarding monogamous marriage offer a similar example of changes in the fitness pressures humans faced, and these cultural practices can have a number of biological effects. First, monogamy provides a kind of societal-level testosterone-suppression program. In normatively monogamous societies, testosterone falls in men when they marry and again when they interact face-to-face with their children, but, in polygynous societies, testosterone stays high after marriage, since men face different sorts of reproductive demands.[69] And as we saw in chapter 5, monogamy is associated with reductions in violence and crime, perhaps in part because of this decline in testosterone. Second, and distinctly, the introduction of religious rules beginning in late antiquity in Europe and extending through the medieval period prohibiting cousin marriage—which obliged families to marry more broadly—may have reduced the reliance on kin networks for support and contributed to the emergence of formal state institutions.[70] But these cultural rules regarding consanguineous marriages might have also affected the kinds of genetic variants passed down to future generations because the offspring of nonconsanguineous marriages are simply more likely to survive.[71]

As if all these effects of culture on human evolution were not enough, mathematical models of gene-culture coevolution suggest that culture may be a further factor explaining the origins of cooperation in our species—though cooperation does not *require* culture in order to emerge (it is seen in many acultural animals, for instance). As we have seen, we humans are unusual in our ability to cooperate with unrelated individuals. And the existence of friendship networks is part of the way the challenge of cooperation is solved in our species. But *cultural* norms of reciprocity are another way of coping with the challenges of sustaining cooperation in groups. Cultural norms regarding punishment, altruism, and reciprocation can clearly support cooperation in circumstances where it would otherwise fail.[72] In short, culture further supports several crucial elements of the social suite—accentuating practices related to identity, friendship, in-group bias, cooperation, and learning—even as culture itself depends on the other elements of the social suite.

### *Pressure on Our Genes*

Careful modern analyses suggest that perhaps a few hundred out of the roughly twenty thousand genes in the human genome have been evolving at an increasing rate over the past ten thousand to forty thousand years in response to high-impact cultural changes (such as animal husbandry, urbanization, and marriage rules).[73] Many of these changes in gene variants are quite recent, having occurred over the span of mere millennia. But work in this area is still very active, and it will require more research for scientists to be certain of the connections between cultural changes and specific genetic responses.

Since human culture is cumulative and human populations are growing, the selection pressure exerted specifically by culture is stronger now than it was when our culture was limited to cave paintings and stone tools in small groups. However, our increased cultural repertoire may actually be *decreasing* the impact of selection pressures originating in the biological and physical aspects of our environment. How could culture dampen the ordinary force of ecologically based selection on our genes? Imagine that the temperature of the environment keeps going up and down. This would mean that, in some time periods, genetic variants that were helpful in coping with hot temperatures would be selected for, and in others, selection would favor variants that were useful in cold weather. But if humans had cultural adaptations—like modifying caves or fabricating air conditioners to stay cool, or making clothes or building fires to stay warm—they would effectively be counteracting these selection forces (this is known as counteractive niche construction). The temperature changes humans experienced would be less consequential because of these cultural adaptations, and therefore the force of selection would be weaker.

When people chose to migrate to the high plateaus of Tibet, they showed some genetic changes (in a gene known as *EPAS1*) to cope with a low-oxygen environment (one that was wholly different than that faced by the Bajau underwater). This evolutionary adaptation took place over the past three thousand years.[74] They had no choice, of course, because there was no available cultural way to increase the supply of oxygen.[75] And the process was similar to the lactase-persistence example we considered earlier. However, while the genes for coping with low oxygen were evolving relatively rapidly in response to the migration to this elevation, other genes, perhaps related to coping with the cold, might *not* have needed to adapt because the people had cultural ways of coping with this aspect of their new environment: clothing and buildings.

The power of culture may mean that our species retains genetic problems that might otherwise have disappeared.[76] For instance, the invention of glasses may, over time, be contributing to making humans more myopic by allowing genes associated with nearsightedness to persist and by failing to preferentially weed out myopic individuals (like me) who would otherwise have fallen off cliffs or been eaten by lions.[77] In a starker example, modern medical advances in the treatment of sickle-cell disease may increase the prevalence of genes for abnormal hemoglobin. When I was a doctor in training in the early 1980s, patients with homozygous sickle-cell disease would typically die by the time they were teenagers. I remember meeting one patient in his early thirties who was described as the longest surviving patient ever seen at Children's Hospital in Boston. Our patients were mostly children admitted to the hospital with agonizing sickle-cell crises involving the occlusion of blood vessels in various parts of their bodies by misshapen red blood cells; typically, they would die of renal failure or infection during late adolescence. Because of medical advances, however, survival into adulthood of these patients is now common; the median survival now exceeds forty-two years for men and forty-eight years for women.[78] Patients can live long enough to start families and pass on their genes, which in turn increases the number of patients carrying this abnormal hemoglobin.[79] There is even more hope, though. In the near future, we will likely have effective gene therapy (another modern cultural innovation!) to cure people suffering from sickle-cell anemia and similar genetic diseases, allowing them to live normal lives, though their sperm and eggs would likely still carry the mutation to pass on.

### *Biology and Culture in Conversation*

Human biology and human culture have always been in conversation, and not just on the genetic level. Rising socioeconomic status with industrial development results in better nutrition and in people becoming taller (a biological effect of a cultural development), and taller people require higher ceilings in their homes (a cultural effect of a biological development). Anyone marveling at the small size of beds in colonial-era houses at historical sites knows this firsthand. Or consider a completely different example: placebos. Their efficacy powerfully illustrates how cultural beliefs, about the generic utility of medication, can affect physiology.[80] Reciprocally, a biological phenomenon like the Black Death epidemic may induce large-scale cultural changes—for instance, in political structures or religious beliefs.

But as we have seen, culture can prompt changes at the *genetic* level too. And it is hard to know where this would stop. We can speculate that there may be genetic variants that result in changes in underlying tastes and behaviors that favor living in cities, saving for the future, consuming opioids, or participating in large online social networks. Or there might be genetic variants that favor living in a democratic society or among computers. If religious beliefs encourage fertility, maybe we will all be more religious in the future.[81]

These ideas have been difficult for me to accept because, unfortunately, this also means that particular ways of living may create advantages for some but not all members of our species across time. Certain populations may gradually acquire certain advantages, and there may be positive or negative feedback loops between genetics and culture. Maybe some people really are better able to cope with modernity than others. The idea that the way we choose to shape our world modifies which of our descendants survive is as troubling to me as it is amazing.

From a scientific perspective, however, work on gene-culture coevolution offers the exciting potential of a unifying framework that might bring social and biological analyses of human nature under one roof. Cultural evolution and genetic evolution should perhaps not be treated as separate at all, since our capacity for culture is indeed something that we evolved to have. The answer to the question "Is it nature or nurture?" is simply "Yes."

Once cultural transmission became possible, our species faced circumstances favoring brains that were capable of learning and teaching, including brains that could respect norms, copy models, or transmit information. In turn, as humans came increasingly to rely on the knowledge of others, they became even friendlier and kinder so that they could interact reasonably peacefully within groups and so that they could take full advantage of culture in order to survive. Culture builds on, and reinforces, our evolved capacity for making a good society. Culture is an emergent property of human groups, a new property of the whole not manifested in the parts themselves. And it arises from humans having the brains and social systems that allow for retaining and exchanging ideas.[82]

Human culture also cumulates. This means that brains and social systems capable of coping with more and more stuff are increasingly advantaged across time. And it also means that the force that culture has been applying to our evolution has been increasing over the past ten thousand to forty thousand years.[83] Once humans evolved to be capable of teaching and learning, they developed a parallel evolutionary strand, cultural evolution, side by side with genetic evolution. These two strands intersect repeatedly in many places and times. Each leaves its mark on the other.

CHAPTER 12

# Natural and Social Laws

Our evolution has shaped not just our bodies and minds, but also our societies. A related idea—about a connection between our bodies and our societies, albeit a metaphorical one—has an ancient heritage.[1] For instance, in 494 BCE, a throng of Roman citizens and soldiers fled Rome in response to draconian debt regulations and the growing power asymmetry between the upper and lower classes. The Roman senate commissioned the Consul Menenius Agrippa, himself of plebeian origins, to talk to the disgruntled citizens, assuage their concerns, and shepherd them back into the city. Agrippa's subsequent speech (as recounted by the Roman historian Livy writing five hundred years later) vividly described the relationship between the various parts of a human body as a metaphor for the interdependence between the various parts of Roman society. Seeing that the stomach appeared to do nothing but profit from the labors of the harder-working parts, the hands, mouth, and teeth

> entered into a conspiracy; the hands were not to bring food to the mouth, the mouth was not to accept it when offered, the teeth were not to masticate it. Whilst, in their resentment, they were anxious to coerce the belly by starving it, the members themselves wasted away.... Then it became evident that the belly rendered no idle service, and the nourishment

> it received was no greater than that which it bestowed by returning to all parts of the body this blood by which we live and are strong, equally distributed into the veins, after being matured by the digestion of the food.[2]

This society-as-body metaphor occurs repeatedly.[3] The Hindu tradition deploys a similar metaphor as early as 1500 BCE in the *Rigveda*, which ordered the castes in terms of body parts.[4] In *The Republic*, in about 360 BCE, Plato claims that societies have a tripartite class structure that mirrors the structure of people's souls and that both states and people are a complex whole made up of several distinct parts, each of which has its own proper role.[5] Plato believed that the most just and ordered state was one whose organization resembled that of an individual in whom these parts were in balance and working together in harmony. And in his letter to the Corinthians in about 53 CE, Saint Paul stressed the importance of interdependence within a community (noting "there are many parts, yet one body").[6]

Perhaps the most famous use of the metaphor was by philosopher Thomas Hobbes in *Leviathan* in 1651, in which he advances his ideas about the "body politic."[7] Hobbes imagines society as a single, embodied entity that he characterizes as the Leviathan:

> [The Leviathan] is but an artificial man, though of greater stature and strength than the natural, for whose protection and defence it was intended; and in which the "sovereignty" is an artificial "soul," as giving life and motion to the whole body; the "magistrates" and other "officers" of judicature and execution, artificial "joints"; "reward" and "punishment," by which fastened to the seat of the sovereignty, every joint and member is moved to perform his duty, are the "nerves," that do the same in the body natural.[8]

**Figure 12.1: Frontispiece from *Leviathan* by Thomas Hobbes (1651)**

This embodied form of society, which Hobbes dubs the Leviathan, represents his ideal model of government, with citizens bound and brought together by the decree and authority of a single powerful king.

Indeed, the famous frontispiece of *Leviathan* (figure 12.1) is a striking visual metaphor of the commonwealth reconfigured into the mold of a single human body.[9] After lengthy discussion with Hobbes, the Parisian Abraham Bosse made the etching of a giant crowned figure emerging from the landscape, clutching a sword and a crook. Above the image of the Leviathan is the passage from Job: *Non est potestas Super Terram quae Comparetur ei*—"There is no power on earth that compares to yours."[10] The torso and arms of the figure are composed of over three hundred blurred, miniature people, all facing inward, with only the giant's head having visible features. This image illustrates Hobbes's conception of the sovereign as an artificial person built up from the consent of many natural persons: "A multitude of men, are made one person, when they are by one man, or one person, represented; so that it be done with the consent of every one of that multitude in particular."[11] Hobbes conceived of the state as "but an artificial man."

Modern understandings of this ancient conundrum of how individuals come together to make a whole society do nothing so simplistic as mapping sectors of society to human body parts. Nor do they reify hierarchy or vindicate state power in quite the same way. But as we have seen, they do place a similar emphasis on groups of people coming together, albeit on the basis of social networks and cooperation. And while we may not have analogized so explicitly to the digestive tract or nervous system of the human body, our analysis has deployed biology to understand the blueprint that provides the foundation of human society.

### *Humans Apart from Nature*

The claim that our species' genes play a role in our society has been controversial, regardless of the persistence of the society-as-body trope. Of course, the main purpose of the metaphors was to show that society must have a certain ideal order to function properly; the metaphors rarely pointed to something organic. But arguing that humans and their behavior are bounded and shaped by biology has historically been contentious in philosophy, social science, and public discourse. It's one thing to see society as a metaphorical body; it's quite another to see society as being driven by the very same processes that guide our physical bodies.

The temptation to see humans as separate from nature and from natural forces has been powerful and long-standing. It was accentuated—in Western thought—by medieval religious ideas: humans were seen as something special, set apart from animals and the rest of the natural world. But this hardly seems an appropriate way to approach the natural world today. In fact, since the nineteenth century, there has been an accelerating movement in philosophy, science, and policy to reintegrate humans into the natural world. We see this in the animal rights movement, in debates about climate change, even in the way that scientists increasingly discover human traits in other animals.[12] As we have noted, elephants have friends. Dolphins have cooperation. Chimpanzees have culture. In fact, every time we break down a distinction between us and other animals—whether it is to avoid eating animals "with faces" (as some vegans express it), to analogize animal diseases to our own (as medical researchers do all the time), or to recognize human qualities of cooperation or friendship in other animals—we reposition ourselves as subject to the same natural laws as all other creatures.

The fundamental changes wrought by the agricultural revolution were mirrored by parallel changes in the way humans thought about nature and their place in it. The domestication of animals and plants implied human mastery over (or at least stewardship of) the natural world. Urbanization and the transformation of natural landscapes, the construction of religious monuments, and, of course, the invention of writing, commerce, and technology—all of these developments served to distance people from a natural world that must have seemed increasingly wild and dangerous, a place to be restrained and controlled. In time, dominating nature or willfully separating from it came to be seen as the source of a good life.

One of the earliest written records, *The Epic of Gilgamesh*, which was inscribed on clay tablets around 1800 BCE in Mesopotamia, addressed our species' place in nature. The first tablet tells of "the Coming of Enkidu," whom the gods create to rival the authority of the demigod Gilgamesh. Enkidu epitomizes all things natural—he has long matted hair, dwells in the wilderness with animals, eats grass, and knows nothing of farming. Conversely, Gilgamesh embodies everything cultivated and civilized, acting as "the shepherd of the city, wise, comely, and resolute," and building city walls to physically demarcate that which is human from that which is not.[13] Other religious strains of thought that arose after humans took up agriculture and built cities generally embraced and expanded this human-versus-nature dichotomy.[14] The Old Testament, for example, is heavily laced with references to humankind's standing above the natural world.[15]

Western philosophers have also come up with all sorts of reasons to pull us away from our natural roots. Aristotle's 350 BCE text *Politics* appeared at a time when human involvement in a specifically political community reached an early crescendo. The notion of *civitas*—that humans achieve their fullest meaning when engaged in political life—powerfully illustrated the way in which humans saw themselves as distinct from nature.[16] Aristotle pinpointed our species' "rational principle" and "endowment with speech" as further intrinsic features that elevated humans above the rest of the animal kingdom.

Although the ascendency of humans was built into early Christian thought, it was Saint Thomas Aquinas's theological writings in the thirteenth century that helped to permanently ingrain this separation between humans and the natural world in Christian doctrine.[17] For Aquinas, it was the soul that separated humans from other animals. In his *Summa Theologica* and *Summa Contra Gentiles,*

Aquinas proposed a hierarchical relationship between humans and nature, arguing that God created the natural world for human dominion.[18]

For centuries, things remained this way. Other traditions of Western thought built on these ideas. Some emphasized humans' sensory/empirical and cognitive/reflective abilities that other creatures lacked. Early forms of empirical and inductive science ("natural philosophy"), first advocated by polymath Francis Bacon in 1620, were predicated in part on the belief that the natural world existed for humans to study using these special faculties, further reinforcing the separation of humans from nature. The Scientific Revolution (building on the work of towering seventeenth-century figures like Galileo and Newton) had complex effects on our relationship with nature. In some ways, it supported an anthropocentric abuse of the natural world, not just a separation from it, because of two simultaneous changes. First, the rise of scientific inquiry stripped nature of its spiritual essence. And second, it fostered the belief that humans ought to assert dominance over the natural world through science.[19]

Philosopher John Locke, also in the seventeenth century, argued that humans, in order to effectively protect property and lead ethical lives, had to create a contract to *exit* a state of nature and come together into a political society. In the same period, philosopher René Descartes propounded notions of dualism, the idea that the human mind and body were distinct spheres and, by extension, that animals were incapable of reason.[20] And in the eighteenth century, philosopher Immanuel Kant characterized humans, who wield agency and reason, as distinctively moral beings.[21] In contrast, philosopher David Hume separated humans from the natural world based not on our capacity to reason about or observe nature but on our capacity to *sympathize.*

Then, the emergence of new industrial technologies in the eighteenth century had two effects in tension with each other. On the one hand, these new technologies represented even greater human mastery over nature, and many thinkers saw this as morally good. On the other hand, some thinkers saw this dominance as troubling and potentially threatening to something pure about nature. Finally, by the middle of the eighteenth century, the long-standing intellectual framework of separation had started to face significant resistance. The competing view was that humans ought to admire and even reside directly within nature and be seen as part of it. Philosopher Jean-Jacques Rousseau, a contemporary of Hume and the father of Romanticism, held views opposed to Hume. He paved the way for subsequent nineteenth-century thinkers, such as Ralph Waldo Emerson and Henry David Thoreau, who embraced the natural world in their philosophy of transcendentalism, as we saw in chapter 3. In *Nature*, published in 1836, for example, Emerson identifies qualities that render nature not only intrinsically valuable but also foundational to human ethics, imbuing nature with a transcendent spirituality on par with humans'.[22] Of course, it was against this backdrop in 1871 that Darwin published *The Descent of Man*, which, among many other advances, argued that the forces of natural selection, so prevalent among other animals, also applied to humans themselves and even their behavior.[23]

In this tradition, nature is also seen as the source of a good life, as we saw with Thoreau's description in *Walden* and in the rural communitarian movements. This appreciation of the intrinsic qualities of the natural world is closely tied to contemporaneous environmental changes. Sparked by enhancements to the steam engine in the 1760s during the Industrial Revolution, mechanized forms of work progressively replaced human and animal labor as the driving force behind societal development in the nineteenth century all the way through the development of tractors in the twentieth century. These changes prompted concerns that humans' relationship with nature was being profoundly altered. Transcendentalism championed the restorative beauty and spiritual unity of nature.[24]

Nevertheless, the separation of humans from the rest of the natural world was an ever-present perspective. And this is still reflected in the modern scientific approach to the world, which has generally—with some exciting exceptions, such as evolutionary psychology, behavior genetics, and social neuroscience—kept the social and biological sciences apart.[25] As a physician and sociologist, I work at the border between these scientific areas, and I can vouch for how strictly it is patrolled. Most people still think of humans as special. But, to me, humans are not so special. And paradoxically, as I've argued, our closeness to animals actually reveals our common humanity.

### *Contentious Claims*

The claim that the social suite is founded on human evolutionary biology and is therefore a universal feature of our societies would be seen as reflecting *positivism, reductionism, essentialism,* or *determinism* (all of which I will define below) by some critics. In some quarters, these are considered epithets; in others, they are even viewed as sins. But really, these critiques merely reflect philosophical stances related to different ways one can understand the world.

One common source of all these philosophically grounded objections is that critics can rightly point to the "contingent" nature of social reality, correctly noting that much of what we observe about social life depends on the particular details under consideration or on the setting, the historical period, the power dynamics and institutional forms that happen to be in place, and even the nature of the observer (whether the observer is male or female or rich or poor, for example).[26]

If social reality is so variable and so mutable, and possibly even so unobservable, how can we study it in the ways ordinarily employed by the natural sciences? Some critics go so far as to state that the social world is never the same from one moment to the next, arguing that it is like Heraclitus's river—you cannot step in the same one twice. One colleague of mine believes that the existence of even one example of, say, polyamory anywhere in the world necessarily negates the claim of a biological basis for pair-bonding. But it is sometimes the exceptions to typical human social behavior that prove the rule, as we saw with the extraordinary cultural machinations that the Na of the Himalayas deploy to suppress romantic attachments. Some social scientists take this intransigence to an extreme and claim that it is not possible to subject social phenomena to scientific inquiry at all, that they may be only "interpreted" and never really explained! To me, this turns science into theology and, in its most extreme form, is a claim that must be strongly resisted.

Of course, it is true that many givens in the social sciences have been overturned even in my lifetime, such as claims about girls' lower-than-average performance in math.[27] Even more dramatically, many things seen as settled matters by nineteenth-century social scientists have been totally discarded—for instance, the perverse justifications for the abhorrent institution of slavery. After all, it was medical doctors in the nineteenth century who came up with the "diagnosis" of *drapetomania*, which is the "insatiable desire of a slave to escape."[28]

It is not just the social sciences that are vulnerable to revision. New thinking and discoveries have upended many scientific claims, such as the number of chromosomes in human cells, the composition of the core of the Earth, the existence of extrasolar planets, the health risks of various nutrients, the efficacy of anti-cancer treatments, and so on.[29] But the provisional nature of scientific discovery does not mean—cannot mean—that it is simply impossible to observe any objective reality. Over time, items of belief become formalized into hypotheses and then, after sustained testing and much experimental evidence, get widely accepted as facts. Cold, hard facts. The social sciences, like the natural sciences, advance. Their previous errors are not sufficient grounds for their present rejection.

Especially in the social sciences, we need to determine whether it is the world that is transforming or just our understanding of it. For instance, just because the *manner* in which we understand certain core aspects of society is updated (for example, if we invent new statistical methods or develop new theories and discard old ones) does not mean that those same *aspects* of society are somehow new. Some of these changes are even to be expected; the contingency we see in the social life of our species is in fact a contingency we evolved to be universally capable of manifesting. One of the ways humans differ from other primates, for instance, is in the variety of mating practices we adopt, albeit grafted onto the core practice of pair-bonding, as we saw.

Finally, just because *some* aspects of social life depend on circumstances and are mutable does not mean that others (such as the social suite) are not constant. Critics of the idea that genes shape our social lives have created a series of straw-man arguments—like the needlessly binary nature-versus-nurture debates—that no longer fit our contemporary understanding of the human condition.

## *Four Isms*

Why is there such resistance to the integration of biology and human behavior? Four strains of thought have contributed to this.

*Positivism* asserts that truth can be known only via scientific study involving the application of logic and mathematics to the natural world in a manner that is verifiable and reproducible. This position can come in for harsh criticism from some quarters when scientists like me advance the claim that social phenomena can indeed be understood *scientifically*, via the discovery of general rules or laws that operate in the social world just as they do in the physical and biological worlds. In my view, both an extreme overconfidence in the completeness of scientific insight and the wholesale rejection of positivism are problematic postures.

The positivist stance that social phenomena are scientifically intelligible was first advanced in modern times in the middle of the nineteenth century by philosopher Auguste Comte.[30] Around the same time, Émile Durkheim, one of the founders of sociology, was also making the case for positivism.[31] His approach to social phenomena stressed that they were, after all, a part of the natural world and so could be approached via a scientific method stressing objectivity and rationality.

But the debate about how to understand social life has ancient roots and can be traced at least as far back as Plato, who analyzed the differing worldviews of poetry and philosophy (which was at the time an approximation of science).[32] Echoes of this debate are still heard today in the endless dialogue between the humanities and the sciences regarding how the world may best be comprehended. Some thinkers argue that the internal states of humans cannot be examined scientifically at all and must instead be understood nonscientifically via intuitive, interpretative, or even religious methods. Even some scientists devoted to strong empiricism adopt this view. B. F. Skinner, the leading twentieth-century advocate of behaviorism and the author of *Walden Two*, famously reasoned that internal mental states are unobservable and unquantifiable subjectivities and thus belong outside the purview of objective scientific scrutiny, in contrast to observable (individual and collective) behaviors.[33] Some philosophers and theologians continue to embrace the age-old dualistic separation between the material world and the mental world.[34] The underlying claim is that we cannot use science to fully understand the soul or even feelings, thoughts, morals, or beauty. While the issue of the soul is a matter unto itself, feelings, thoughts, morals, and even beauty—and their evolutionary origins—are, in fact, yielding increasingly to science in the twenty-first century with techniques as diverse as MRI imaging and behavior genetics.[35]

It is easy to caricature positivism and claim, as did twentieth-century physicist Werner Heisenberg (famous for his uncertainty principle), that positivism is concerned with only the most boring parts of the world and ignores the much greater and more interesting quantity of things that are scientifically unknowable.[36] Positivism can also be legitimately criticized for skirting real problems of observer bias and the abuse of science for ideological or fraudulent ends, which can result in scientific "facts" that may not be facts at all. The early-twentieth-century eugenics movement and the barbaric history of experimentation on humans are obvious examples.[37] The conduct of science is a human activity, fraught with all the imperfections that bedevil anything humans do. But the existence of limitations in scientific practitioners does not mean that the world is not ultimately knowable.

Moreover, in a defense of positivism, I take it as a given that some observation is better than no observation. Just because our tools are limited and sometimes (or even often) fail does not mean they serve no useful purpose. Ever since Galileo used his rudimentary telescope to find moons around Jupiter, spots on the sun, and craters on the moon, people have discovered quite a lot with tools that seem primitive by today's standards. Science is an iterative process. We cannot go back and skip over the earlier, imperfect invention of a terrestrial telescope to today's space-based telescopes, even if the former had limitations imposed by light pollution and atmospheric interference. It would be like saying that the paper maps many people used to navigate for centuries served no purpose now that we have GPS.

Those examining the role of genetics in human affairs are often accused of another sin, of reducing complex or higher-level phenomena to their parts. *Reductionism* assumes that the whole is merely the sum of its parts and nothing more. An effort to reduce society to a set of universal features or to a smaller set of rules partially encoded in our species' genes—like the social suite—is seen as problematic. Such an effect is seen as a gross oversimplification of something that is inherently complex and irreducible.

Reductionism neglects the reality of *emergence*, which is the process by which wholes can have properties that are not present in the parts. An example of emergence we have seen is the differing properties of collections of carbon atoms depending on whether they are arranged as graphite or diamond. Other natural expressions of emergence are stunning. Perhaps the most impressive is that carbon, hydrogen, oxygen, nitrogen, sulfur, phosphorus, iron, and a few other elements, when combined, yield life itself. And life has properties not present in or (given current science) reducible to these constituent parts. Another moving example is the way that consciousness can emerge from the pattern of connections among neurons in the brain. In these examples, there is a transcendent synergy among the parts in these complex systems.

Physicist Philip Anderson, in his discussion of emergence, offered a piercing critique of reductionist thinking: "The ability to reduce everything to simple fundamental laws does not imply the ability to start from those laws and reconstruct the universe."[38] This seeming paradox stems from what Anderson calls "the twin difficulties of complexity and scale." As systems expand along these axes—as physical entities aggregate into chemical ones, as chemical entities aggregate into biological ones, as biological entities aggregate into social ones—they acquire new, emergent properties, and they require entirely new approaches, concepts, and laws to comprehend.

I have indeed been arguing in part that social order can be understood with some reductionistic reference to biological principles and social facts. Yet at the same time, I have also been arguing for the importance of holism (which is the opposite of reductionism) and for the role of emergence in our social lives. I have spent most of my career showing how, when it comes to society, the whole comes to be greater than the sum of its parts, with distinctive features not present in the components. Our examination of social life has therefore embraced *both* reductionism and holism. Collections of people have qualities that transcend the sum of each individual's traits. And accepting an evolutionary foundation for collective phenomena—even if this is a reductionistic enterprise—actually allows us to see *how* emergent properties like cooperation and social networks can arise at all.

For the past few centuries, the thrust of science has been to break things down into ever smaller bits in the pursuit of understanding. This has been a huge success. We can now understand matter by breaking it down to atoms, then protons and electrons and neutrons, then quarks and gluons, and so on. Organisms can be similarly broken into smaller units, from tissues to cells to organelles and then to proteins and amino acids and DNA. The opposite approach does not come naturally. Putting the parts back together into a whole in order to understand what is happening is usually harder. And so, holism typically comes later in the development of a scientist or in the development of a scientific discipline. But putting-back-together is no less a crucial or productive part of science. In short, I am arguing that our focus on the genetic foundations of society is not only, or merely, reductionistic; on the contrary, it sets the stage for a truly holistic understanding of our social lives.

Another critique sometimes leveled at the ideas we have explored is that they are essentialist. *Essentialism* is the position that things in the material world (including people and society) have a fundamental set of properties that are necessary for their identity. Like positivism, essentialism can be traced back at least to Plato; he expounded his theory of "forms," showing that, while every chair is different (for example, dining-room chairs, outdoor chaises, three-legged stools, Barcaloungers, ejection seats, and so on), they all share something in common, the *essence* of being a chair. Essentialism looks beyond particular examples of things to deduce their essence, which does not change.[39] In this regard, I have been arguing that societies share some universal, essential properties. But many critics, as we saw in chapter 1, do not like this claim because it would seem to devalue the role of culture and somehow neglect the incredible variety of ways humans live. Nevertheless, I think it is possible to embrace the reality of essentialism when it comes to society without abandoning the recognition that there is also much that is quite variable about social life wrapped around the social suite—and even, as we have seen, facilitated by it.

Finally, *determinism* is the position that the state of any system is completely fixed by its preceding state or states. From this perspective, I have been, in a sense, claiming that society can be meaningfully determined by human genes. In strict "causal determinism," no events are self-caused, and no events have no causes. In the extreme, every observable feature of the natural world can be causally traced back to the big bang![40] This stance has obvious implications for the notion of free will or the concept of prediction in social systems.[41] But criticism of an all-powerful biological determinism is sometimes used as a cudgel to crush the more limited (and reasonable) claim that biology can prime—even if it does not completely govern—the flow of certain human behaviors. This is my argument.

Unlike essentialism with respect to human nature, which explicitly and implicitly tips toward nature in the nature/nurture dichotomy, the concept of determinism is agnostic on whether it's our genes, our environments, or our cultures that "determine" our behaviors. Of course, as we have seen, the reality is that all of these are important. Still, despite this neutrality, critics often suggest that determinism is more nefarious, more socially problematic, in its biological form.[42] But why would it be inherently more problematic to conclude that human actions or social life depend on genes as opposed to any other fundamental factor? There are indeed plenty of nongenetic features—things that cannot be changed, like one's birth order, preschool education, or childhood trauma—that also influence behavior.

In obvious ways, the notion of a blueprint for an inevitable society is a kind of determinism. But this is complex. A crucial part of our species' genetic inheritance is, ironically, the capacity *not* to be wholly tethered to our biology! Considering the astounding range of habitats in which *Homo sapiens* has survived and thrived and the evolution, in our species, of the capacity for culture, there has never been a more elastic creature on earth. We have an evolved capacity (using cooperation, friendship, and social learning) to very deftly respond to the myriad circumstances we face, to manifest a kind of bounded flexibility shaped by our genes.

### *Aversion to Genetics*

Whenever a new study finds that genes are relevant to some behavioral phenomenon (from personality to depression to resilience to violence), a hectoring little asterisk is often slapped onto the conclusions.[43] Genes are not destiny, we are told; they neither determine nor essentially define our fate. One reason that people are admonished to downplay the role of genetics as an explanation for human nature is the fear of the abuse that it might inspire if this claim went uncontested. Many sensible people want the evidence of the role of genetics in human behavior to go away because they are convinced that "grave dangers await."[44] Alarm bells go off. Could this lead to an apologia for the vile eugenics of the past?[45] Could this be used to support ongoing bigotry in the present? Misguided and poisonous biological arguments have been used to promote racism, misogyny, child abuse, colonialism, and other destructive ideas in many times and places.

But when observers downplay the role of genetics in human affairs, they create a different kind of problem, one that forces people to ignore what is plainly in front of their eyes. And this can lead to missed opportunities for ameliorating the human predicament. First, accepting the genetic predicates of human behavior helps us all understand why so many social problems recur with such maddening frequency. I once had a conversation with an eminent social scientist who denied that biology played any role whatsoever in social affairs. When I asked whether he believed that genetics might have some role in criminality, he said, "Absolutely not." And when I gently pointed out that 93 percent of incarcerated criminals were men and that scientists have a lot of evidence, for example, regarding the role of testosterone in aggression in nonhuman primates (for instance, among chimpanzees, 92 percent of attackers and 73 percent of victims are male), he seemed befuddled.[46]

Concerns about eugenics and discrimination are obviously profoundly legitimate, but they are not justifications for persistent, willful ignorance of the scientific underpinnings of our social lives. Acceptance of this evidence alone does not inexorably lead to discrimination. That requires a further ugly moral and political overlay. Public policy is better served by rejecting that overlay.

In fact, I would even argue that the best way to avoid morally bad outcomes is to be fully aware of the scientific reality—while also acknowledging that this reality might be revised with further research. Doing this will allow us to come up with the most humane policies.[47] All else being equal, it serves us better to adopt practices that take account of our species' evolved nature. In this way, we could avoid the harm imposed on individual and social welfare by misguided prohibitions and repression, whether these are enforced through customs or through laws (such as the grotesque separation of refugee children from their parents). It is neither good nor sustainable to treat humans as the mere "puppets of social arrangements," to quote sociologist and philosopher Erich Fromm.[48] Conversely, if and when we do indeed decide to push back against our evolved nature, we can be more conscious of the serious challenges we will face and plan accordingly and thoughtfully.

The universals of the social suite—shaped by natural selection and encoded in our genes—are not only facts, but also sources of our happiness. They are essential to the ability to judge which social arrangements are good for humans in the first place. As Fromm argues, "In fact if man were nothing but the reflex of culture patterns, no social order could be criticized or judged from the standpoint of man's welfare since there would be no concept of 'man.'"[49] And there would be no concept of human welfare either.

For too long, many people have perpetuated a false dichotomy whereby genetic explanations for human behavior are viewed as antediluvian and social explanations are viewed as progressive. But there is a further problem with people putting their heads in the sand when it comes to human evolution, and that is the overcorrection that results. Choosing cultural over genetic explanations for human affairs is not more forgiving. After all, culture has played a huge role in slavery, pogroms, and the Inquisition. Why should the social determinants of human affairs be considered any better—morally or scientifically—than the genetic determinants? In fact, belief in the sociological mutability of human beings has, in my judgment, done more harm to people through the ages than the belief in their genetic immutability. There is a long history, for example, of denying any biological basis for homosexuality and seeing it as a lifestyle choice under the control of the individual, a choice that others then may respond to with opprobrium, oppression, and violence.[50]

Efforts in social engineering, sponsored by leaders such as Joseph Stalin, Mao Zedong, and Pol Pot, have killed countless millions of people, and they are often driven by a false belief that fundamental, genetically encoded, and universal aspects of human behavior and social order can simply be swept aside. Stalin, for example, advocated creating a "revolutionary authority" that could be used "to abolish by force the old system of productive relations and establish the new system. The spontaneous process of development gives place to the conscious action of men, peaceful development to violent upheaval, evolution to revolution."[51] Stalin wrote and thought much about the science of controlling and manipulating human nature, and, as we saw in his enthusiastic support of Lysenkoism (and also Pavlovian psychology), he felt that behavior in animals was completely determined by the environment, which in turn offered the prospect of fundamentally controlling it.[52] This did not work out well. Historians estimate that at least three million people (and probably more than nine million) died because of Stalin; this included roughly eight hundred thousand executions, over 1.7 million deaths in the Gulag, and hundreds of thousands of deaths related to the resettlement of ethnic minorities.[53] A notable aspect of Mao's philosophy was an analogous confidence in the malleability of human behavior, on both individual and collective levels. Mao felt that the state must directly intervene to shape the beliefs and actions of human beings because the transformation of society "depends entirely on the consciousness, the wills, and the activities of men."[54] Mao did not think highly of notions of an innate, shared human nature.[55]

There is a long and sordid history of using genetics in ways that divide and alienate people from each other. Some have responded to this by ignoring the empirical evidence about the evolutionary origins of human behavior and social organization in the hope that the evidence would simply disappear. But the fact that the truth could be dangerous—if misunderstood, misapplied, or combined with faulty moral premises—does not mean it should be suppressed.

I believe that a better path forward is to examine our shared evolutionary heritage in order to identify the ancient roots of human similarity. The thing about genes is this: we all have them. And at least 99 percent of the DNA in all humans is *exactly* the same. A scientific understanding of human beings actually fosters the cause of justice by identifying the deep sources of our common humanity. The underpinnings of society that we have come to understand—the social suite that is our blueprint—have to do with our genetic similarities, not our differences.

## *The Natural and the Good*

How can we say that the blueprint is good? Many of the elements of the social suite—individuality, love, friendship, cooperation, learning, and so on—seem manifestly pleasing and, indeed, good. But philosopher G. E. Moore argued a hundred years ago that goodness cannot be equated with pleasingness. Furthermore, he went on to state—overly nihilistically, in my view—that goodness cannot be defined because it is not a natural property at all. He coined the term *naturalistic fallacy* to explore these ideas, a concept that, in its contemporary usage, refers to the claim that just because something is natural does not make it good. Maternal death during delivery is natural, in the sense that it is not uncommon without the "artificial" intervention of modern medicine, but no one would characterize it as good. Similarly, veganism, surgery, and nation-states are all unnatural, but that does not mean that they are bad or undesirable.

A long-standing problem in moral philosophy relates to the fundamental origin of moral values.[56] Are they independent of humans, naturally woven into the fabric of the universe (or, as some believe, an expression of God's will), or are they simply human creations? If the latter, how can we avoid an endless downward spiral of relativism that would reduce morality to a trivial seesaw between "Boo" and "Hurrah," merely reflecting any particular person's or group's preferences at the moment—no different than judgments about aesthetics or religion? Morality would devolve to plebiscite or individual taste.

Some evolutionary biologists have argued that it is precisely this capacity for moral *deliberation* that has evolved (because it must have been fitness-enhancing in humans) rather than the *content* of any resulting morality per se.[57] Moral deliberations, and even moral tensions, function as a kind of social cement, binding groups together because such deliberation requires awareness of, and appeals to, both group-level norms and the well-being of others.

The topic of the origin of morality is related to another famous dichotomy in moral philosophy, known as the is/ought problem; that is, the distinction between the state the world is in and the state we would want it to be in. Moral judgments implicitly contain commands. *Ought* invites us to do something in a way that *is* does not. This dichotomy also sets morality apart from nature, outside it, because it can prescribe a state of affairs that does not or could not exist—for instance, an impossibly utopian community. However, other philosophers argue that if humans are part of the natural world, then their morals must be too.

Because moral judgments do not describe the world but prescribe it, they are usually felt to be incapable of disproof and thus not scientific. While you can say that the claim that the Earth is flat is either true or false, you cannot make the same objective statement about the claim that killing is wrong. Yet there would seem to be *something* objective about morality. As Hume argued, morality is related to the objective state of affairs we find in the world, though it also contains something more that is "determined by sentiment."[58]

A strange thing happened to moral philosophy after World War II. Philosophers looked at photographs of the concentration camps and felt, strongly, that it could simply not be a case of some people saying "Boo" and others (the Nazi doctor Josef Mengele, perhaps) saying "Hurrah." They felt that surely there must be some basis for adjudicating these positions. Part of the lesson of the Holocaust for many secular moral philosophers seemed to be that the pre–World War II view of morality was too optimistic and that it was in fact not possible to see people as "fundamentally decent chaps," to quote philosopher and novelist Iris Murdoch.[59] The war prompted not only a desire to find an objective basis for morality but also a sense that humans were, in the end, actually quite bad. The latter is not my own view, of course. Quite the contrary. I think that humans are fundamentally good and that, as we have seen, we are pre-wired to make societies that are filled with the sorts of things moral systems see as good. The social blueprint is a nontheological, human-independent source for the good things in life that we value.

Also emerging after World War II was a set of ideas advanced by British philosopher R. M. Hare.[60] He had been taken prisoner in 1942 and survived the long march up the river Kwai and the Japanese camps. Hare retained the perception that people freely choose values, but he argued that they did so *against constraints.* One's morality, slipping and sliding, would eventually have to reach a bedrock of those constraints imposed by something objective, something outside oneself—in other words, by nature. At least parts of morality could be natural, after all.

The argument goes like this: When one understands what it means to be a clock (which is to tell time accurately), one can then be in a position to say whether the function of a clock is good or bad. Similarly, when one understands what it means to be human, one can be in a position to say whether a human experience is good or bad. For instance, we might say that a human being who lacks the capacity for love does not fully participate in being human, which is bad. From this perspective, these natural constraints and definitions can stop an otherwise endless relativistic moral regress.[61] We could say that a society is good when it enhances its members' happiness or survival. Such are the constraints against which both evolution and morality abut. This, too, is an ancient idea, at least as old as Plato and Aristotle.

Philosopher Philippa Foot famously and provocatively said, "In moral philosophy it is useful, I believe, to think about plants."[62] She argued that there is no fundamental difference between a notion of "good," whether the topic is a tree having "good roots" or people being in a "good" state. The roots have a purpose, a logical constraint that they must satisfy, which sets the standard for deciding whether they are good or not. For instance, humans, animals, and plants are all living things. In all three cases, we can speak of them as being healthy or unhealthy, say, or excellent or defective examples of their kind. This means that we can characterize the qualities that are conducive to their being healthy or excellent and so on. We can speak of human virtues such as kindness and bravery the same way. These virtues are "natural excellences" and their opposites are "natural defects."[63] Foot explained that "moral action is rational action" and that morality can be determined by the constraints imposed by the nature of our species. In our case, *rational* means that it is good when humans live socially, as we are naturally compelled to do. Morality, insofar as it relates to the making of a good society, which is what makes it possible for us to be fully human, is guided by our evolutionary past. In answering the question of what is a good society, therefore, we must ask, *What kind of society is good for us?* What do we need from, and what can we offer to, one another? Here, again, the blueprint provides an answer.

There is a long tradition, epitomized by psychologist Abraham Maslow, of ordering human needs in terms of those that are necessary for survival and that motivate and animate human actions.[64] Maslow arranged these needs from most basic to most advanced, from "physiological needs," to "safety," "belonging," "esteem," and finally "self-actualization." In a simplified version of his theory, one moves in one direction through these needs so that, in a particular setting or across a life course, one satisfies physiological needs before safety needs, safety needs before esteem needs, and so on. In essence, Maslow's argument was that "man does not live by bread alone."

For instance, at the first level, we have a need for food, sleep, shelter, and sex. Once these needs are satisfied, we move to the next level. Under safety, we have a need for emotional and physical security. Then we have needs for social belonging, such as friendship and love. At the next step, humans have needs for recognition, status, and respect. And finally, Maslow argued, people have a need to realize their full potential, to find meaning by giving to some higher cause outside themselves, and to be "self-actualized." Maslow later put it as follows, slightly expanding his concept: "Transcendence refers to the very highest and most inclusive or holistic levels of human consciousness, behaving and relating, as ends rather than means, to oneself, to significant others, to human beings in general, to other species, to nature, and to the cosmos."[65]

And yet, perhaps this order of needs and motivations should be upended. The shipwrecked crews, for instance, were able to satisfy their needs for food, shelter, and safety precisely because they first honored and attended to their needs for friendship, cooperation, and learning. The latter were predicates for the former. In fact, as a species, humans have evolved to have these higher-order needs, including even for transcendence (which can involve our sense of duty to our social groups and the sense of meaning we derive from such selflessness), precisely in order to allow us to more efficiently satisfy the basic needs of life.

### *Engineering New Social Worlds*

An understanding of, and respect for, our social blueprint is valuable for coping with radical new technologies that give humans entirely new powers to intervene in their social world. Of course, humans have had to cope with disruptive innovations with relevance to social organization in the past. For instance, the invention of cities, with the imposition of a less nomadic existence and a higher population density, affected social interactions. Humans responded to this both as individuals who live in this new way and also as a species, possibly evolving new capacities (such as resistance to the different kinds of urban-facilitated infections).

The invention of technologies that give us the power to communicate faster than a human can travel has also affected our social interactions. Up until the invention of telegraphy—and with the exception of signal fires and carrier pigeons—the speed of any message between two people was limited to the speed of a human messenger traveling on foot, on horseback, or by boat. With the invention of telegraphy and telephony, however, messages could move faster than a human. The invention of the internet was another big step, not so much in speed, although that too, but in volume, breadth, and searchability. We can see the impact of these technologies in everything from the way we teach our children to the way we use our memory (because we have Google-equipped phones in our pockets). The ways in which we interact with one another—certain social niceties—can be supplanted by a computational device. What would it mean to live in a world in which you could wear glasses that recognize another person's face and then bring up his or her online biography so that you no longer have to remember whether an approaching person is friend or foe? Something that was crucial for hundreds of thousands of years in hominids—knowing a person's identity and whether he or she meant you ill or well—can now be delegated to a machine. All of these developments have had, and will have, implications for social interaction, social organization, and social behavior.

Still, none of the foregoing technologies has—so far—fundamentally altered the social suite. One of the ways we know this is that the basic structure of human social networks and the basic outlines of human cooperation are the same throughout the world, as we have seen, even among populations who do not live in cities and who lack access to modern telecommunications and other technologies. To the extent that foragers show certain behaviors in common with contemporary populations (and they do), it means there is something very deep and fundamental about our humanity. However, there are two new technologies that could have a more substantial impact than anything we have created before.

The first radical technology is artificial intelligence (AI). A key difference between technologies humans have heretofore invented and modern AI is that all prior technology was at our service. That technology—whether spears or satellites—was concerned with means directed at ends humans specified. But AI has the potential of specifying its own ends, of having its own "desires."

Even more amazing, to me, is that AI might actually affect our social organization. We will increasingly be adding machines (such as driverless cars) and autonomous agents (such as online bots) to our social systems. These devices are not simply tools we will use to supplement our own efforts (like the face-recognizing glasses), but machines that might act in humanlike ways. At the moment, these devices—such as companion robots that express emotions and carry on simple conversations or online bots that spread misinformation—are still crude. Still, many of these technologies interact with us on a level playing field, as if they were human, in what I call *hybrid systems.*

These hybrid systems of humans and machines offer opportunities for a new world of *social* artificial intelligence. My own lab has experimented with some of the ways that such AI might modify the performance of groups. In one experiment, we added bots to online groups of humans and showed how the bots—even though very simple (equipped only with what we have called "dumb AI")—can make it easier for groups of (intelligent) humans to work together by helping them to overcome friction in their efforts to coordinate their actions.[66] In another experiment, we brought people into the lab and placed them into groups of four (three people and one humanoid robot); the groups were given the challenge of solving a game (involving laying railroad tracks in a virtual world) and the robot was programmed to make some mistakes and—importantly—to vocally acknowledge them (for instance, saying, "Sorry, guys, I made the mistake this round. I know it may be hard to believe, but robots make mistakes too").[67] The presence of this robot willing to admit error modified how the humans interacted among themselves, making them work better together. And consider the impact of driverless cars on the roads. Such vehicles will interact with nearby human-driven cars and will be programmed to drive in a way that reduces the likelihood of a collision.[68] But a human who has driven near a driverless car for a while might wind up driving in a different way even once he or she is no longer near that car. The robotic car might create a cascade of benefits, modifying the behavior not only of drivers with whom it has direct contact but also of others with whom it has not interacted.[69]

Still other advances in artificial intelligence will affect our social lives. One of the most remarkable features of AlphaGo, the software that beat the reigning human champion, Lee Sedol, at the ancient game of go in March of 2016, is not its astonishing ability to learn the game on its own but the fact that, after playing against the machine, Lee Sedol reported that he himself had learned new things from the weird, beautiful, and previously unimagined moves the machine made.[70] That is, interacting with this artificial intelligence changed how Sedol interacted with *other humans*. In the case of the game of go, this was not so troubling. But what if machines take on other teaching functions, affecting the social learning that is a key part of the social suite? This is the stuff of science fiction, but do we want machines to teach our children how and when to be kind? Machines might have rather different ideas about altruism than we do.

The second radical technology that might affect the social suite is the gene-editing tool CRISPR (which stands for "clustered regularly interspaced short palindromic repeats").[71] With such biological techniques, humans can modify their genes and direct their evolution, not just by modifying their culture (as discussed in chapter 11), but more directly, in a more rapid and intentional way, using tools that involve nonrandom mutation and unnatural selection. CRISPR offers the prospect of gene therapy that modifies not only our somatic tissues, but also, possibly, the human germline (that is, the gametes that give rise to subsequent generations), permanently introducing modified genes into our species' gene pool. It's the difference between curing sickle-cell anemia in a patient by modifying the blood-producing cells in the bone marrow, on the one hand, and making it so that the patient's descendants will not suffer from the disease by modifying that individual's gametes, on the other hand.

Of course, the initial applications of germline gene editing and directed evolution are likely to be done by scientists who have good ends in mind. They will start by preventing genetic diseases.[72] Few would object to that. Then they might move on to simply enhancing humans, perhaps by engineering faster versions. But it is not inconceivable that it will be possible to modify humans to make them, say, less empathetic or less friendly, and some people might choose to do so. What would the social systems composed of or involving such individuals look like? We know that the introduction of pets can improve how humans interact with one another, so it is not hard to imagine that these new kinds of humans might have powerful effects on others as well. How social or cooperative will any such human groups be?

We are mechanically and biologically transforming ourselves—and our societies. These developments will oblige us, yet again, to come to grips with how separable from nature humans really are. A new social contract may be required if we are to avoid a dystopian future. This new contract would prescribe that these innovations respect the social suite. Part of me is fearful, of course. But part of me is also optimistic. The reason is that the social suite is so deeply ingrained within us that it is hard to imagine anything being able to advertently or inadvertently modify our blueprint over any realistic time frame.

## *Justifying Our Society*

When we look around the world, we see endless and timeless fear, ignorance, hatred, and violence. From our position, we could also boundlessly catalog the minute details of human groups, highlighting and emphasizing the differences among them. But this pessimistic gaze that separates humans from one another by highlighting evil and by emphasizing difference misses an important underlying unity and overlooks our common humanity. The project of evolutionary sociology in which we have been engaged reveals that humans everywhere are pre-wired to make a particular kind of society—one full of love, friendship, cooperation, and learning.

What accounts for our species' general success in living together in the face of all of our defects and differences? How can we understand the goodness of the social world despite the badness? In theology, this is known as a question of theodicy: How is God to be justified in the face of all the evil in the world? I believe, analogously, that we can focus on what I call *sociodicy*.[73] This is the vindication of our confidence in the virtue of society despite its numerous failures, so obvious to anyone. This is not just idle optimism. Rather, it's a recognition of the fundamental good that lies within us.

It's tempting to look at human history as full of abject misery and dysfunction. One can pick any century or millennium and find it replete with horrors. It is true that there was a dramatic inflection for the better that occurred in the eighteenth century with the arrival of the Enlightenment and its philosophical values and scientific discoveries. Life became longer, richer, freer, and more peaceful.[74] But people do not have to rely solely on such recent historical developments to make the world better. More ancient and more powerful forces are at work, propelling a good society.

Humans have always had both competitive and cooperative impulses, both violent and beneficent tendencies. Like the two strands of the double helix of our DNA, these conflicting impulses are intertwined. We are primed for conflict and hatred but also for love, friendship, and cooperation. If anything, modern societies are just a patina of civilization on top of this evolutionary blueprint.

There is another reason to step off the plateau and look at mountains rather than hills. A key danger of viewing historical forces as more salient than evolutionary ones in explaining human society is that our species' story then becomes more fragile. Giving historical forces primacy may even tempt us to give up and feel that a good social order is unnatural. But the good things we see around us are part of what makes us human in the first place.

We should be humble in the face of temptations to engineer society in opposition to our instincts. Fortunately, we do not need to exercise any such authority in order to have a good life. The arc of our evolutionary history is long. But it bends toward goodness.

# Acknowledgments

I could not be the person I am, nor could I have written this book, were it not for my friends, new and old, some of very long duration, all of them still a deep part of my life, even if just in memory. They are, in order, from the first year of my life: Lisa Frantzis, Dimitris Dimitrelias, Kevin Sheehan, James Billington, Nasi Samiy, Morris Panner, Anne Stack, Ophelia Dahl, Bemy Jelin, Nicolas McConnell, Kathryn Viguerie, Les Viguerie, Renée C. Fox, Curt Langlotz, Mary Leonard, Lucy Twose, Cordelia Dyer, Paul Allison, Diana Young, Bill Brown, Marie Randazzo, Oswaldo Morales, Daniel Gilbert, Gary King, James Fowler, Doug Melton, Mark Pachucki, Winslow Carroll, Adam Glick, Nan Carroll, John Carroll, Bob Stevens, Elizabeth Anderson, Julia Adams, and Hans van Dyk.

I would like to thank my research assistants on this book, many of whom were undergraduates and graduate students at Harvard and Yale, many of whom I got to know initially as students, and all of whom were filled with enthusiasm and intelligence that uplifted me: Joseph Brennan, Kevin Garcia, Libby Henry, Andrew Lea, Sam Southgate, Duncan Tomlin, Namratha Vedire, and Zachary Wood.

Many other people read parts (or even all) of this book or gave me valuable and critical comments that saved me from errors and stimulated my thinking. They were Marcus Alexander, Dorsa Amir, Coren Apicella, Dimitri Christakis, Katrina Christakis, Peter Dewan, Brian

Earp, Felix Elwert, Joseph Henrich, Heather Heying, Viveca Morris, Jonathan Schulz, Samuel Snow, and Margaret Traeger.

Much of my own original research discussed in this book is based on efforts undertaken by the extraordinary personnel in the Human Nature Lab at Yale. Over the past decade, I have had brilliant graduate students and postdoctoral fellows who spearheaded many of the projects featured here, including Feng Fu, Luke Glowacki, Alexander Isakov, David Kim, Akihiro Nishi, Jessica Perkins, and Hirokazu Shirado. Mark McKnight directed the development of our Trellis and Breadboard software platforms that we have used to such powerful effect to map social networks around the world and do online experiments with many thousands of subjects. Liza Nicoll has expertly managed all our data. Rennie Negron has done superhuman work managing our long-standing research effort in Honduras, mapping the social networks of 24,812 people in 176 villages; I am enormously grateful for her many interpersonal and management skills. I thank Kim Kuzina for administrative support. Tom Keegan manages the Human Nature Lab and, for ten years, has never ceased to amaze me with his judgment, creativity, farsightedness, and patience.

My agent, Katinka Matson, solidified the focus of the book proposal. My long-standing editor Tracy Behar is fabulous in every way, from sharpening arguments to showing patience; she improves everything she touches. My copyeditors, Tracy Roe and Barbara Jatkola, saved me from much foolishness. Some of my own research in this book was supported by the Bill and Melinda Gates Foundation, the Robert Wood Johnson Foundation, the John Templeton Foundation, Tata Sons Limited, and the National Institute on Aging.

Sebastian, Lysander, and Eleni, and now Orien, by their admirable actions, remarkable character, and warm affection, have shown me the meaning of a good life. I dedicated this book to my beloved wife, Erika Christakis, but I will take this further opportunity to acknowledge her beautiful heart and mind, which improved this book immeasurably.

# Illustration Credits

All illustrations are courtesy of the author, with the exception of the following:

Page 41: F. E. Raynal, *Wrecked on a Reef, or Twenty Months Among the Auckland Isles* (London: T. Nelson & Sons, 1874).

Page 91: Redrawn from J. C. Johnson, J. S. Boster, and L. A. Palinkas, "Social Roles and the Evolution of Networks in Extreme and Isolated Environments," *Journal of Mathematical Sociology* 27 (2003): 89–121.

Page 102: J. F. von Racknitz, *Ueber den Schachspieler des Herrn von Kempelen* (Leipzig und Dresden: J. G. I. Breitkopf, 1789).

Page 115: Figure courtesy of the Society for Sedimentary Geology, from D. M. Raup, A. Michelson, "Theoretical Morphology of the Coiled Shell," *Science* 147 (1965): 1294–1295.

Page 116: Figure courtesy of Cavan Huang.

Page 142: F. W. Marlowe, "Mate Preferences Among Hadza Hunter-Gatherers," *Human Nature* 15 (2004): 365–376.

Page 206: Photo courtesy of National Geographic Creative.

Page 210: Photos courtesy of John Mitani, from J. C. Mitani, "Male Chimpanzees Form Enduring and Equitable Social Bonds," *Animal Behaviour* 77 (2009): 633–640.

Pages 211, 212: Redrawn from J. Lehmann and C. Boesch, "Sociality of the Dispersing Sex: The Nature of Social Bonds in West African Female Chimpanzees, *Pan troglodytes*," *Animal Behaviour* 77 (2009): 377–387.

Page 224: Figure courtesy of Shermin de Silva, from S. de Silva and G. Wittemyer, "A Comparison of Social Organization in Asian Elephants and African Savannah Elephants," *International Journal of Primatology* 33 (2012): 1125–1141.

Page 227: D. Lusseau, "The Emergent Properties of a Dolphin Social Network," *Proceedings of the Royal Society B* 270 (2003): S186–S188.

Page 243: Figure courtesy of Arthur Aron, from S. Gächter, C. Starmer, and F. Tufano, "Measuring the Closeness of Relationships: A Comprehensive Evaluation of the 'Inclusion of the Other in the Self' Scale," *PLOS ONE* 10 (2015): e0129478.

Page 245: D. J. Hruschka, *Friendship: Development, Ecology, and Evolution of a Relationship* (Berkeley: University of California Press, 2010).

Page 286: Figure courtesy of Cavan Huang.

Page 290: Image courtesy of Michael Sheehan, from M. J. Sheehan and M. W. Nachman, "Morphological and Population Genomic Evidence That Human Faces Have Evolved to Signal Individual Identity," *Nature Communications* 5 (2014): 4800.

Page 292: Redrawn from M. J. Sheehan and M. W. Nachman, "Morphological and Population Genomic Evidence that Human Faces Have Evolved to Signal Individual Identity," *Nature Communications* 5 (2014): 4800.

Page 305: Drawing courtesy of Frans de Waal, from J. M. Plotnik, R. Lair, W. Suphachoksahakun, and F. B. M. de Waal, "Elephants Know When They Need a Helping Trunk in a Cooperative Task," *PNAS: Proceedings of the National Academy of Sciences* 108 (2011): 5116–5121.

Page 322: Redrawn from P. I. Chiyo, C. J. Moss, and S. C. Alberts, "The Influence of Life History Milestone and Association Networks on Crop-Raiding Behavior in Male African Elephants," *PLOS ONE* 7 (2012): e31382.

Page 328: Photo courtesy of Edwin van Leeuwen.

Page 340: Redrawn from J. N. Weber, B. K. Peterson, and H. E. Hoekstra, "Discrete Genetic Modules Are Responsible for Complex Burrow Evolution in *Peromyscus* Mice," *Nature* 493 (2013): 402–405.

Page 376: Adapted from M. A. Kline and R. Boyd, "Population Size Predicts Technological Complexity in Oceania," *Proceedings of the Royal Society B* 277 (2010): 2559–2564.

Plate 2: Redrawn from J. C. Flack, M. Girvan, F. B. M. de Waal, and D. C. Krakauer, "Policing Stabilizes Construction of Social Niches in Primates," *Nature* 439 (2006): 426–429.

Plate 3: Image courtesy of F. Bibi, B. Kraatz, N. Craig, M. Beech, M. Schuster, and A. Hill, "Early Evidence for Complex Social Structure in *Proboscidea* from a Late Miocene Trackway Site in the United Arab Emirates," *Biology Letters* 8 (2012): 670–673.

Plate 6: Image courtesy of A. Whiten, S. Smart, and *Nature* magazine, from A. Whiten et al., "Cultures in Chimpanzees," *Nature* 399 (1999): 682–685.

Plate 7: Figure courtesy of Todd A. Blackledge, from T. A. Blackledge et al., "Reconstructing Web Evolution and Spider Diversification in the Molecular Era," *PNAS: Proceedings of the National Academy of Sciences* 106 (2009): 5229–5234.

Plate 8: Photo courtesy of David Hughes.

# Notes

## *Preface: Our Common Humanity*


1. C. Mackay, *Extraordinary Popular Delusions and the Madness of Crowds* (1841; New York: Farrar, Straus and Giroux, 1932), p. xx.
2. People's Republic of Bangladesh Const. part III, sect. 37; Canadian Charter of Rights and Freedoms sect. 2; Republic of Hungary Const. art. LXIII; Indian Const. art. XIX (1) (b).
3. See, for example, C. Andris, D. Lee, M. J. Hamilton, M. Martino, C. E. Gunning, and J. A. Selden, "The Rise of Partisanship and Super-Cooperators in the U.S. House of Representatives," *PLOS ONE* 10 (2015): e0123507; and E. Saez, "Striking It Richer: The Evolution of Top Incomes in the United States (Updated with 2013 Preliminary Estimates)" (unpublished manuscript, January 25, 2015), https://eml.berkeley.edu/~saez/saez-UStopincomes-2013.pdf.
4. K. E. Steinhauser, N. A. Christakis, E. C. Clipp, M. McNeilly, L. McIntyre, and J. A. Tulsky, "Factors Considered Important at the End of Life by Patients, Family, Physicians, and Other Care Providers," *JAMA* 284 (2000): 2476–2482.
5. M. V. Llosa, "The Culture of Liberty," *Foreign Policy*, November 20, 2009, http://foreignpolicy.com/2009/11/20/the-culture-of-liberty/.
6. Darrell Powers, interview, *Band of Brothers*, episode 9, "Why We Fight," first aired October 28, 2001, on HBO.
7. *The Vietnam War*, episode 4, "'Resolve' (January 1966–June 1967)," a film by Ken Burns and Lynn Novick, first aired September 20, 2017, on PBS.


## *Chapter 1: The Society Within Us*


1. M. Fortes, *Social and Psychological Aspects of Education in Taleland* (London: Oxford University Press, 1938), p. 44.
2. M. Martini, "Peer Interactions in Polynesia: A View from the Marquesas," in J. L. Roopnarine, J. E. Johnson, and F. H. Hoper, eds., *Children's Play in Diverse Cultures*, pp. 73–103 (Albany: State University of New York Press, 1994), p. 74.
3. B. Whiting, J. Whiting, and R. Longabaugh, *Children of Six Cultures: A Psycho-Cultural Analysis* (Cambridge, MA: Harvard University Press, 1975). See also C. P. Edwards, "Children's Play in Cross-Cultural Perspective: A New Look at the Six Culture Study,"

in F. F. McMahon, D. E. Lytle, and B. Sutton-Smith, eds., *Play: An Interdisciplinary Synthesis* (Lanham, MD: University Press of America, 2005), pp. 81–96; and D. F. Lancy, *The Anthropology of Childhood: Cherubs, Chattel, and Changelings* (Cambridge: Cambridge University Press, 2008). See also E. Christakis, *The Importance of Being Little: What Preschoolers Really Need from Grownups* (New York: Viking, 2016).

4. J. Huizinga, *Homo Ludens: A Study of the Play Element in Culture* (Boston: Beacon Press, 1950): p. 1.
5. Y. Dunham, A. S. Baron, and S. Carey, "Consequences of 'Minimal' Group Affiliations in Children," *Child Development* 82 (2011): 793–811. In these experiments, the individuals in the groups were, crucially, not in competition with the other groups. The size of these effects was roughly half as big as the preference children expressed for their own gender, though the gender preference was mainly evinced by little girls favoring other girls (boys favored other boys and girls equally).
6. Three-month-olds already prefer faces of their own race. D. Kelly et al., "Three-Month-Olds, but Not Newborns, Prefer Own-Race Faces," *Developmental Science* 8 (2005): F31–F36. Five-month-olds prefer their native language and shun foreign accents. K. D. Kinzler, E. Dupoux, and E. S. Spelke, "The Native Language of Social Cognition," *PNAS: Proceedings of the National Academy of Sciences* 104 (2007): 12577–12580.
7. P. Bloom, *Just Babies: The Origins of Good and Evil* (New York: Crown, 2013).
8. J. K. Hamlin, K. Wynn, and P. Bloom, "3-Month-Olds Show a Negativity Bias in Their Social Evaluations," *Developmental Science* 13 (2010): 923–929.
9. Y. J. Choi and Y. Luo, "13-Month-Olds' Understanding of Social Interactions," *Psychological Science* 26 (2015): 274–283.
10. F. Warneken and M. Tomasello, "Altruistic Helping in Human Infants and Young Chimpanzees," *Science* 311 (2006): 1301–1303.
11. For an expansive treatment of this phenomenon, which Frank White has called "the overview effect," see F. White, *The Overview Effect: Space Exploration and Human Evolution*, 3rd ed. (Reston, VA: American Institute of Aeronautics and Astronautics, 2014). The quotes from Aleksandrov and Williams are widely attributed online, but I could not find the original sources. Two similar quotes, with sources, include the following: "You develop an instant global consciousness, a people orientation, an intense dissatisfaction with the state of the world, a compulsion to do something about it. From out there on the moon, international politics looks so petty." Edgar Mitchell (Apollo 14 astronaut), "Edgar Mitchell's Strange Voyage," *People*, April 8, 1974. "When you're finally up at the moon looking back on Earth, all those differences and nationalistic traits are pretty well going to blend, and you're going to get a concept that maybe this really is one world and why the hell can't we learn to live together like decent people." Frank Borman (Apollo 8 astronaut), "Christmas Journey," *Newsweek*, December 23, 1968.
12. D. Keltner and J. Haidt, "Approaching Awe, a Moral, Spiritual, and Aesthetic Emotion," *Cognition and Emotion* 17 (2003): 297–314. Of course, we may also be awestruck by beautiful music, deep scientific theories, or even a charismatic leader, not just by nature.

13. For instance, chimpanzees appear to have goose bumps during thunderstorms. J. Marchant, "Awesome Awe: The Emotion That Gives Us Superpowers," *New Scientist,* July 26, 2017.
14. The right to have a name is so universal that it was even codified by the UN. See Office of the United Nations High Commissioner for Human Rights, *Convention on the Rights of the Child, Adopted and opened for signature, ratification and accession by General Assembly resolution 44/25 of 20 November 1989 entry into force 2 September 1990, in accordance with article 49.* A very few societies (e.g., the Machiguenga), do not have personal names but use other sorts of descriptors to uniquely identify people.
15. J. Fajans, *Work and Play Among the Baining of Papua New Guinea* (Chicago: University of Chicago Press, 1997).
16. The first person to use this terminology may have been Charles Darwin. In a letter to J. D. Hooker in 1857, he noted that "those who make many species are the 'splitters,' and those who make few are the 'lumpers.'" C. Darwin and F. Darwin, *The Life and Letters of Charles Darwin,* vol. 2 (London: John Murray, 1887), day 153. These terms were introduced more widely in G. G. Simpson, "The Principles of Classification and a Classification of Mammals," *Bulletin of the American Museum of Natural History* 85 (1945): 22–24.
17. D. M. Buss, "Human Nature and Culture: An Evolutionary Psychological Perspective," *Journal of Personality* 69 (2001): 955–978.
18. C. Geertz, *The Interpretation of Cultures: Selected Essays* (New York: Basic Books, 1973), pp. 40–41.
19. S. Pinker, *The Blank Slate: The Modern Denial of Human Nature* (New York: Penguin, 2002).
20. See D. E. Brown, *Human Universals* (New York: McGraw-Hill, 1991), pp. 58–59, 66–67. Another influential midcentury essay, by Clyde Kluckhorn, also proposed the possibility of biological and psychological explanations for cultural universals in addition to the possibilities of shared social interactions and shared environmental contexts. C. C. Kluckhorn, "Universal Categories of Culture," in A. L. Kroeber, ed., *Anthropology Today* (Chicago: University of Chicago Press, 1953), pp. 507–523.
21. G. P. Murdock, "The Common Denominator of Cultures," in R. Linton, ed., *The Science of Man in a World of Crisis* (New York: Columbia University Press, 1945), pp. 123–142.
22. Brown, *Human Universals,* p. 50.
23. Ibid., p. 47.
24. P. Turchin et al., "Quantitative Historical Analysis Uncovers a Single Dimension of Complexity that Structures Global Variation in Human Social Organization," *PNAS: Proceedings of the National Academy of Sciences* 115 (2018): E144—E151.
25. P. Ekman, "Facial Expressions," in T. Dalgleish and M. Power, eds., *Handbook of Cognition and Emotion* (Chichester, UK: John Wiley and Sons, 1999), pp. 301–320. See also G. A. Bryant et al., "The Perception of Spontaneous and Volitional Laughter Across 21 Societies," *Psychological Science* 29 (2018): 1515–1525. Of course, again, some links can be severed by very strong cultural overlays, as in decoupling smiles from happiness (as happens in some cultures). Human personality structures are also likely universal. See

R. R. McCrae and P. T. Costa Jr., "Personality Trait Structure as a Human Universal," *American Psychologist* 52 (1997): 509–516; and S. Yamagata et al., "Is the Genetic Structure of Human Personality Universal? A Cross-Cultural Twin Study from North America, Europe, and Asia," *Journal of Personality and Social Psychology* 90 (2006): 987–998.

26. C. Chen, C. Crivelli, O. G. B. Garrod, P. G. Schyns, J. M. Fernandez-Dols, and R. E. Jack, "Distinct Facial Expressions Represent Pain and Pleasure Across Cultures," *PNAS: Proceedings of the National Academy of Sciences* 115 (2018): E10013–E10021.
27. N. Chomsky, *Syntactic Structures* (Berlin: Mouton de Gruyter, 1957); S. Pinker, *The Language Instinct: How the Mind Creates Language* (New York: Harper Perennial, 1995).
28. P. E. Savage, S. Brown, E. Sakai, and T. E. Currie, "Statistical Universals Reveal the Structures and Functions of Human Music," *PNAS: Proceedings of the National Academy of Sciences* 112 (2015): 8987–8992.
29. See, for example, R. Heinsohn, C. N. Zdenek, R. B. Cunningham, J. A. Endler, and N. E. Langmore, "Tool Assisted Rhythmic Drumming in Palm Cockatoos Shares Elements of Human Instrumental Music," *Science Advances* 3 (2017): e1602399.
30. E. O. Wilson, *The Social Conquest of Earth* (New York: Liveright, 2013).
31. These traits are supported by still others that are expressed on a more individual level, such as a need for transcendence or a sense of purpose; a capacity to make or appreciate art and music; and a desire to tell or hear stories.
32. Other scholars are less troubled by this metaphor. See, for example, R. Plomin, *Blueprint: How DNA Makes Us Who We Are* (Cambridge, MA: MIT Press, 2018), which explores how our genes predict our psychological strengths and weaknesses. With respect to my use of the metaphor of a blueprint, it's worth noting that I take our capacity for culture (which is evolutionarily enabled) to be part of what specifies our social order. But, more particularly, the social suite, not our DNA per se, is the blueprint for a good society.
33. Causes of possible intergroup variation include adaptation to environment, neutral drift, reproductive isolation of human populations, and founder effects. There are also broader genetic differences in populations of humans that, generally, track the continents of origin. See L. B. Jorde and S. P. Wooding, "Genetic Variation, Classification, and 'Race,'" *Nature Genetics* 36 (2004): 528–533.
34. A. Quamrul and O. Galor, "The Out-of-Africa Hypothesis, Human Genetic Diversity, and Comparative Development," *American Economic Review* 103 (2013): 1–46.

### *Chapter 2: Unintentional Communities*

1. *Castaway 2000,* produced by C. Kelley, BBC One, 2000. In 2016, *Eden,* a similar reality survival show, focused on a failed community in an isolated area of Scotland. See Sam Knight, "Reality TV's Wildest Disaster: 'Eden' Aspired to Remake Society Altogether. What Could Go Wrong?," *New Yorker,* September 4, 2017.
2. R. Copsey, "How *Castaway* Made My Life Hell," *Guardian,* August 11, 2010.
3. J. Kibble-White, "This is What Happens to Make Reality TV," *Off The Telly,* November 2004.
4. Copsey, "How *Castaway* Made My Life Hell."

5. G. Martin, "Return to Castaway Island: The Cast of Britain's First Reality TV Programme Reunite," *Daily Mail*, July 17, 2010.
6. Copsey, "How *Castaway* Made My Life Hell."
7. R. Shattuck, *The Forbidden Experiment: The Story of the Wild Boy of Aveyron* (New York: Farrar, Straus and Giroux, 1980).
8. K. Steel, "Feral and Isolated Children from Herodotus to Akbar to Hesse: Heroes, Thinkers, and Friends of Wolves" (presentation, CUNY Brooklyn College, April 11, 2016), https://academicworks.cuny.edu/gc_pubs/216/.
9. H. Fast, "The First Men," *Magazine for Science Fiction and Fantasy*, February 1960.
10. Demarcating what is and is not "science" is difficult, and philosophers of science have begun to think about science less in terms of specific methods and more in terms of an underlying process that the sociologist Robert K. Merton has called "organized skepticism." R. K. Merton, *The Sociology of Science: Theoretical and Empirical Investigations* (Chicago: University of Chicago Press, 1973).
11. Earnings were reduced by about 15 percent (among white men) for at least ten years. This loss of earnings roughly reflects the fact that while they were serving in the military for two years, they were failing to get relevant labor market experience. J. D. Angrist, "Lifetime Earnings and the Vietnam Era Draft Lottery: Evidence from Social Security Administrative Records," *American Economic Review* 80 (1990): 313–336. Similar natural experiments have taken advantage of people winning monetary lotteries to evaluate the link between wealth and health, trying to sort out whether wealthy people become healthy, or healthy people become wealthy (it's both). J. Gardner and A. J. Oswald, "Money and Mental Wellbeing: A Longitudinal Study of Medium-Sized Lottery Wins," *Journal of Health Economics* 26 (2007): 49–60.
12. A. Banerjee and L. Iyer, "Colonial Land Tenure, Electoral Competition, and Public Goods in India," in J. Diamond and J. A. Robinson, eds., *Natural Experiments of History* (Cambridge, MA: Belknap Press, 2010), pp. 185–220.
13. D. Acemoglu, D. Cantoni, S. Johnson, and J. A. Robinson, "From Ancien Régime to Capitalism: The Spread of the French Revolution as a Natural Experiment," in J. Diamond and J. A. Robinson, eds., *Natural Experiments of History* (Cambridge, MA: Belknap Press, 2010), pp. 221–256.
14. A. Duncan, *The Mariner's Chronicle Containing Narratives of the Most Remarkable Disasters at Sea, Such as Shipwrecks, Storms, Fires and Famines* (New Haven, CT: G. W. Gorton, 1834). See also M. Gibbs, "Maritime Archaeology and Behavior During Crisis: The Wreck of the VOC Ship *Batavia* (1629)," in R. Torrence and J. Grattan, eds., *Natural Disasters and Cultural Change* (Abingdon, UK: Routledge, 2002), pp. 66–86.
15. J. Lichfield, "Shipwrecked and Abandoned: The Story of the Slave Crusoes," *Independent*, February 4, 2007.
16. C. A. Dard, J. G. des Odonais, and P.-R. de Brisson, *Perils and Captivity: Comprising the sufferings of the Picard family after the shipwreck of the Medusa, in the year 1816; Narrative of the captivity of M. de Brisson, in the year 1785; Voyage of Madame Godin along the river of the Amazons, in the year 1770*, trans. P. Maxwell (Edinburgh: Constable; London: Thomas Hurst, 1827); P. Viaud, *The Shipwreck and Adventures of Monsieur Pierre Viaud* (London:

T. Davies, 1771). Cannibalism is left out of the much shorter account of Monsieur Viaud's adventures published in the anonymous compendium *Tales of Shipwreck and Peril at Sea* (London: Burns and Lambert, 1858).

17. Regarding the case of the Andes survivors, see P. P. Read, *Alive: The Story of the Andes Survivors* (New York: J. B. Lippincott, 1974).
18. According to historian Keith Huntress, the earliest collection of shipwreck narratives was published in London in 1675, in the form of *Mr. James Janeway's Legacy to His Friends, Containing Twenty-Seven Famous Instances of God's Providence in and About Sea-Dangers and Deliverances.* K. Huntress, *Narratives of Shipwrecks and Disasters* (Ames: Iowa State University Press, 1974).
19. M. Nash, *The Sydney Cove Shipwreck Survivors Camp,* Flinders University Maritime Archaeology Monograph Series, no. 2 (Adelaide: Flinders University Department of Archaeology, 2006).
20. That is, while we may know about wrecks, we often do not know about how the survivors were able to rebuild a society. For example, the *Tamaris,* a French brig bound for New Caledonia, was wrecked on the Crozet Islands in 1887, leaving the thirteen-man crew stranded on the uninhabited, cold, windy, and treeless island of Île aux Cochons. Desperate for any help, the men attached a note pleading for rescue to the leg of a large seabird, a note that, astonishingly, was found seven months later in Fremantle, Western Australia, more than four thousand miles away. The men, however, were never located. "The Crozet Islands," *Adelaide Express and Telegraph,* March 21, 1889.
21. To estimate the total number of shipwrecks that occurred between the years 1500 and 1900, I (along with my capable research assistants) made use of the data set regarding over 176,000 wrecks compiled on the website Wrecksite.com (https://www.wrecksite.eu) as of 2016. This site attempts to catalog all ships lost worldwide, including causes other than shipwreck (e.g., foundering, fire, naval battle, or scuttling). I filtered entries by cause of wreck, disregarding those ships that did not actually strike shore, and also eliminated all wrecks occurring outside 1500 to 1900 CE. These constraints yielded a total of over 8,100 wrecks. This includes all wrecks that hit land, including those catastrophic wrecks in which all passengers immediately perished. I was primarily concerned with shipwrecks that may have produced survival colonies on land. From the population of 8,100 wrecks, I used information in the database to select all those wrecks where a survival colony of nineteen or more people resided on land for at least sixty days, and where at least one person survived to tell the tale. I found twenty such wrecks and reviewed all first-person accounts available from these wrecks. An intriguing wreck that I did not include was the *Jamaica Sloop,* from 1711, which involved sixteen survivors who encamped for four months, as discussed in Duncan, *The Mariner's Chronicle,* pp. 242–275.

    I could not find any examples from Asia. For instance, in a comprehensive sample of twenty-four cases of Japanese shipwreck survivors who drifted to the coasts of Southeast Asia in the seventeenth, eighteenth, and nineteenth centuries, just three had nineteen or more people who made it to shore, and I did not include these wrecks because the subjects made rapid contact with locals and were repatriated back to Japan. S. F. Liu, "Shipwreck Salvage and Survivors' Repatriation Networks of the East

Asian Rim in the Qing Dynasty," in F. Kayoko, M. Shiro, and A. Reid, eds., *Offshore Asia: Maritime Interactions in Eastern Asia Before Steamships* (Singapore: ISEAS, 2013), pp. 211–235. A search of the few naval histories of China that exist and of some searchable primary sources using the Chinese words for "shipwreck" by historian Zvi Ben-Dor Benite reveals only a handful of recorded cases, and in all cases, the sailors were rescued, often within days (Ben-Dor Benite, personal communication, July 14, 2018).

22. M. Gibbs, "The Archeology of Crisis: Shipwreck Survivor Camps in Australasia," *Historical Archeology* 37 (2003): 128–145.
23. F. E. Woods, *Divine Providence: The Wreck and Rescue of the Julia Ann* (Springville, UT: Cedar Fort, 2014), p. 58.
24. Ibid., p. 48.
25. Ibid., pp. 61–62.
26. J. G. Lockhart, *Blenden Hall: The True Story of a Shipwreck, a Casting Away, and Life on a Desert Island* (New York: D. Appleton, 1930).
27. Quoted in ibid., pp. 153–154. The son was also named Alexander M. Greig.
28. Interestingly, the HMS *Beagle* itself, on its third voyage (with Darwin no longer on board—he had been on the second voyage), stopped at the wreck site in 1842. S. Harris and H. McKenny, "Preservation Island, Furneaux Group: Two Hundred Years of Vegetation Change," *Papers and Proceedings of the Royal Society of Tasmania* 133, no. 1 (1999): 85–101.
29. "Supercargo William Clark's Account," in M. Nash, *Sydney Cove: The History and Archaeology of an Eighteenth-Century Shipwreck* (Hobart, Australia: Navarine, 2009), p. 235.
30. Ibid., p. 237. Italics added.
31. Ibid., p. 238.
32. "Governor Hunter's Account" (from a letter dated August 15, 1797), in ibid., p. 243.
33. Mr. Webb, "A Journal of the Proceedings of the *Doddington* East Indiaman," in B. Plaisted, ed., *A Journal from Calcutta to England, in the Year, 1750. To Which Are Added, Directions by E. Eliot, for Passing over the Little Desart from Busserah. With a Journal of the Proceedings of the Doddington East-Indiaman,* 2nd ed. (London: T. Kinnersly, 1758), p. 238.
34. Tensions between crew and passengers and between men and women, as well as differential survival, have been studied in other maritime disasters. See B. S. Frey, D. A. Savage, and B. Torgler, "Interaction of Natural Survival Instincts and Internalized Social Norms Exploring the *Titanic* and *Lusitania* Disasters," *PNAS: Proceedings of the National Academy of Sciences* 107 (2010): 4862–4865. For an examination of chivalrous norms with respect to the treatment of women and children, exploiting a sample of eighteen maritime disasters and fifteen thousand people, see M. Elinder and O. Erixson, "Gender, Social Norms, and Survival in Maritime Disasters," *PNAS: Proceedings of the National Academy of Sciences* 109 (2012): 13220–13224. The authors observed that women had a significant survival disadvantage compared with men and that passengers had a significant disadvantage compared to captains and crew. They also concluded that the best way to describe behavior in maritime disasters was "every man for himself."
35. Two months into their stay, on a more careful exploration of the island, they also discovered evidence of prior castaways on Bird Island. The *Doddington* was carrying gold

and silver; the wreck was discovered by divers and looted two hundred years later. J. Shaw, "Clive of India's Gold Comes Up for Sale After Legal Settlement," *Independent*, August 27, 2000.

36. A prior effort to get help had not ended well. On September 3, three men set out on a risky mission to the mainland. As the small boat approached land, it was overset by surf, and one of them drowned. The two who made it ashore with the boat soon encountered hostile local people. The men were stripped of their clothes and strongly encouraged to leave. They took the hint and returned to Bird Island, barely alive.
37. Webb, "Proceedings of the *Doddington*," p. 268.
38. Ibid., p. 269.
39. The men had salvaged a treasure chest from the ship shortly after landfall, and on September 28, they discovered that it had been broken open and two-thirds of it had been hidden elsewhere. It was not possible to discern who had done this.
40. G. Dalgarno, "Letter from the Captain," *Otago Witness* (Dunedin, New Zealand), October 28, 1865.
41. Even after rescue, the officers could not be magnanimous to the seaman (Holding) whose inventive efforts on the island had actually saved them. After the rescue, Captain Dalgarno seemed pleased that, while he and Smith were entertained by the rescue ship's officers in finer accommodations, Holding was relegated to his proper station "among his forecastle equals." J. Druett, *Island of the Lost: Shipwrecked at the Edge of the World* (Chapel Hill, NC: Algonquin Books, 2007), p. 201. Class stratification had been the doom of the *Invercauld* crew on the island. The resourceful Holding, however, died in Canada in 1933, at the age of eighty-six. An account of the *Invercauld* authored by Holding before he died was found by his great-granddaughter and published in 1997: M. F. Allen, *Wake of the Invercauld* (Auckland: Exisle Press, 1997). Captain Raynal also included a since-lost account of the *Invercauld* written by Captain Dalgarno as an appendix to his own book: F. E. Raynal, *Wrecked on a Reef, or Twenty Months Among the Auckland Isles* (London: T. Nelson and Sons, 1874).
42. "Captain and Mate," *Otago Witness* (Dunedin, New Zealand), October 28, 1865. Lest we judge Dalgarno too harshly, in the foregoing account he noted, perhaps self-servingly, that when his boat sank, "everything in and belonging to the vessel was lost, including a medal presented to me in the year 1862 by the United States Government, for saving the lives of the crew of a water-logged vessel, and a telescope presented to me by the British Government in the same year, for saving the lives of the crew of a British vessel... both of which I valued most highly, as mementos of service which I had been enabled to perform to brother seamen."
43. Some contemporary research suggests that demographic diversity may be beneficial to group performance, subject to certain provisos. See E. Smith and Y. Hou, "Redundant Heterogeneity and Group Performance," *Organization Science* 26 (2014): 37–51.
44. Raynal's account, *Wrecked on a Reef,* was originally published in French. Musgrave's was published as T. Musgrave, *Castaway on the Auckland Isles: A Narrative of the Wreck of the 'Grafton' and the Escape of the Crew After Twenty Months Suffering* (London: Lockwood, 1866).
45. Musgrave, *Castaway,* p. ix.
46. Raynal, *Wrecked on a Reef,* p. 82.

47. Ibid., pp. 159–160.
48. Ibid., p. 152.
49. Druett, *Island of the Lost,* pp. 163–164.
50. Musgrave, *Castaway,* p. 129. See also A. W. Eden, *Islands of Despair* (London: Andrew Melrose, 1955), p. 101.
51. W. H. Norman and T. Musgrave, *Journals of the Voyage and Proceedings of the HMCS "Victoria" in Search of Shipwrecked People at the Auckland and Other Islands* (Melbourne: F. F. Bailliere, 1866), p. 28.
52. Druett, *Island of the Lost,* p. 248.
53. Ibid., p. 280.
54. S. Sheppard, "Physical Isolation and Failed Socialization on Pitcairn Island: A Warning for the Future?," *Journal of New Zealand and Pacific Studies* 2 (2014): 21–38; D. T. Coenen, "Of Pitcairn's Island and American Constitutional Theory," *William and Mary Law Review* 38 (1997): 649–675.
55. Four other loyalists were also later set free by the mutineers.
56. R. B. Nicolson, *The Pitcairners* (Honolulu: University of Hawaii Press, 1997).
57. T. Lummis, *Life and Death in Eden: Pitcairn Island and the Bounty Mutineers* (Farnham, UK: Ashgate, 1997), p. 46.
58. R. W. Kirk, *Pitcairn Island, the Bounty Mutineers, and Their Descendants: A History* (Jefferson, NC: McFarland, 2008).
59. H. L. Shapiro, *The Pitcairn Islanders* (formerly *"The Heritage of the Bounty")* (New York: Simon and Schuster, 1968), p. 54.
60. Sheppard, "Physical Isolation."
61. Ibid, p. 31.
62. Teehuteatuaonoa [Jenny], "Account of the Mutineers of the Ship *Bounty,* and Their Descendants at Pitcairn's Island," *Sydney Gazette,* July 17, 1819.
63. Lummis, *Life and Death in Eden,* p. 63.
64. Teehuteatuaonoa, "Account of the Mutineers."
65. Ibid.
66. Lummis, *Life and Death in Eden,* p. 69.
67. *Pitcairn Island Encyclopedia,* s.v. "Pitcairn Islands Study Center: Folger, Mayhew," with text taken from S. Wahlroos, *Mutiny and Romance in the South Seas: A Companion to the Bounty Adventure* (Salem, MA: Salem House, 1989). Capitalization modernized. Visiting the island community another forty-two years later, Walter Brodie, a New Zealand sailor who was temporarily stranded on Pitcairn, similarly marveled at the community's warmth and hospitality. W. Brodie, *Pitcairn's Island and the Islanders in 1850. Together with Extracts from His Private Journal and a Few Hints Upon California: Also, the Reports of All the Commanders of H.M. Ships That Have Touched at the Above Island Since 1800* (London: Whittaker, 1851), pp. 30–32.
68. For a summary of the trial, see "Six Found Guilty in Pitcairn Sex Offences Trial," *Guardian,* October 25, 2004.
69. J. Diamond, *Collapse: How Societies Choose to Fail or Succeed* (New York: Penguin, 2005).
70. M. Weber, *The Vocation Lectures,* ed. D. S. Owen and T. B. Strong, trans. R. Livingstone (Indianapolis: Hackett, 2004).

71. "Shackleton's Voyage of Endurance," *NOVA,* season 29, episode 6, first aired March 26, 2002, on PBS. The original version of the ad in the *Times* (UK) has never been located by historians, leading a growing body of researchers to think that it may have been fabricated. Some organizations even offer a monetary reward for the original ad.
72. M. T. Fisher and J. Fisher, *Shackleton* (London: Barrie, 1957); R. Huntford, *Shackleton* (New York: Carroll and Graf, 1998).
73. Fisher and Fisher, *Shackleton,* p. 340.
74. Ibid., p. 345.
75. F. Hurley, *The Diaries of Frank Hurley, 1912–1941,* ed. R. Dixon and C. Lee (London: Anthem Press, 2011), p. 24.
76. Fisher and Fisher, *Shackleton,* p. 345. Italics added.
77. Geographers Jared Diamond and Barry Rolett could not go back in time and, on a vast scale, experimentally assign inhabitants to sixty-nine different Polynesian islands in order to figure out why Easter Island was deforested and others were not. But they assumed that similar people settled these islands more or less at random, and they concluded, from this natural experiment, that the deforestation was due to geographic factors (such as windborne volcanic ash and rainfall) more than to various behaviors later adopted by the settlers. J. Diamond, "Intra-Island and Inter-Island Comparisons," in J. Diamond and J. A. Robinson, eds., *Natural Experiments of History* (Cambridge, MA: Belknap Press, 2010), pp. 120–141. See also Diamond, *Collapse.*
78. P. V. Kirch, "Controlled Comparison and Polynesian Cultural Evolution," in J. Diamond and J. A. Robinson, eds., *Natural Experiments of History* (Cambridge, MA: Belknap Press, 2010), p. 35.
79. M. D. Sahlins, *Social Stratification in Polynesia* (Seattle: University of Washington Press, 1958).
80. Kirch, "Controlled Comparison," pp. 27–28.
81. Irrigation probably allowed landed elites in arid areas to monopolize both water and arable land, as well as to oppose democratic rule. See J. S. Bentzen, N. Kaarsen, and A. M. Wingender, "Irrigation and Autocracy," *Journal of the European Economic Association* 15 (2017): 1–53. See also A. Sharma, S. Varma, and D. Joshi, "Social Equity Impacts of Increased Water for Irrigation," in U. A. Amarasinghe and B. R. Sharma, eds., *Strategic Analyses of the National River Linking Project (NRLP) of India, Series 2. Proceedings of the Workshop on Analyses of Hydrological, Social and Ecological Issues of the NRLP* (Colombo, Sri Lanka: International Water Management Institute, 2008).
82. For several other factors that contribute to a society manifesting human sacrifice or cannibalism, see B. Schutt, *Cannibalism: A Perfectly Natural History* (Chapel Hill, NC: Algonquin Books, 2017).
83. Liu, "Shipwreck Salvage."

### *Chapter 3: Intentional Communities*

1. H. D. Thoreau, *A Week on the Concord and Merrimack Rivers; Walden, or Life in the Woods; The Maine Woods; Cape Cod,* ed. R. F. Sayre (New York: Literary Classics of the United States, 1985), p. 105.
2. Thoreau, *Walden,* p. 84.

3. Ibid., p. 99.
4. Ibid., p. 102. Thoreau was not impressed with the human capacity for cooperation either: "The only cooperation which is commonly possible is exceedingly partial and superficial; and what little true cooperation there is, is as if it were not, being a harmony inaudible to men." Ibid., p. 55.
5. Ibid., p. 128.
6. M. Meltzer, *Henry David Thoreau: A Biography* (Minneapolis: Twenty-First Century Books, 2007).
7. H. D. Thoreau, *Walden and Civil Disobedience: Complete Texts with Introduction, Historical Contexts, Critical Essays* (Boston: Houghton Mifflin, 2000).
8. F. Tönnies, *Community and Society* [originally published as *Gemeinschaft und Gesellschaft*], ed. and trans. C. P. Loomis (East Lansing: Michigan State University Press, 1957). M. Weber, *Economy and Society* [originally published as *Wirtschaft und Gesellschaft*], ed. and trans. G. Roth and C. Wittich (Berkeley: University of California Press, 1978).
9. B. Zablocki, *Alienation and Charisma: A Study of Contemporary American Communes* (New York: Free Press, 1980).
10. D. E. Pitzer, *America's Communal Utopias* (Chapel Hill: University of North Carolina Press, 1997).
11. T. More, *Utopia: Written in Latin by Sir Thomas More, Chancellor of England; Translated into English,* trans. G. Burnet (London: printed for R. Chiswell, 1684).
12. Pitzer, *America's Communal Utopias,* p. 5.
13. C. Nordhoff, *The Communistic Societies of the United States, from Personal Visit and Observation* (New York: Harper and Brothers, 1875). See also J. H. Noyes, *History of American Socialisms* (Philadelphia: J. B. Lippincott, 1870); and A. F. Tyler, *Freedom's Ferment: Phases of American Social History from the Colonial Period to the Outbreak of the Civil War* (New York: Harper and Row, 1944).
14. E. Green, "Seeking an Escape Hatch from Trump's America," *Atlantic,* January 15, 2017.
15. There were perhaps ten thousand such communities of varying size across the United States in the 1960s. By 1995, there were only five hundred groups active in North America, according to one census. Pitzer, *America's Communal Utopias,* p. 12.
16. A. de Tocqueville, *Democracy in America* [originally published as *De la démocratie en Amérique*], trans. H. Reeve (London: Saunders and Otley, 1838).
17. R. W. Emerson, cited in E. K. Spann, *Brotherly Tomorrows: Movements for a Cooperative Society in America, 1820–1920* (New York: Columbia University Press, 1989), p. 52.
18. A. R. Schultz and H. A. Pochmann, "George Ripley: Unitarian, Transcendentalist, or Infidel?," *American Literature* 14 (1942): 1–19.
19. J. Myerson, "Two Unpublished Reminiscences of Brook Farm," *New England Quarterly* 48 (1975): 253–260.
20. George Ripley, cited in Spann, *Brotherly Tomorrows,* p. 56.
21. Myerson, "Two Unpublished Reminiscences."
22. Ibid., p. 256.
23. S. F. Delano, *Brook Farm: The Dark Side of Utopia* (Cambridge, MA: Harvard University Press, 2004), pp. 60–76.

24. R. Francis, "The Ideology of Brook Farm," *Studies in the American Renaissance* (1977): 1–48.
25. Ibid., p. 11.
26. J. Haidt, *The Righteous Mind: Why Good People Are Divided by Politics and Religion* (New York: Pantheon, 2012), chap. 10.
27. Francis, "Ideology of Brook Farm," pp. 14–15.
28. A. E. Russell, *Home Life of the Brook Farm Association* (Boston: Little, Brown, 1900), p. 15.
29. Myerson, "Two Unpublished Reminiscences," p. 259.
30. C. A. Dana, cited in Francis, "Ideology of Brook Farm," p. 8.
31. Russell, *Home Life,* p. 24.
32. Myerson, "Two Unpublished Reminiscences," p. 256.
33. Fourier's ideas had to do with the laws of symmetry and seriality in the natural world and with idealized dwellings for self-sufficient communities of 1,620 people, sorted by personality type and occupation, which he called "phalanxes." Fourier had mostly progressive ideas about the equality of women and the instruction of children and about homosexuality and casual sex (which were permitted in his theory). C. Fourier, *Theory of Social Organization* (New York: C. P. Somerby, 1876). Fourier's thinking inspired perhaps thirty utopian communities in the United States.
34. Russell, *Home Life,* p. 2.
35. Ibid., p. 134.
36. C. A. Russell, "The Rise and Decline of the Shakers," *New York History* 49 (1968): 29–55.
37. For a contemporaneous account of the reception of the Shakers, see V. Rathbun, *An Account of the Matter, Form, and Manner of a New and Strange Religion, Taught and Propagated by a Number of Europeans Living in a Place Called Nisqueunia, in the State of New-York* (Providence, RI: Bennett Wheeler, 1781).
38. W. S. Bainbridge, "Shaker Demographics 1840–1900: An Example of the Use of U.S. Census Enumeration Schedules," *Journal for the Scientific Study of Religion* 21 (1982): 352–365.
39. Ibid.
40. S. J. Stein, *Shaker Experience in America* (New Haven, CT: Yale University Press, 1992).
41. M. M. Cosgel and J. E. Murray, "Productivity of a Commune: The Shakers, 1850–1880," *Journal of Economic History* 58 (1998): 494–510.
42. Stein, *Shaker Experience in America,* pp. 149–154.
43. Bainbridge, "Shaker Demographics."
44. Russell, "Rise and Decline," p. 46.
45. Ibid.
46. Population size was always a key consideration in survival. Just as small founding colonies of animals on an island are more susceptible to extinction, so too with small founding groups for communal living. The Amish, Hutterites, and Mormons (all religiously inspired) have endured so long in part because of their larger family size and in part because of their ongoing interactions with the outside world. They also, it bears mentioning, avoid alcohol. It may seem astonishing that alcohol can wreak such havoc on human social organization, yet, from varied sources—not only case histories of shipwrecks but also modern longitudinal research such as the eighty-year-old

Harvard Study of Adult Development—we have abundant evidence for this too. P. Hoehnle, "Community in Transition: Amana's Great Change, 1931–1933," *Annals of Iowa* 60 (2001): 1–34; R. Janzen and M. Stanton, *The Hutterites in North America* (Baltimore: Johns Hopkins University Press, 2010). Regarding alcohol, see, for example, G. E. Vaillant, *Aging Well: Surprising Guideposts to a Happier Life from the Landmark Harvard Study of Adult Development* (Boston: Little, Brown, 2003).

47. M. Palgi and S. Reinharz, eds., *One Hundred Years of Kibbutz Life: A Century of Crises and Reinvention* (New Brunswick, NJ: Transaction, 2014), p. 2.
48. B. Beit-Hallahmi and A. I. Rabin, "The Kibbutz as a Social Experiment and as a Child-Rearing Laboratory," *American Psychologist* 32 (1977): 533.
49. D. Lieberman and T. Lobel, "Kinship on the Kibbutz: Co-Residence Duration Predicts Altruism, Personal Sexual Aversions and Moral Attitudes Among Communally Reared Peers," *Evolution and Human Behavior* 33 (2012): 26–34.
50. O. Aviezer, M. H. Van IJzendoorn, A. Sagi, and C. Schuengel, "'Children of the Dream' Revisited: 70 Years of Collective Early Child Care in Israeli Kibbutzim," *Psychological Bulletin* 116 (1994): 99–116.
51. Plato, *The Republic,* trans. B. Jowett (New York; Vintage Books, 1991), bk. 5.
52. This idea is explored in J. Rawls, *A Theory of Justice* (Cambridge, MA: Harvard University Press, 1971). For a more recent summary of the debate, see A. L. Alstott, "Is the Family at Odds with Equality? The Legal Implications of Equality for Children," *Southern California Law Review* 82, no. 1 (2008): 1–43.
53. One of my former graduate students, Peter Dewan, grew up in a commune known as the Fort Hill Community or the Lyman Family, which was active beginning in the 1960s and where he was reared collectively. Dewan confirmed that lack of sexual interest in his peers, and he describes a special intimacy with a larger group of sibling surrogates: "From my perspective, as one of those children, this was a success, in that I have a large kin group, and we care for each other and even each other's children. Unfortunately, we don't have a word for this type of kinship. It is different than blood, in that I have always recognized a special closeness to my blood relatives, as have all of the other children of the community. However, my mother, when she had my brother eighteen years after I was born, deeply regretted the childhood years that she spent apart from me, sometimes as long as three years and thousands of miles apart. She determined that she would not do that again, and it is something that I have never considered. Interestingly, although this feeling is common among the mothers, many of the children never developed strong relationships with their fathers, and the fathers never seemed to mind that much. From this perspective, I would say that collective child rearing failed. Nobody continued it, and many regretted it." Peter Dewan, personal communication, August 31, 2017. For more about this group, see R. L. Levey, "Friendly Fifty on Fort Hill—Better Way for People," *Boston Globe,* December 12, 1967; and D. Johnston, "Once-Notorious '60s Commune Evolves into Respectability," *Los Angeles Times,* August 4, 1985.
54. H. Barry and L. M. Paxton, "Infancy and Early Childhood: Cross-Cultural Codes," *Ethnology* 10 (1971): 466–508.
55. S. Mintz, *Huck's Raft: A History of American Childhood* (Cambridge, MA: Belknap Press, 2004).

56. L. Tiger and J. Shepher, *Women in the Kibbutz* (New York: Harcourt Brace Jovanovich, 1975).
57. One study found that only 59 percent of kibbutz children were securely attached to their mothers, compared with 75 percent of Israeli infants in day care. This difference seems to be due to communal sleeping arrangements resulting in a lack of parental caregiving during nighttime hours, as well as limited daytime contact. A lack of attachment to a primary caregiver has many negative effects. Aviezer et al., " 'Children of the Dream' Revisited."
58. A. Sagi, M. E. Lamb, R. Shoham, R. Dvir, and K. S. Lewkowicz, "Parent-Infant Interaction in Families on Israeli Kibbutzim," *International Journal of Behavioral Development* 8 (1985): 273–284.
59. E. Ben-Rafael, *Crisis and Transformation: The Kibbutz at Century's End* (Albany: State University of New York Press, 1997), p. 62.
60. Aviezer et al., " 'Children of the Dream' Revisited."
61. The aversion to sex with peers was observed as early as the 1970s and has been subject to investigation ever since. See, for example, J. Shepher, "Mate Selection Among Second Generation Kibbutz Adolescents and Adults: Incest Avoidance and Negative Imprinting," *Archives of Sexual Behavior* 1 (1971): 293–307. For a review, see E. Shor, "The Westermarck Hypothesis and the Israeli Kibbutzim: Reconciling Contrasting Evidence," *Archives of Sexual Behavior* 44 (2015): 1–12. See also Lieberman and Lobel, "Kinship on the Kibbutz," which found that duration of co-residence predicted stronger altruistic motivations toward one another and also more moralistic attitudes toward sex between third parties who were childhood peers.
62. Aviezer et al., " 'Children of the Dream' Revisited," p. 113.
63. Palgi and Reinharz, *One Hundred Years.*
64. R. Abramitzky, "Lessons from the Kibbutz on the Equality-Incentives Trade-Off," *Journal of Economic Perspectives* 25 (2011): 185–207. Economists who studied the crises that beset kibbutzim during this period note a significant "brain drain" problem, with highly skilled and qualified residents more likely to leave and individuals with lower wages more likely to join. See also R. Abramitzky, "The Limits of Equality: Insights from the Israeli Kibbutz," *Quarterly Journal of Economics* 123 (2008): 1111–1159.
65. Tiger and Shepher, *Women in the Kibbutz,* p. 14.
66. B. J. Ruffle and R. Sosis, "Cooperation and the In-Group–Out-Group Bias: A Field Test on Israeli Kibbutz Members and City Residents," *Journal of Economic Behavior and Organization* 60 (2006): 147–163.
67. They also stated that, despite their knowledge of the relative intractability of sex roles, they were "surprised at how the major innovations in kibbutz women's lives have failed to stimulate the expected new social patterns." Mothers sought closer connections to their offspring due to a "species-wide attraction between mothers and their young" that was biologically and not simply culturally encoded. Tiger and Shepher, *Women in the Kibbutz,* pp. 6, 272. See also L. Tiger and R. Fox, *The Imperial Animal* (New York: Transaction, 1971); and M. E. Spiro, *Gender and Culture: Kibbutz Women Revisited* (New York: Transaction, 1979).

68. B. F. Skinner, *Walden Two* (New York: Macmillan, 1948). In 1985, Skinner published a kind of coda, recounting what happened after the novel ended, in the voice of Professor Burris, one of the characters. B. F. Skinner, "News from Nowhere, 1984," *Behavior Analyst* 8 (1985): 5–14.
69. Skinner, *Walden Two,* p. 194. Skinner also had other ideas that did not quite pan out as intended, including using pigeons to guide missiles during World War II and marketing his famous Skinner box for the training of children other than his own (upon whom he famously experimented). B. F. Skinner, *The Shaping of a Behaviorist* (New York: Knopf, 1979).
70. Skinner, *Shaping of a Behaviorist,* p. 292.
71. D. E. Altus and E. K. Morris, "B. F. Skinner's Utopian Vision: Behind and Beyond *Walden Two,*" *Behavior Analyst* 32 (2009): 319–335. When first published, *Walden Two* sold about seven hundred copies per year.
72. Skinner, *Walden Two,* p. 22.
73. J. K. Jessup, "Utopia Bulletin," *Fortune* (October 1948): 191–198, cited in Altus and Morris, "B. F. Skinner's Utopian Vision," p. 321.
74. H. Kuhlmann, *Living* Walden Two (Urbana: University of Illinois Press, 2005); E. K. Morris, N. G. Smith, and D. E. Altus, "B. F. Skinner's Contributions to Applied Behavior Analysis," *Behavior Analyst* 28 (2005): 99–131; Altus and Morris, "B. F. Skinner's Utopian Vision."
75. Kuhlmann, *Living* Walden Two, p. 92. See also T. Jones, "The Other American Dream," *Washington Post Magazine,* November 15, 1998. The compound at Twin Oaks would eventually encompass 450 acres.
76. Kuhlmann, *Living* Walden Two, p. 102.
77. Ibid., p. 98.
78. I. Komar, *Living the Dream: A Documentary Study of Twin Oaks Community* (Norwood, PA: Norwood Editions, 1983), pp. 99–101.
79. Kuhlmann, *Living* Walden Two, p. 101.
80. D. Ruth, "The Evolution of Work Organization at Twin Oaks," *Communities: Journal of Cooperative Living* 35 (1975): 58–60. See also H. Kuhlmann, "The Illusion of Permanence: Work Motivation and Membership Turnover at Twin Oaks Community," in B. Goodwin, ed., *The Philosophy of Utopia: A Special Issue of Critical Review of International Social and Political Philosophy* (London: Frank Cass, 2001), pp. 157–171.
81. Regarding the impact of high turnover in social groups, see H. Shirado, F. Fu, J. H. Fowler, and N. A. Christakis, "Quality Versus Quantity of Social Ties in Experimental Cooperative Networks," *Nature Communications* 4 (2013): 2814.
82. Komar, *Living the Dream,* p. 72. See also Jones, "The Other American Dream."
83. L. Rohter, "Isolated Desert Community Lives by Skinner's Precepts," *New York Times,* November 7, 1989.
84. Comunidad Los Horcones, "News from Now-Here, 1986: A Response to 'News from Nowhere, 1984,'" *Behavior Analyst* 9 (1986): 129–132.
85. See, for example, F. S. Keller, "Goodbye Teacher..." *Journal of Applied Behavior Analysis* 1 (1968): 79–89.
86. Kuhlmann, *Living* Walden Two, p. 190.

87. Ibid., p. 145.
88. Zablocki, *Alienation and Charisma*. There were between 800 and 1,725 communes in the chosen urban areas. These cities had a combined population of 25.6 million people, which corresponds to as many as 7 communes per 100,000 people.
89. S. Vaisey, "Structure, Culture, and Community: The Search for Belonging in 50 Urban Communes," *American Sociological Review* 72 (2007): 851–873; A. A. Aidala and B. D. Zablocki, "The Communes of the 1970s: Who Joined and Why?," *Marriage and Family Review* 17 (1991): 87–116. The average size of 13.4 is obtained by dividing the total census of 804 adults (age fifteen and older) by 60 in data from Aidala and Zablocki. Vaisey gives a lower average, of 10.4, likely because he used fifty communes and because he restricted his analysis to respondents who provided data of diverse sorts.
90. Zablocki, *Alienation and Charisma,* p. 44.
91. Ibid., p. 96.
92. Aidala and Zablocki, "The Communes of the 1970s," p. 112.
93. Ibid., p. 108.
94. D. French and E. French, *Working Communally: Patterns and Possibilities* (New York: Russell Sage Foundation, 1975), p. 89.
95. Zablocki, *Alienation and Charisma,* p. 319.
96. Ibid., p. 124. This was from one of the sixty rural communes. Regarding leadership in these communes, see also S. L. Carlton-Ford, "Ritual, Collective Effervescence, and Self-Esteem," *Sociological Quarterly* 33 (1992): 365–387; and J. L. Martin, "Is Power Sexy?," *American Journal of Sociology* 111 (2005): 408–446.
97. Zablocki, *Alienation and Charisma,* pp. 127, 153.
98. Ibid., pp. 115–118. The values for these behaviors reported in Aidala and Zablocki, "The Communes of the 1970s," are somewhat different, likely because the latter report focuses only on the urban communes. Part of the decline in some of these nonconventional behaviors no doubt had to do with the aging of the members, and with the winding down of the Vietnam War and civil rights movement, and not simply to being in the commune.
99. Vaisey, "Structure, Culture, and Community."
100. A. A. Harrison, Y. A. Clearwater, and C. P. McKay, eds., *From Antarctica to Outer Space: Life in Isolation and Confinement* (New York: Springer-Verlag, 1991).
101. This is a diary entry of Edward Wilson, cited in D. J. Lugg, "The Adaptation of a Small Group to Life on an Isolated Antarctic Station," in O. G. Edholm and E. K. E. Gunderson, eds., *Polar Human Biology* (Chichester UK: William Heinemann, 1973), pp. 401–409.
102. A. Lansing, *Endurance: Shackleton's Incredible Voyage* (New York: McGraw-Hill, 1959), p. 51.
103. E. K. E. Gunderson, "Psychological Studies in Antarctica: A Review," in O. G. Edholm and E. K. E. Gunderson, eds., *Polar Human Biology* (Chichester UK: William Heinemann, 1973), pp. 352–361.
104. Even though the winter-over groups are new each year, the stations still have a distinctive culture, history, and set of traditions—as well as meteorological and material exigencies—that shape behavior. For example, winter-over crews think of the summer workers as "tourists," wear specially designed patches and clothing, and have

traditions such as the "Three Hundred Club," in which crew members at the South Pole station sit in a 200-degree sauna and then run outside into 100-below weather wearing nothing more than their boots. S. K. Narula, "On Getting Naked in Antarctica," *Atlantic,* January 7, 2014.

105. In fact, conflict between naval crews and scientists had a long history during the age of European exploration, including on Captain Cook's vessels. B. Finney, "Scientists and Seamen," in A. A. Harrison, Y. A. Clearwater, and C. P. McKay, eds., *From Antarctica to Outer Space: Life in Isolation and Confinement* (New York: Springer-Verlag, 1991), pp. 89–101. Upon returning from such long voyages, scientists were known to congratulate one another for their "narrow escape from naval servitude," as Harvard botanist Asa Gray wrote to his geologist friend James Dwight Dana. Cited in W. R. Stanton, *The Great United States Exploring Expedition of 1838–1842* (Berkeley: University of California Press, 1975), p. 137.

106. One early study of groups that wintered over in 1969–1971 found that as many as 72 percent of subjects suffered some symptoms during the winter. R. E. Strange and W. J. Klein, "Emotional and Social Adjustment of Recent U.S. Winter-Over Parties in Isolated Antarctic Station," in O. G. Edholm and E.K.E. Gunderson, eds., *Polar Human Biology* (Chichester UK: William Heinemann, 1973), pp. 410–416. But a more recent study using psychometric testing to evaluate seventy-eight men who wintered over in 1977 found no increase in depression. D. C. Oliver, "Psychological Effects of Isolation and Confinement of a Winter-Over Group at McMurdo Station, Antarctica," in A. A. Harrison, Y. A. Clearwater, and C. P. McKay, eds., *From Antarctica to Outer Space: Life in Isolation and Confinement* (New York: Springer-Verlag, 1991), pp. 217–227. See also L. A. Palinkas, "Going to Extremes: The Cultural Context of Stress, Illness, and Coping in Antarctica," *Social Science and Medicine* 35 (1992): 651–664.

107. P. E. Cornelius, "Life in Antarctica," in A. A. Harrison, Y. A. Clearwater, and C. P. McKay, eds., *From Antarctica to Outer Space: Life in Isolation and Confinement* (New York: Springer-Verlag, 1991), p. 10.

108. Gunderson, "Psychological Studies in Antarctica," p. 357.

109. J. C. Johnson, J. S. Boster, and L. A. Palinkas, "Social Roles and the Evolution of Networks in Extreme and Isolated Environments," *Journal of Mathematical Sociology* 27 (2003): 89–121.

110. P. V. Marsden, "Core Discussion Networks of Americans," *American Sociological Review* 52 (1987): 122–131; H. B. Shakya, N. A. Christakis, and J. H. Fowler, "An Exploratory Comparison of Name Generator Content: Data from Rural India," *Social Networks* 48 (2017): 157–168.

111. M. C. Pachucki, E. J. Ozer, A. Barrat, and D. Cattuto, "Mental Health and Social Networks in Early Adolescence: A Dynamic Study of Objectively Measured Social Interaction Behaviors," *Social Science and Medicine* 125 (2015): 40–50; M. Salathe, M. Kazandjieva, J. W. Lee, P. Levis, M. W. Feldman, and J. H. Jones, "A High-Resolution Human Contact Network for Infectious Disease Transmission," *PNAS: Proceedings of the National Academy of Sciences* 107 (2010): 22020–22025; J. P. Onnela, B. N. Waber, A. Pentland, S. Schnorf, and D. Lazer, "Using Sociometers to Quantify Social Interaction Patterns," *Scientific Reports* 4 (2014): 5604.

112. W. M. Smith, "Observations over the Lifetime of a Small Isolated Group: Structure, Danger, Boredom, and Vision," *Psychological Reports* 19 (1966): 475–514.
113. Strictly, a network is a hyperdimensional object; it's usually not two-dimensional or even three-dimensional. This description of networks as buttons and strings is taken from N. A. Christakis and J. H. Fowler, *Connected: The Surprising Power of Our Social Networks and How They Shape Our Lives* (New York: Little, Brown, 2009).
114. In a different study of winter-over crews, when they were still all-male (1967, 1968, and 1969), altercations involving "strong verbal disagreements, an exchange of threats, or the exchange of physical blows" occurred every year, often between beakers and trades. K. Natani, J. T. Shurley, and A. T. Joern, "Inter-Personal Relationships, Job Satisfaction, and Subjective Feelings of Competence: Their Influence upon Adaptation to Antarctic Isolation," in O. G. Edholm and E. K. E. Gunderson, eds., *Polar Human Biology,* (Chichester UK: William Heinemann, 1973), pp. 384–400.
115. Palinkas, "Going to Extremes."
116. One study of ninety-three people from New Zealand who had wintered over during one of five years in the late 1960s found that 40 percent spontaneously noted the importance of "singing and games." A. J. W. Taylor, "The Adaptation of New Zealand Research Personnel in the Antarctic," in O. G. Edholm and E. K. E. Gunderson, eds., *Polar Human Biology* (Chichester UK: William Heinemann, 1973), pp. 417–429.
117. M. Weber, "Science as Vocation," in *From Max Weber: Essays in Sociology,* ed. and trans. H. H. Gerth and C. Wright Mills (Oxford: Routledge, 1991), p. 155.

### *Chapter 4: Artificial Communities*

1. T. Standage, *The Turk: The Life and Times of the Famous Eighteenth-Century Chess-Playing Machine* (London: Walker Books, 2002).
2. H. Reese and N. Heath, "Inside Amazon's Clickworker Platform," TechRepublic, 2016, https://www.techrepublic.com/article/inside-amazons-clickworker-platform-how-half-a-million-people-are-training-ai-for-pennies-per-task/.
3. J. Bohannon, "Psychologists Grow Increasingly Dependent on Online Research Subjects," *Science,* June 7, 2016.
4. J. J. Horton, D. G. Rand, and R. J. Zeckhauser, "The Online Laboratory: Conducting Experiments in a Real Labor Market," *Experimental Economics* 14 (2011): 399–425; E. Snowberg and L. Yariv, "Testing the Waters: Behavior Across Participant Pools" (working paper no. 24781, National Bureau of Economic Research, June 2018).
5. M. Zelditch, "Can You Really Study an Army in the Laboratory?," in A. Etzioni, ed., *Complex Organizations,* 2nd ed. (New York: Holt, Rinehart, and Winston, 1969) pp. 528–539.
6. D. Rand, S. Arbesman, and N. A. Christakis, "Dynamic Social Networks Promote Cooperation in Experiments with Humans," *PNAS: Proceedings of the National Academy of Sciences* 108 (2011): 19193–19198.
7. D. G. Rand, M. Nowak, J. H. Fowler, and N. A. Christakis, "Static Network Structure Can Stabilize Human Cooperation," *PNAS: Proceedings of the National Academy of Sciences* 111 (2014): 17093–17098.

8. H. Shirado, F. Fu, J. H. Fowler, and N. A. Christakis, "Quality Versus Quantity of Social Ties in Experimental Cooperative Networks," *Nature Communications* 4 (2013): 2814.
9. Rand et al., "Static Network Structure."
10. A. Nishi, H. Shirado, D. G. Rand, and N. A. Christakis, "Inequality and Visibility of Wealth in Experimental Social Networks," *Nature* 526 (2015): 426–429.
11. Regarding *Second Life,* see T. Boellstorff, *Coming of Age in Second Life: An Anthropologist Explores the Virtually Human* (Princeton, NJ: Princeton University Press, 2008). Social interactions in online games are also reviewed in N. A. Christakis and J. H. Fowler, *Connected: The Surprising Power of Our Social Networks and How They Shape Our Lives* (New York: Little, Brown, 2009).
12. K. McKeand, "Blizzard Says World of Warcraft 10.1 Million Subscriber Statement Was a 'Misquote or Misunderstanding,'" *PCGamesN,* October 5, 2016.
13. N. Ducheneaut, N. Yee, E. Nickell, and R. J. Moore, "The Life and Death of Online Gaming Communities: A Look at Guilds in World of Warcraft," *Proceedings of the SIGCHI Conference on Human Factors in Computing Systems* (New York: ACM, 2007), pp. 839–848.
14. H. Cole and M. D. Griffiths, "Social Interactions in Massively Multiplayer Online Role-Playing Games," *CyberPsychology and Behavior* 10 (2007): 575–583.
15. P. W. Eastwick and W. L. Gardner, "Is It a Game? Evidence for Social Influence in the Virtual World," *Social Influence* 1 (2008): 1–15.
16. N. Yee, J. N. Bailenson, M. Urbanek, F. Chang, and D. Merget, "The Unbearable Likeness of Being Digital: The Persistence of Nonverbal Social Norms in Online Virtual Environments," *CyberPsychology and Behavior* 10 (2007): 115–121.
17. E. K. Yuen, J. D. Herbert, E. M. Forman, E. M. Goetter, R. Comer, and J. C. Bradley, "Treatment of Social Anxiety Disorder Using Online Virtual Environments in Second Life," *Behavior Therapy* 44 (2013): 51–61.
18. M. Szell, R. Lambiotte, and S. Thurner, "Multirelational Organization of Large-Scale Social Networks in an Online World," *PNAS: Proceedings of the National Academy of Sciences* 107 (2010): 13636–13641.
19. D. M. Raup, "Geometric Analysis of Shell Coiling: General Problems," *Journal of Paleontology* 40 (1966): 1178–1190.
20. D. M. Raup and A. Michelson, "Theoretical Morphology of the Coiled Shell," *Science* 147 (1965): 1294–1295.
21. Raup actually had a fourth parameter, which characterized the "shape" of the "generating curve," or the shape of the aperture of the shell.
22. Subsequent studies dealt with some of the limitations and oversights of Raup's model. For example, some scholars addressed the conceptual problem of shelled organisms that change their parameters over the course of their development. Other scholars noted that Raup's three parameters were not entirely independent of one another (an unintended oversight). This is concerning because it might mean that the reason parts of the Raup cube are unoccupied is that they are mathematically impossible, not biologically implausible. Nevertheless, follow-up work by ecologist Bernard Tursch developed a more complex model with ten (rather than three) parameters, and it still shows a shell morphospace that is only partially filled by known organisms.

B. Tursch, "Spiral Growth: The 'Museum of All Shells' Revisited," *Journal of Molluscan Studies* 63 (1997): 547–554. Tursch also argues that the final shape of a shell is largely determined by its starting conditions, which, being starting conditions, are presumably largely genetically encoded within the organism.

23. R. Dawkins, *Climbing Mount Improbable* (New York: W. W. Norton, 1996).
24. R. D. K. Thomas and W. E. Reif, "The Skeleton Space: A Finite Set of Organic Designs," *Evolution* 47 (1993): 341–360.
25. S. Wolfram, *A New Kind of Science* (Champaign, IL: Wolfram Media, 2002).
26. G. L. Stebbins, "Natural Selection and the Differentiation of Angiosperm Families," *Evolution* 5 (1951): 299–324.
27. There is a similar explanation for why constricting snakes do not coil around their prey with their backs facing the prey: it's physically impossible for their spines to bend that way, so this behavior is just not seen. D. E. Willard, "Constricting Methods of Snakes," *Copeia* 2 (1977): 379–382.
28. The so-called Gravner-Griffeath model generates dazzling snowflake shapes while keeping the parameter for number of sides fixed at six. This model is based on seven parameters; by allowing them to vary, it defines a world of all possible snowflakes. Many of these mathematically foreseen possible shapes do, in fact, arise. But some of them (for instance, very thin, cross-like snowflakes) might be so unstable as not to last. M. Krzywinski and J. Lever, "In Silico Flurries: Computing a World of Snowflakes," *Scientific American,* December 23, 2017.
29. Another, more complicated issue relates to what is known as the "fitness landscape." Here, the problem is not necessarily that there isn't enough genetic availability per se, but that most organisms are sitting at the top of "adaptive peaks" (even if those peaks may be suboptimal), and shifts between peaks are less common or very difficult to achieve (and happen only through genetic drift or by traversing "adaptive ridges").
30. M. LaBarbera, "Why the Wheels Won't Go," *American Naturalist* 121 (1983): 395–408.
31. J. Hsu, "Walking Military Robots Stumble Toward Future," *Discover,* December 31, 2015.
32. Dawkins, *Climbing Mount Improbable,* p. 222.
33. R. I. M. Dunbar, "Neocortex Size as a Constraint on Group Size in Primates," *Journal of Human Evolution* 22 (1992): 469–493.
34. J. Henrich, R. Boyd, S. Bowles, C. Camerer, E. Fehr, and H. Gintis, eds., *Foundations of Human Sociality: Economic Experiments and Ethnographic Evidence from Fifteen Small-Scale Societies* (Oxford: Oxford University Press, 2004).
35. L. Cronk, *That Complex Whole: Culture and the Evolution of Human Behavior* (Boulder, CO: Westview Press, 1999), p. 21.
36. J. Sawyer and R. A. Levine, "Cultural Dimensions: A Factor Analysis of the World Ethnographic Sample," *American Anthropologist* 68 (1966): 708–731. It is also worth noting that there is path specificity in cultures too, and, just as biology can lead species down specific paths and into dead ends, so can history lead a culture to a set of practices that, while theoretically changeable, are not actually mutable.
37. D. Brown, *Human Universals* (New York: McGraw-Hill, 1991), pp. 130–141.

38. Such radically alien social worlds are uncommon. This may have to do not so much with a failure of imagination on the part of the authors, but rather with their need to conform to constraints imposed by what readers expect from a story. This is analogous, in many ways, to the issue of what evolution *may* yield, as opposed to what it *does* yield given environmental constraints, as discussed earlier.
39. B. M. Stableford, "The Sociology of Science Fiction" (PhD diss., University of York, UK, 1978).
40. H. G. Wells, *The Time Machine* (London: William Heinemann, 1895).
41. A. Huxley, *Brave New World* (London: Chatto and Windus, 1932); R. A. Heinlein, *Orphans of the Sky* (New York: G. P. Putnam's Sons, 1964); G. Orwell, *Nineteen Eighty-Four* (London: Secker and Warburg, 1949).
42. C. P. Gilman, *Herland* (New York: Pantheon, 1979). Whereas Gilman's *Herland* envisions a utopia with hyper-cooperativeness as its defining trait, William Golding's dystopian novel *Lord of the Flies* explores the opposite end of the spectrum, imagining the striking descent of self-governance into savagery. W. Golding, *Lord of the Flies* (New York: Penguin, 1954).
43. Gilman, *Herland,* p. 60. Societal roles in Herland are, in fact, somewhat differentiated, with women of exceptional wisdom and nobility occupying the village temples.
44. The women credit this diversity to their systems of education and child-rearing, claiming that "so much divergence without cross-fertilization" is due partly to "the careful education, which followed each slight tendency to differ, and partly to the law of mutation." Gilman, *Herland,* p. 77. In this way, though described as a single "unit, a conscious group," the women each still have a separate sense of self and distinctive quirks.
45. L. Lowry, *The Giver* (New York: Random House, 1993).
46. R. Kipling, R. Jarrell, and E. Bishop, eds., *The Best Short Stories of Rudyard Kipling* (Garden City, NY: Hanover House, 1961).
47. Cronk, *That Complex Whole,* p. 33.
48. J. Tooby and L. Cosmides, "The Psychological Foundation of Culture," in J. H. Barkow, L. Cosmides, and J. Tooby, eds., *The Adapted Mind: Evolutionary Psychology and the Generation of Culture* (Oxford: Oxford University Press, 1992), pp. 19–136.
49. Our species has a very high level of genetic similarity—mostly due to founder effects from small migratory groups—compared with species like chimpanzees. This likely also imposes constraints on the societal forms we make.

### *Chapter 5: First Comes Love*

1. H. A. Junod, *Life of a South African Tribe,* vol. 1 (London: Macmillan, 1927), pp. 353–354.
2. W. R. Jankowiak, S. L. Volsche, and J. R. Garcia, "Is the Romantic-Sexual Kiss a Near Human Universal?," *American Anthropologist* 117 (2015): 535–539. For a short, early review of this, see I. Eibl-Eibesfeldt, *Love and Hate: The Natural History of Behavior Patterns* (New York: Holt, Rinehart and Winston, 1971), p. 129. Thankfully, kissing one's children *does* appear to be universal, so far as I can tell.

3. E. W. Hopkins, "The Sniff-Kiss in Ancient India," *Journal of the American Oriental Society* 28 (1907): 120–134.
4. Jankowiak, Volsche, and Garcia, "Romantic-Sexual Kiss."
5. Yet kissing is extremely common in Arctic foragers. Also, societies with more complex social systems involving social stratification (e.g., industrialized societies with distinct social classes) evince romantic kissing more often than egalitarian societies (e.g., foragers). It is unclear why this is the case; it may relate to factors such as improved oral hygiene or an emphasis on formalized emotional displays in more complex societies.
6. F. B. M. de Waal, "The First Kiss: Foundations of Conflict Resolution Research in Animals," in F. Aureli and F. B. M. de Waal, eds., *Natural Conflict Resolution* (Berkeley: University of California Press, 2000), pp. 15–33; R. Wlodarksi and R. I. M. Dunbar, "Examining the Possible Functions of Kissing in Romantic Relationships," *Archives of Sexual Behavior* 42 (2013): 1415–1423.
7. For instance, Bronislaw Malinowski's classic (if unfortunately titled) 1929 account *The Sexual Life of Savages in Northwestern Melanesia* describes the Trobrianders' bemused impression of kissing. B. Malinowski, *The Sexual Life of Savages in Northwestern Melanesia* (New York: Halcyon House, 1929), p. 331.
8. C. Wagley, *Welcome of Tears: The Tapirapé Indians of Central Brazil* (New York: Oxford University Press, 1977), p. 158.
9. Wagley's informants also denied the practice of cunnilingus, though they did provide accounts of fellatio (which was not uncommon) and homosexuality. Homosexual men were deemed especially good candidates to accompany hunting parties that would be absent from camp a long time, and there did not appear to be much opprobrium associated with homosexual behavior.
10. E. E. Evans-Pritchard, *Kinship and Marriage Among the Nuer* (Oxford: Clarendon Press, 1901). See also G. H. Herdt, *Same Sex, Different Cultures: Gays and Lesbians Across Cultures* (Boulder, CO: Westview Press, 1997).
11. T. A. Kohler et al., "Greater Post-Neolithic Wealth Disparities in Eurasia Than in North America and Mesoamerica," *Nature* 551 (2017): 619–622.
12. E. D. Gould, O. Moav, and A. Simhon, "The Mystery of Monogamy," *American Economic Review* 98 (2008): 333–357; J. Henrich, R. Boyd, and P. J. Richerson, "The Puzzle of Monogamous Marriage," *Philosophical Transactions of the Royal Society B* 367 (2012): 657–669.
13. S. J. Gould and E. S. Vrba, "Exaptation—a Missing Term in the Science of Form," *Paleobiology* 8 (1982): 4–15.
14. In the anthropological sense, there are at least three fundamental axes along which societies can be organized with respect to sex: patriarchy versus matriarchy (where the power and decision-making are situated); patrilocal versus matrilocal (whether people live with or near their fathers or their mothers); and patrilineal versus matrilineal (whether ancestry and property follow the paternal or maternal line).
15. J. Henrich, S. J. Heine, and A. Norezayan, "The Weirdest People in the World?," *Behavioral and Brain Sciences* 33 (2010): 61–135.

16. D. R. White et al., "Rethinking Polygyny: Co-Wives, Codes, and Cultural Systems," *Current Anthropology* 29 (1988): 529–572.
17. UN Department of Economic and Social Affairs, *Population Facts*, December 2011, http://www.un.org/en/development/desa/population/publications/pdf/popfacts/PopFacts_2011-1.pdf.
18. 1 Kings 11:3 (New International Version).
19. G. M. Williams, *Handbook of Hindu Mythology* (Oxford: Oxford University Press, 2003), p. 188.
20. K. MacDonald, "The Establishment and Maintenance of Socially Imposed Monogamy in Western Europe," *Politics and the Life Sciences* 14 (1995): 3–23; W. Scheidel, "A Peculiar Institution? Greco-Roman Monogamy in Global Context," *History of the Family* 14 (2009): 280–291; D. Herlihy, "Biology and History: The Triumph of Monogamy," *Journal of Interdisciplinary History* 25 (1995): 571–583.
21. L. Betzig, "Roman Polygyny," *Ethology and Sociobiology* 13 (1992): 309–349.
22. Gould, Moav, and Simhon, "Mystery of Monogamy"; L. Betzig, "Medieval Monogamy," *Journal of Family History* 20 (1995): 181–216.
23. This numerical illustration is adapted from Henrich, Boyd, and Richerson, "Puzzle of Monogamous Marriage."
24. Ibid.
25. Having a lot of unpartnered, willing-to-be-violent young men could also be useful in ecological situations (perhaps of great resource scarcity) in which war between groups is inevitable or helpful to the group. The balance between intragroup and intergroup violence as well as the relative scarcity of resources are likely important, if complex, factors in guiding the emergence of monogamy.
26. Henrich, Boyd, and Richerson, "Puzzle of Monogamous Marriage," p. 660.
27. T. Hesketh and Z. W. Xing, "Abnormal Sex Ratios in Human Populations," *PNAS: Proceedings of the National Academy of Sciences* 103 (2006): 13271–13275; T. Hesketh, L. Lu, and Z. W. Xing, "The Consequences of Son Preference and Sex-Selective Abortion in China and Other Asian Countries," *Canadian Medical Association Journal* 183 (2011): 1374–1377; L. Jin, F. Elwert, J. Freese, and N. A. Christakis, "Preliminary Evidence Regarding the Hypothesis That the Sex Ratio at Sexual Maturity May Affect Longevity in Men," *Demography* 47 (2010): 579–586.
28. MacDonald, "Establishment and Maintenance"; Scheidel, "Peculiar Institution"; Herlihy, "Biology and History."
29. A. Korotayev and D. Bondarenko, "Polygyny and Democracy: A Cross-Cultural Comparison," *Cross-Cultural Research* 34 (2000): 190–208; R. McDermott and J. Cowden, "Polygyny and Violence Against Women," *Emory Law Journal* 64 (2015): 1767–1814.
30. Henrich, Boyd, and Richerson, "Puzzle of Monogamous Marriage." As always when talking about human behavior, it is important to avoid the assumption that every individual in a given society has the same sexual proclivities or the same desires for partners, whatever the cultural norm.
31. F. W. Marlowe, *The Hadza: Hunter-Gatherers of Tanzania* (Berkeley: University of California Press, 2010).

32. R. Sear and F. W. Marlowe, "How Universal Are Human Mate Choices? Size Doesn't Matter When Hadza Foragers Are Choosing a Mate," *Biology Letters* 5 (2009): 606–609.
33. F. W. Marlowe, "Mate Preferences Among Hadza Hunter-Gatherers," *Human Nature* 15 (2004): 365–376.
34. C. L. Apicella, A. N. Crittenden, and V. A. Tobolsky, "Hunter-Gatherer Males Are More Risk-Seeking Than Females, Even in Late Childhood," *Evolution and Human Behavior* 38 (2017): 592–603.
35. A. Little, C. L. Apicella, and F. W. Marlowe, "Preferences for Symmetry in Human Faces in Two Cultures: Data from the UK and the Hadza, an Isolated Group of Hunter-Gatherers," *Proceedings of the Royal Society B* 274 (2007): 3113–3117; C. L. Apicella, A. C. Little, and F. W. Marlowe, "Facial Averageness and Attractiveness in an Isolated Population of Hunter-Gatherers," *Perception* 36 (2007): 1813–1820; C. L. Apicella and D. R. Feinberg, "Voice Pitch Alters Mate-Choice-Relevant Perception in Hunter-Gatherers," *Proceedings of the Royal Society B* 276 (2009): 1077–1082; F. W. Marlowe, C. L. Apicella, and D. Reed, "Men's Preferences for Women's Profile Waist-to-Hip Ratio in Two Societies," *Evolution and Human Behavior* 26 (2005): 458–468. See also D. M. Buss and M. Barnes, "Preferences in Human Mate Selection," *Journal of Personality and Social Psychology* 50 (1986): 559–570.
36. F. W. Marlowe, "Mate Preferences Among Hadza Hunter-Gatherers," p. 374.
37. C. L. Apicella, "Upper Body Strength Predicts Hunting Reputation and Reproductive Success in Hadza Hunter-Gatherers," *Evolution and Human Behavior* 35 (2014): 508–518.
38. K. Hawkes, J. O'Connell, and N. G. Blurton Jones, "Hunting and Nuclear Families: Some Lessons from the Hadza About Men's Work," *Current Anthropology* 42 (2001): 681–709.
39. F. W. Marlowe, "A Critical Period for Provisioning by Hadza Men: Implications for Pair-Bonding," *Evolution and Human Behavior* 24 (2003): 217–229.
40. Ibid., pp. 224–225. Marlowe notes that among men who have stepchildren at home, their efforts are attenuated, in keeping with provisioning theory.
41. R. Dyson-Hudson, D. Meekers, and N. Dyson-Hudson, "Children of the Dancing Ground, Children of the House: Costs and Benefits of Marriage Rules (South Turkana, Kenya)," *Journal of Anthropological Research* 54 (1998): 19–47.
42. P. H. Gulliver, *A Preliminary Survey of the Turkana,* Communications from the School of African Studies, n.s., no. 26 (Cape Town: University of Cape Town, 1951), p. 199.
43. Generally speaking, dowry societies, where the bride's family makes a payment to the groom's family (i.e., the other way around compared with bride-wealth), are more likely to be monogamous, patrilineal, and endogamous (marrying within a group).
44. Gulliver, *Preliminary Survey,* p. 206.
45. From one point of view, the exchange of bride-wealth is adaptive, because, during times of drought, men cannot assemble the requisite amount, and so marriages (and, quite likely, births) are postponed to a time when the food supply might again be adequate (though it is unclear whether the Turkana themselves are aware of the possible benefits of such cyclicity).

46. Gulliver, *Preliminary Survey*, p. 199.
47. Ibid., pp. 198–199. This classic study of Turkana marriage, conducted in 1951, reported no instance in which this trait (fertility) was noted as essential in a partner. As many as 50 percent of firstborn children are conceived prior to marriage, according to Dyson-Hudson, Meekers, and Dyson-Hudson, "Children of the Dancing Ground," p. 26.
48. Ibid. Among males, puberty (which is known as *abu akoun*, or "testes come") is usually at age 17. Age at first marriage for women is 22.4 (± 5.2) and for men is 32.6 (±7.2).
49. Ibid.
50. R. Dyson-Hudson and D. Meekers, "Universality of African Marriage Reconsidered: Evidence from Turkana Males," *Ethnology* 35 (1996): 301–320. My grandfather Nicholas C. Christakis, described in the preface, told me a related story about his own childhood. He was orphaned as a teenager in about 1910 in Greece and had several older sisters. Greece had a dowry system, and the norm was that he had to earn the money to provide dowries for all his sisters before he himself could get married, which took him until age thirty-five.
51. Ibid.
52. According to Utah officials, as many as one thousand teenage boys were separated from their parents and thrown out of their communities by the polygamous Mormon sect FLDS (Fundamentalist Church of Jesus Christ of Latter-Day Saints). They were often left to fend for themselves. FLDS officials described them as "delinquents," but Utah authorities claimed they were thrown out primarily to make younger girls available as plural wives for older, more powerful men in the sect. J. Borger, "The Lost Boys, Thrown Out of US Sect So That Older Men Can Marry More Wives," *Guardian*, June 13, 2005.
53. P. W. Leslie, R. Dyson-Hudson, and P. H. Fry, "Population Replacement and Persistence," in M. A. Little and P. W. Leslie, eds., *Turkana Herders of the Dry Savanna* (Oxford: Oxford University Press, 1999), pp. 281–301. However, tests of androgen levels in older Turkana men present a complex picture, in part because the practice of polygyny keeps testosterone relatively high into older age.
54. W. Jankowiak, M. Sudakov, and B. C. Wilreker, "Co-Wife Conflict and Cooperation," *Ethnology* 44 (2005): 81–98.
55. B. I. Strassmann, "Polygyny as a Risk Factor for Child Mortality Among the Dogon," *Current Anthropology* 38 (1997): 688–695.
56. Interestingly, it's not only in polygynous households that we find nonrelatedness a risk factor for violence. Empirical investigations in monogamous societies demonstrate that lower degrees of relatedness among household members are associated with higher rates of abuse, neglect, and homicide. Living with genetically unrelated adults is the single biggest risk factor for a child to be abused or murdered. Conforming to familiar fairy-tale tropes, stepmothers are more than twice as likely as birth mothers to kill children, and children living with an unrelated parent are more than ten times as likely to die "accidentally." M. Daly and M. Wilson, "Discriminative Parental Solicitude: A Biological Perspective," *Journal of Marriage and Family* 42 (1980): 277–288; M. Daly and M. Wilson, *The Truth About Cinderella: A Darwinian View of Parental*

*Love* (New Haven, CT: Yale University Press, 1999); V. A. Weekes-Shackelford and T. K. Shackelford, "Methods of Filicide: Stepparents and Genetic Parents Kill Differently," *Violence Victims* 19 (2004): 75–81; K. Gibson, "Differential Parental Investment in Families with Both Adopted and Genetic Children," *Evolution and Human Behavior* 30 (2009): 184–189.

57. D. MacDougall and J. MacDougall, *A Wife Among Wives,* documentary film (1981). Lorang continues: "But if she accepted a man chosen for her, she would be blessed by her parents and live happily with her husband. Then her father might give her more animals. They would help support her. That's the Turkana way."
58. Ibid.
59. D. MacDougall and J. MacDougall, *The Wedding Camels,* documentary film (1980).
60. S. Beckerman and P. Valentine, eds., *Cultures of Multiple Fathers: The Theory and Practice of Partible Paternity in Lowland South America* (Gainesville: University Press of Florida, 2002).
61. Wagley, *Welcome of Tears,* p. 134.
62. R. M. Ellsworth, D. H. Bailey, K. R. Hill, A. M. Hurtado, and R. S. Walker, "Relatedness, Co-Residence, and Shared Fatherhood Among Aché Foragers of Paraguay," *Current Anthropology* 55 (2014): 647–653.
63. K. G. Anderson, "How Well Does Paternity Confidence Match Actual Paternity?," *Current Anthropology* 47 (2006): 513–520.
64. G. J. Wyckoff, W. Want, and D. I. Wu, "Rapid Evolution of Male Reproductive Genes in the Descent of Man," *Nature* 403 (2000): 304–309. This has also led to all sorts of adaptations in humans, including the claim that the shape of the human penis reflects its role as a "semen displacement device," as assessed with experiments involving artificial penises and vaginas. G. G. Gallup et al., "The Human Penis as a Semen Displacement Device," *Evolution and Human Behavior* 24 (2003): 277–289. See also L. W Simmons, R. C. Firman, G. Rhodes, and M. Peters, "Human Sperm Competition: Testis Size, Sperm Production, and Rates of Extrapair Copulations," *Animal Behaviour* 68 (2004): 297–302.
65. S. Beckerman and P. Valentine, "The Concept of Partible Paternity Among Native South Americans," in S. Beckerman and P. Valentine, eds., *Cultures of Multiple Fathers: The Theory and Practice of Partible Paternity in Lowland South America* (Gainesville: University Press of Florida, 2002), pp. 1–13.
66. On the other hand, the exact opposite may be the case for polygynous households, where evidence suggests that child survival might actually be reduced by the practice of having multiple wives (e.g., because of competition among the mothers for resources for their children). E. Smith-Greenaway and J. Trinitapoli, "Polygynous Contexts, Family Structure, and Infant Mortality in Sub-Saharan Africa," *Demography* 51 (2014): 341–366.
67. S. B. Hrdy, *Mother Nature: A History of Mothers, Infants, and Natural Selection* (New York: Pantheon, 1999).
68. C. Hua, *A Society Without Fathers or Husbands: The Na of China* (New York: Zone Books, 2001), p. 22. See also C. K. Shih, *Quest for Harmony: The Moso Traditions of Sexual Unions and Family Life* (Stanford, CA: Stanford University Press, 2010).

69. Hua has to be especially careful to mention the "genitor" of the child, because to say "father" would evoke in non-Na readers a sense of roles and obligations that are simply not present among the Na. Hua notes that some women and children are able to identify the genitor. The role of genitors in the lives of their biological offspring might also be increasing. A quantitative study in 2008 of 140 respondents discovered that biological fathers did contribute time and money to their offspring (to a non-trivial degree) and that this was associated with positive childhood outcomes. S. M. Mattison, B. Scelza, and T. Blumenfield, "Paternal Investment and the Positive Effects of Fathers Among the Matrilineal Mosuo of Southwest China," *American Anthropologist* 116 (2014): 591–610.
70. Hua, *Society Without Fathers,* p. 226.
71. Ibid., p. 119.
72. Ibid., p. 127.
73. Ibid., p. 205.
74. Ibid., p. 232.
75. Ibid., p. 187. I have lightly paraphrased the sort of conversation that Hua sketches here. If the man wants to refuse the woman's overture, the last response might simply be, according to Hua, "I don't want to come."
76. Ibid., p. 197.
77. Ibid., p. 237. Italics added.
78. Cohabitation is uncommon. In one sample of data from 1963, only 10 percent of people were living in a cohabiting relationship. Ibid., p. 273.
79. Ibid., p. 408.
80. Ibid., p. 249.
81. N. K. Choudhri, *The Complete Guide to Divorce Law,* 1st ed. (New York: Kensington, 2004).
82. Hua, *Society Without Fathers,* p. 446. Incidentally, Hua also concludes that the Na do not, in fact, have a family in any traditional sense of the word as used by anthropologists, and that their matrilineal households do not meet the definitions of this concept. I do wonder how men love children in these households. Do they feel the same warmth toward their sisters' children as they would, in our society, feel toward their own?
83. V. Safronova, "Dating Experts Explain Polyamory and Open Relationships," *New York Times,* October 26, 2016.
84. Hua, *Society Without Fathers,* p. 447.
85. This 5 percent figure comes from a survey by the International Institute for Population Sciences and the Population Council, cited in G. Harris, "Websites in India Put a Bit of Choice into Arranged Marriages," *New York Times,* April 24, 2015. Based on a sample from the urban middle class of Mumbai, 8 percent of marriages in the parents' generation and about 30 percent in the current generation are not arranged. See D. Mathur, "What's Love Got to Do with It? Parental Involvement and Spouse Choice in Urban India" (paper, November 7, 2007), http://dx.doi.org/10.2139/ssrn.1655998.

86. J. Marie, "What It's Really Like to Have an Arranged Marriage," *Cosmopolitan,* November 25, 2014.
87. "Seven Couples in Arranged Marriage Reveal When They 'Actually Fell in Love' with Each Other," *Times of India,* November 14, 2017.
88. R. Epstein, M. Pandit, and M. Thakar, "How Love Emerges in Arranged Marriages: Two Cross-Cultural Studies," *Journal of Comparative Family Studies* 43 (2013): 341–360.
89. For examples from some small studies of convenience samples, see J. Madathil and J. M. Benshoff, "Importance of Marital Characteristics and Marital Satisfaction: A Comparison of Asian Indians in Arranged Marriages and Americans in Marriages of Choice," *Family Journal* 16 (2008): 222–230; J. E. Myers, J. Madathil, and L. R. Tingle, "Marriage Satisfaction and Wellness in India and the United States: A Preliminary Comparison of Arranged Marriages and Marriages of Choice," *Journal of Counseling and Development* 83 (2005): 183–190; and P. Yelsma and K. Athappilly, "Marital Satisfaction and Communication Practices: Comparisons Among Indian and American Couples," *Journal of Comparative Family Studies* 19 (1988): 37–54.
90. P. C. Regan, S. Lakhanpal, and C. Anguiano, "Relationship Outcomes in Indian-American Love-Based and Arranged Marriages," *Psychological Reports* 110 (2012): 915–924.
91. Epstein, Pandit, and Thakar, "How Love Emerges." Divorce rates in arranged marriages are usually substantially lower than in marriages of choice, although, again, this outcome is hard to disentangle from the fact that arranged marriages typically take place in societies with other strong cultural disincentives, and even legal barriers, to divorce.
92. D. M. Buss et al., "International Preferences in Selecting Mates: A Study of 37 Cultures," *Journal of Cross-Cultural Psychology* 21 (1990): 5–47. Love ranked lower on the list in some countries, such as China and Nigeria. This survey also found that people everywhere place tremendous value on the partner characteristics of dependability, emotional stability, kindness, and intelligence—not too dissimilar from the ratings offered by the Turkana or the Hadza. Though culture had a discernible effect on what was considered desirable in a partner, there was much more similarity than difference worldwide. The factor with the greatest variation was chastity, which was highly valued in China, India, Indonesia, and Iran but was deemed irrelevant in Sweden, Finland, and Germany. Conversely, an "exciting personality" was a highly desirable spousal trait in France, Japan, Brazil, Spain, Ireland, and the United States but was deemed less important in China, India, and Iran.

### *Chapter 6: Animal Attraction*

1. L. A. McGraw and L. J. Young, "The Prairie Vole: An Emerging Model Organism for Understanding the Social Brain," *Trends in Neuroscience* 33 (2010): 103–109.
2. T. Pizzuto and L. Getz, "Female Prairie Voles (*Microtus ochrogaster*) Fail to Form a New Pair After Loss of Mate," *Behavioural Processes* 43 (1998): 79–86.
3. Animals can also practice genetic or "true" monogamy—total sexual exclusivity in which all the pair's offspring are genetically related to them. Genetic monogamy is seen in some deep-sea fishes, where (given how hard it is for a male and female to find

each other) the much smaller male latches onto and can even become incorporated within the female's body, as a mere source of sperm. Genetic monogamy can also describe species in which "extra-pair matings" are exceptionally rare. It is important not to conflate the notion of a pair-bond with dyadic living in a species; that is, animals might form long-term mating unions and live apart from their mates. For a primate example, see M. Huck, E. Fernandez-Duque, P. Babb, and T. Schurr, "Correlates of Genetic Monogamy in Socially Monogamous Mammals: Insights from Azara's Owl Monkeys," *Proceedings of the Royal Society B* 281 (2014): 20140195.

4. We also see close pair-bonds in some same-sex animal pairs, such as in some species of penguins. B. Bagemihl, *Biological Exuberance: Animal Homosexuality and Natural Diversity* (New York: St. Martin's, 1999).
5. B. B. Smuts, "Social Relationships and Life Histories of Primates," in M. E. Morbeck, A. Galloway, and A. Zihlman, eds., *The Evolving Female* (Princeton, NJ: Princeton University Press, 1997), pp. 60–68.
6. B. Chapais, *Primeval Kinship: How Pair-Bonding Gave Birth to Human Society* (Cambridge, MA: Harvard University Press, 2008).
7. In other taxa, 16 percent of carnivores (such as dogs, cats, and bears) and only 3 percent of ungulates (such as pigs, deer, and hippopotamuses) are monogamous. Surprisingly, monogamy is entirely absent among whales, one of the most social of species. D. Lukas and T. H. Clutton-Brock, "The Evolution of Social Monogamy in Mammals," *Science* 341 (2013): 526–530.
8. The origin of mammals and the timing are still under investigation. See N. M. Foley, M. S. Springer, and E. C. Teeling, "Mammal Madness: Is the Mammal Tree of Life Not Yet Resolved?," *Philosophical Transactions of the Royal Society B* 371 (2016): 20150140.
9. Lukas and Clutton-Brock, "Evolution of Social Monogamy."
10. J. C. Mitani, "The Behavioral Regulation of Monogamy in Gibbons (*Hylobates muelleri*)," *Behavioral Ecology and Sociobiology* 15 (1984): 225–229.
11. S. Shultz, C. Opie, and Q. D. Atkinson, "Stepwise Evolution of Stable Sociality in Primates," *Nature* 479 (2011): 219–222. Changes in dietary patterns among our ancestors might have resulted in larger female foraging ranges, which would have limited the ability of a male to guard more than one female. There are a number of possible causes of the initial shift from solitary to social living in primates. Living in groups would have reduced the risk of predation that followed a prior evolutionary shift from a nocturnal lifestyle to a diurnal one. Moreover, once a lineage becomes social, it remains social; there were no reversions to solitary lifestyle patterns.
12. W. D. Lassek and S. J. C. Gaulin, "Costs and Benefits of Fat-Free Muscle Mass in Men: Relationship to Mating Success, Dietary Requirements, and Native Immunity," *Evolution and Human Behavior* 30 (2009): 322–328; A. Sell, L. S. E. Hone, and N. Pound, "The Importance of Physical Strength to Human Males," *Human Nature* 23 (2012): 30–44.
13. J. M. Plavcan, "Sexual Dimorphism in Primate Evolution," *American Journal of Physical Anthropology* 116 (2002): 25–53; J. M. Plavcan and C. P. van Schaik, "Intrasexual Competition and Body Weight Dimorphism in Anthropoid Primates," *American Journal of Physical Anthropology* 103 (1997): 37–68.

14. In a letter from Thomas Jefferson to John Adams, dated October 28, 1813, Jefferson also notes this: "Formerly, bodily powers gave place among the aristoi. But since the invention of gunpowder has armed the weak as well as the strong with missile death, bodily strength, like beauty, good humor, politeness and other accomplishments, has become but an auxiliary ground of distinction." *The Adams-Jefferson Letters: The Complete Correspondence Between Thomas Jefferson and Abigail and John Adams,* ed. L. J. Cappon, vol. 2 (Chapel Hill: University of North Carolina Press, 1959), pp. 387–392.
15. C. L. Apicella, "Upper Body Strength Predicts Hunting Reputation and Reproductive Success in Hadza Hunter-Gatherers," *Evolution and Human Behavior* 35 (2014): 508–518.
16. S. Gavrilets, "Human Origins and the Transition from Promiscuity to Pair-Bonding," *PNAS: Proceedings of the National Academy of Sciences* 109 (2012): 9923–9928.
17. A. Fuentes, "Patterns and Trends in Primate Pair Bonds," *International Journal of Primatology* 23 (2002): 953–978. See also C. Opie, Q. D. Atkinson, R. I. M. Dunbar, and S. Shultz, "Male Infanticide Leads to Social Monogamy in Primates," *PNAS: Proceedings of the National Academy of Sciences* 110 (2013): 13328–13332.
18. R. O. Prum, "Aesthetic Evolution by Mate Choice: Darwin's Really Dangerous Idea," *Philosophical Transactions of the Royal Society B* 367 (2012): 2253–2265.
19. Gavrilets, "Human Origins."
20. This term is used in formal writing and is often attributed to the famous evolutionary biologist John Maynard Smith (though I have been unable to find the original reference). An early use of this expression can be seen in J. Cherfas, "The Games Animals Play," *New Scientist* 75 (1977): 672–674. This is also known as *kleptogyny* (theft of females).
21. Gavrilets, "Human Origins."
22. The maximum reliably recorded number of offspring a single woman has ever produced is sixty-nine (a Russian woman in the eighteenth century). M. M. Clay, *Quadruplets and Higher Multiple Births* (Auckland: MacKeith Press, 1989), p. 96. The practice of contraception or abortion would, of course, reduce fertility. Forager populations often have very long birth intervals and longer periods of breastfeeding.
23. T. Zerjal et al., "The Genetic Legacy of the Mongols," *American Journal of Human Genetics* 72 (2003): 717–721.
24. J. Henrich, R. Boyd, and P. J. Richerson, "The Puzzle of Monogamous Marriage," *Philosophical Transactions of the Royal Society B* 367 (2012): 657–669.
25. Interest in different strategies may vary with the ovulatory cycle. M. G. Haselton and S. W. Gangestad, "Conditional Expression of Women's Desires and Men's Mate Guarding Across the Ovulatory Cycle," *Hormones and Behavior* 49 (2006): 509–518. See also D. M. Buss, "Sex Differences in Human Mate Preferences: Evolutionary Hypotheses Testing in 37 Cultures," *Behavioral and Brain Sciences* 12 (1989): 1–49.
26. E. Turkheimer, "Three Laws of Behavior Genetics and What They Mean," *Current Directions in Psychological Science* 9 (2000): 160–164.
27. T. J. C. Polderman et al., "Meta-Analysis of the Heritability of Human Traits Based on Fifty Years of Twin Studies," *Nature Genetics* 47 (2015): 702–729.
28. J. Wu, H. Xiao, H. Sun, L. Zou, and L. Q. Zhu, "Role of Dopamine Receptors in ADHD: A Systematic Meta-Analysis," *Molecular Neurobiology* 45 (2012): 605–620; C. Chen, M.

Burton, E. Greenberger, and J. Dmitrieva, "Population Migration and the Variation of Dopamine D4 Receptor (DRD4) Allele Frequencies Around the Globe," *Evolution and Human Behavior* 20 (1999): 309–324; R. P. Ebstein et al., "Dopamine D4 Receptor (D4DR) Exon III Polymorphism Associated with the Human Personality Trait of Novelty Seeking," *Nature Genetics* 12 (1996): 78–80; J. Benjamin, L. Li, C. Patterson, B. D. Greenberg, D. L. Murphy, and D. H. Hamer, "Population and Familial Association Between the D4 Dopamine Receptor Gene and Measures of Novelty Seeking," *Nature Genetics* 12 (1996): 81–84; M. R. Munafo, B. Yalcin, S. A. Willis-Owen, and J. Flint, "Association of the Dopamine D4 Receptor (DRD4) Gene and Approach-Related Personality Traits: Meta-Analysis and New Data," *Biological Psychiatry* 63 (2008): 197–206.

29. See, for example, J. N. Rosenquist, S. F. Lehrer, A. J. O'Malley, A. M. Zaslavsky, J. W. Smoller, and N. A. Christakis, "Cohort of Birth Modifies the Association Between FTO Genotype and BMI," *PNAS: Proceedings of the National Academy of Sciences* 112 (2015): 354–359.
30. Typically, variation between animals or species reflects changes in the receptors for the hormone or transmitter rather than changes in the level or structure of the hormone. The variation can include the strength with which the receptors bind the hormone, the number of such receptors, the ways the receptors transmit information related to such binding, or the location of the receptors at the cellular or neuroanatomical level.
31. L. J. Young and Z. Wang, "The Neurobiology of Pair-Bonding," *Nature Neuroscience* 7 (2004): 1048–1054.
32. E. A. Hammock and L. J. Young, "Variation in the Vasopressin V1a Receptor Promoter and Expression: Implications for Inter- and Intraspecific Variation in Social Behaviour," *European Journal of Neuroscience* 16 (2002): 399–402.
33. M. Nagasawa et al., "Oxytocin-Gaze Positive Loop and the Coevolution of Human-Dog Bonds," *Science* 348 (2015): 333–336.
34. P. T. Ellison and P. B. Gray, eds., *Endocrinology of Social Relationships* (Cambridge, MA: Harvard University Press, 2009); Z. R. Donaldson and L. J. Young, "Oxytocin, Vasopressin, and the Neurogenetics of Sociality," *Science* 322 (2008): 900–904.
35. For instance, in a simplified taxonomy, there is the explorer (which is driven primarily by the dopamine system), the builder (serotonin), the director (testosterone), and the negotiator (estrogen). H. Fisher, *Why Him? Why Her?* (New York: Henry Holt, 2009).
36. M. M. Lim et al., "Enhanced Partner Preference in a Promiscuous Species by Manipulating the Expression of a Single Gene," *Nature* 429 (2004): 754–757.
37. The situation is actually more complicated than this. Subsequent work by another group showed that there were other, nonmonogamous vole species that also had similar vasopressin receptors. It seems that the precise sequence of the DNA controlling the receptor expression may be important. This is an ongoing area of research. See S. Fink, L. Excoffier, and G. Heckel, "Mammalian Monogamy Is Not Controlled by a Single Gene," *PNAS: Proceedings of the National Academy of Sciences* 103 (2006): 10956–10960; and McGraw and Young, "Prairie Vole."

38. A. Bendesky et al., "The Genetic Basis of Parental Care Evolution in Monogamous Mice," *Nature* 544 (2017): 434–439.
39. H. Walum et al., "Genetic Variation in the Vasopressin Receptor 1a Gene (*AVPR1A*) Associates with Pair-Bonding Behavior in Humans," *PNAS: Proceedings of the National Academy of Sciences* 105 (2008): 14153–14156.
40. Z. M. Prichard, A. J. Mackinnon, A. F. Jorm, and S. Easteal, "*AVPR1A* and *OXTR* Polymorphisms Are Associated with Sexual and Reproductive Behavioral Phenotypes in Humans," *Human Mutation* 28 (2007): 1150; T. H. Wassink et al., "Examination of AVPR1a as an Autism Susceptibility Gene," *Molecular Psychiatry* 9 (2004): 968–972; N. Yirmiya et al., "Association Between the Arginine Vasopressin 1a Receptor (AVPR1a) Gene and Autism in a Family-Based Study: Mediation by Socialization Skills," *Molecular Psychiatry* 11 (2006): 488–494; A. Knafo et al., "Individual Differences in Allocation of Funds in the Dictator Game Associated with Length of the Arginine Vasopressin 1a Receptor RS3 Promoter Region and Correlation Between RS3 Length and Hippocampal mRNA," *Genes, Brain and Behavior* 7 (2007): 266–275. Other experiments have shown that partner preference is not just regulated genetically; it is also regulated *epigenetically*. This refers to a set of biological processes that affect how genes are expressed through processes outside the genetic sequence itself—like a set of biological on/off switches. H. Wang, F. Duclot, Y. Liu, Z. Wang, and M. Kabbaj, "Histone Deacetylase Inhibitors Facilitate Partner Preference Formation in Female Prairie Voles," *Nature Neuroscience* 16 (2013): 919–924. Many other genes coding for many other structural and physiological aspects of our bodies surely play similar roles in our mating and social behavior. G. E. Robinson, R. D. Fernald, and D. F. Clayton, "Genes and Behavior," *Science* 322 (2008): 896–900.
41. D. Pissonnier, J. C. Thiery, C. Fabre-Nys, P. Poindron, and E. B. Keverne, "The Importance of Olfactory Bulb Noradrenalin for Maternal Recognition in Sheep," *Physiology and Behavior* 35 (1985): 361–363.
42. A. Bartels and S. Zeki, "The Neural Correlates of Maternal and Romantic Love," *NeuroImage* 21 (2004): 1155–1166.
43. G. B. Wislocki, "Size, Weight, and Histology of the Testes in the Gorilla," *Journal of Mammalogy* 23 (1942): 281–287. There are several theories about penis size in humans. One is that the penis functions as a kind of display, like a lion's mane. Another theory posits that females among our hominid ancestors might have mated with many males in sequence and that males with longer penises were able to deposit sperm in the vagina closer to its ultimate destination. Still another theory is that large penises facilitate female orgasms and that this resulted in ancestral females preferentially mating with males with bigger penises. L. J. Young and B. Alexander, *The Chemistry Between Us: Love, Sex, and the Science of Attraction* (New York: Penguin, 2012).
44. For example, one study found that neither oral sex nor masturbation gave women the same feeling of overall satisfaction with their relationship, or the feeling of being close to their partner, that vaginal sex created. S. Brody and R. M. Costa, "Satisfaction (Sexual, Life, Relationship, and Mental Health) Is Associated Directly with Penile-Vaginal Intercourse, but Inversely with Other Sexual Behavior Frequencies,"

*Journal of Sexual Medicine* 6 (2009): 1947–1954; S. Brody, "The Relative Health Benefits of Different Sexual Activities," *Journal of Sexual Medicine* 7 (2010): 1336–1361.

45. Experiments with prairie voles, for example, demonstrate that virgin males are generally indifferent to one another, but once they have mated, they will fight off other males who come near their mates, and this effect (of having a partner preference and of fighting off other males) could be blocked by the administration of antagonists to certain hormones (e.g., to vasopressin). J. T. Winslow, N. Hastings, C. S. Carter, C. R. Harbaugh, and T. R. Insel, "A Role for Central Vasopressin in Pair-Bonding in Monogamous Prairie Voles," *Nature* 365 (1993): 545–548.
46. Spouses can have an "opposites attract" dynamic too. Examples include big spenders marrying big savers (see, e.g., S. I. Rick, D. A. Small, and E. J. Finkel, "Fatal [Fiscal] Attraction: Spendthrifts and Tightwads in Marriage," *Journal of Marketing Research* 48 [2011]: 228–237); followers marrying leaders (see, e.g., C. D. Dryer and L. M. Horowitz, "When Do Opposites Attract? Interpersonal Complementarity Versus Similarity," *Journal of Personality and Social Psychology* 72 [1997]: 592–603); dissimilar HLA types choosing each other (see, e.g., R. Chaix, C. Chao, and P. Donnelly, "Is Mate Choice in Humans MHC-Dependent?," *PLOS Genetics* 4 [2008]: e1000184); or even couples who are well matched for sadomasochistic sexual interests (see, e.g., B. L. Stiles and R. E. Clark, "BDSM: A Subcultural Analysis of Sacrifices and Delights," *Deviant Behavior* 32 [2011]: 158–189).
47. R. A. Fisher, "The Correlation Between Relatives on the Supposition of Mendelian Inheritance," *Transactions of the Royal Society of Edinburgh* 52 (1918): 399–433; S. Wright, "Systems of Mating. III: Assortative Mating Based on Somatic Resemblance," *Genetics* 6 (1920): 144–161.
48. B. D. Neff and T. E. Pitcher, "Genetic Quality and Sexual Selection: An Integrated Framework for Good Genes and Compatible Genes," *Molecular Ecology* 14 (2005): 19–38; H. L. Mays Jr. and G. E. Hill, "Choosing Mates: Good Genes Versus Genes That Are a Good Fit," *Trends in Ecology and Evolution* 19 (2004): 554–559; M. Andersson and L. W. Simmons, "Sexual Selection and Mate Choice," *Trends in Ecology and Evolution* 21 (2006): 296–302; A. G. Jones and N. L. Ratterman, "Mate Choice and Sexual Selection: What Have We Learned Since Darwin?," *PNAS: Proceedings of the National Academy of Sciences* 106 (2009): 10001–10008.
49. F. de Waal and S. Gavrilets, "Monogamy with a Purpose," *PNAS: Proceedings of the National Academy of Sciences* 110 (2013): 15167–15168; Lukas and Clutton-Brock, "Evolution of Social Monogamy"; G. Stulp, A. P. Buunk, R. Kurzban, and S. Verhulst, "The Height of Choosiness: Mutual Mate Choice for Stature Results in Suboptimal Pair Formation for Both Sexes," *Animal Behaviour* 86 (2013): 37–46; S. A. Baldauf, H. Kullmann, S. H. Schroth, T. Thunken, and T. C. Bakker, "You Can't Always Get What You Want: Size Assortative Mating by Mutual Mate Choice as a Resolution of Sexual Conflict," *BMC Evolutionary Biology* 9 (2009): 129.
50. R. J. H. Russell, P. A. Wells, and J. P. Rushton, "Evidence for Genetic Similarity Detection in Human Marriage," *Ethology and Sociobiology* 6 (1985): 183–187.
51. T. Antal, H. Ohtsuki, J. Wakeley, P. D. Taylor, and M. A. Nowak, "Evolution of Cooperation by Phenotypic Similarity," *PNAS: Proceedings of the National Academy of Sciences*

106 (2009): 8597–8600; M. A. Nowak, "Five Rules for the Evolution of Cooperation," *Science* 314 (2006): 1560–1563; F. Fu, M. A. Nowak, N. A. Christakis, and J. H. Fowler, "The Evolution of Homophily," *Scientific Reports* 2 (2012): 845.

52. In this case, genotypic assortative mating might turn an otherwise neutral mutant genotype into a kind of "good gene." However, finding a mate with a minor allele might be costly when the mutant allele is not common and especially when it is not initially advantageous.
53. Jones and Ratterman, "Mate Choice"; Y. Jiang, D. I. Bolnick, and M. Kirkpatrick, "Assortative Mating in Animals," *American Naturalist* 181 (2013): E125–E138; Russell, Wells, and Rushton, "Evidence for Genetic Similarity."
54. R. Laurent and R. Chaix, "MHC-Dependent Mate Choice in Humans: Why Genomic Patterns from the HapMap European American Dataset Support the Hypothesis," *Bioessays* 34 (2012): 267–271. There may also be disassortativity for similarity to one's own face. L. M. DeBruine et al., "Opposite-Sex Siblings Decrease Attraction, but Not Prosocial Attributions, to Self-Resembling Opposite-Sex Faces," *PNAS: Proceedings of the National Academy of Sciences* 108 (2011): 11710–11714.
55. J. Havlicek and S. C. Roberts, "MHC-Correlated Mate Choice in Humans: A Review," *Psychoneuroendocrinology* 34 (2009): 497–512. Interestingly, some of the studies of odor preference and MHC have shown that women using oral contraceptives do not evince the same effects. To the extent that contraceptive use interferes with "natural" partner choices, it is possible that this might result in a woman subsequently finding her partner less attractive later in any marriage, and hence face a higher risk of divorce. In principle, it should be possible to test this idea epidemiologically.
56. C. Wedekind, T. Seebeck, F. Bettens, and A. J. Paepke, "MHC-Dependent Mate Preferences in Humans," *Proceedings of the Royal Society B* 260 (1995): 245–249. See also C. Wedekind and S. Füri, "Body Odour Preferences in Men and Women: Do They Aim for Specific MHC Combinations or Simply Heterozygosity?," *Proceedings of the Royal Society B* 264 (1997): 1471–1479.
57. While it is clear that MHC genes play a role in body odor, it is still unclear how, exactly, they do so. M. Milinski, I. Croy, T. Hummel, and T. Boehm, "Major Histocompatibility Complex Peptide Ligands as Olfactory Cues in Human Body Odour Assessment," *Proceedings of the Royal Society B* 280 (2013): 20122889.
58. Another way that HLA might affect partner choice is through face preference. Humans clearly place a lot of emphasis on the faces of prospective mates, and there is some suggestive evidence that HLA heterozygosity is associated with faces that are judged as more attractive, at least in men. But there is a further hurdle for a man to overcome to be seen as desirable by a female partner. Women rated the faces of men whose HLA genome was similar to theirs as more attractive than HLA-dissimilar men. So, ideally, a man should carry two copies of HLA genes that differ, but these copies should *resemble* the two copies of his prospective mate if she is to find him optimal. At least, that may be the case with visual cues. Note that this imperative to resemble a mate is the *opposite* of the response to olfactory cues. The possibly opposing forces at play with respect to olfaction and vision may indicate that natural selection equipped us with the capacity to choose mates with an *optimal* level of dissimilarity in

HLA, rather than a maximal level. S. C. Roberts et al., "MHC-Assortative Facial Preferences in Humans," *Biology Letters* 1 (2005): 400–403.

59. To my knowledge, no one has studied HLA dissimilarity in homosexual couples. If results similar to heterosexual unions were found, it might mean that HLA preferences are generic and not tightly coupled to reproduction per se. But if these results were not found, it might mean that these phenomena are still linked to the biology of reproduction. Either result would be interesting.
60. C. Ober, L. R. Weitkamp, N. Cox, H. Dytch, D. Kostyu, and S. Elias, "HLA and Mate Choice in Humans," *American Journal of Human Genetics* 61 (1997): 497–504; P. W. Hedrick and F. L. Black, "HLA and Mate Selection: No Evidence in South Amerindians," *American Journal of Human Genetics* 61 (1997): 505–511.
61. T. Bereczkei, P. Gyuris, and G. E. Weiseld, "Sexual Imprinting in Human Mate Choice," *Proceedings of the Royal Society B* 271 (2004): 1129–1134.
62. T. J. C. Polderman et al., "Meta-Analysis of the Heritability of Human Traits Based on Fifty Years of Twin Studies," *Nature Genetics* 47 (2015): 702–709.
63. R. S. Herz and M. Inzlicht, "Sex Differences in Response to Physical and Social Factors Involved in Human Mate Selection: The Importance of Smell for Women," *Evolution and Human Behavior* 23 (2002): 359–364.
64. R. McDermott, D. Tingley, and P. K. Hatemi, "Assortative Mating on Ideology Could Operate Through Olfactory Cues," *American Journal of Political Science* 58 (2014): 997–1005.
65. A. Nishi, J. H. Fowler, and N. A. Christakis, "Assortative Mating at Loci Under Recent Natural Selection in Humans" (unpublished manuscript, 2012). Several small-scale studies have explored the extent to which humans preferentially mate with people they resemble genetically and what this might mean for evolution. See R. Sebro, T. J. Hoffman, C. Lange, J. J. Rogus, and N. J. Risch, "Testing for Non-Random Mating: Evidence for Ancestry-Related Assortative Mating in the Framingham Heart Study," *Genetic Epidemiology* 34 (2010): 674–679; and R. Laurent, B. Toupance, and R. Chaix, "Non-Random Mate Choice in Humans: Insights from a Genome Scan," *Molecular Ecology* 21 (2012): 587–596.
66. In our analysis, we did not find substantial evidence for disassortative mating in the HLA regions. Nishi, Fowler, and Christakis, "Assortative Mating." This was in keeping with some prior studies. See, for example, Chaix, Chao, and Donnelly, "Mate Choice"; and A. Derti, C. Cenik, P. Kraft, and F. P. Roth, "Absence of Evidence for MHC-Dependent Mate Selection Within HapMap Populations," *PLOS Genetics* 6 (2010): e1000925.
67. The former relates to the disadvantages of inbreeding. The latter relates to the fact that sets of genes in our genome often are coadapted and work together; reproducing with people who are *too* dissimilar from us may, therefore, break down the synergistic way such genes interact, again resulting in fewer surviving offspring. A. Helgason, S. Palsson, D. F. Gudbjartsson, T. Kristjansson, and K. Stefansson, "An Association Between the Kinship and Fertility of Human Couples," *Science* 319 (2008): 813–816.
68. K. R. Hill et al., "Co-Residence Patterns in Hunter-Gatherer Societies Show Unique Human Social Structure," *Science* 331 (2011): 1286–1289; C. L. Apicella, F. W. Marlowe,

J. H. Fowler, and N. A. Christakis, "Social Networks and Cooperation in Hunter-Gatherers," *Nature* 481 (2012): 497–501; M. Sikora et al., "Ancient Genomes Show Social and Reproductive Behavior of Early Upper Paleolithic Foragers," *Science* 358 (2017): 659–662.

69. M. Dyble et al., "Sex Equality Can Explain the Unique Structure of Hunter-Gatherer Bands," *Science* 348 (2015): 796–798.
70. Chapais, *Primeval Kinship*, p. 179.
71. As we saw, pair-bonding also directed male energy away from competition for mates and toward competition to be a better provider. And since pair-bonding made it possible to recognize male kin, it may have dramatically expanded the efficiency of mechanisms individuals used to favor their kin, thus also facilitating the emergence of within-group coalitions and alliances. See Gavrilets, "Human Origins"; Chapais, *Primeval Kinship;* M. Mesterton-Gibbons, S. Gavrilets, J. Gravner, and E. Akcay, "Models of Coalition or Alliance Formation," *Journal of Theoretical Biology* 274 (2011): 187–204. Once family groups and sharing were established, this set the stage for a new type of cultural selection to operate, and for the further emergence of cooperation with non-kin. P. Richerson et al., "Cultural Group Selection Plays an Essential Role in Explaining Human Cooperation: A Sketch of the Evidence," *Behavioral and Brain Sciences* 39 (2014): e30.

### *Chapter 7: Animal Friends*

1. J. van Lawick–Goodall, *In the Shadow of Man* (Boston: Houghton Mifflin, 1971), p. 268.
2. The way we reach out to other animals says a lot about our own capacity to befriend one another, in my view. The abuse of animals also reflects vile parts of our individual or collective nature. For instance, people who abuse animals are very likely to abuse people. See, for example, R. Lockwood and G. R. Hodge, "The Tangled Web of Animal Abuse: The Links Between Cruelty to Animals and Human Violence," in R. Lockwood and F. Ascione, eds., *Cruelty to Animals and Interpersonal Violence* (West Lafayette, IN: Purdue University Press, 1998), pp. 77–82.
3. J. O'Neill, *Prodigal Genius: The Life of Nikola Tesla* (New York: Cosimo, 2006), p. 312.
4. M. Seifer, *Wizard: The Life and Times of Nikola Tesla; Biography of a Genius* (Secaucus, NJ: Birch Lane Press, 1996), p. 414.
5. O'Neill, *Prodigal Genius*, p. 316.
6. C. M. Parkes, B. Benjamin, and R. G. Fitzgerald, "Broken Heart: A Statistical Study of Increased Mortality Among Widowers," *British Medical Journal* 1, no. 5646 (1969): 740–743; F. Elwert and N. A. Christakis, "The Effect of Widowhood on Mortality by the Causes of Death of Both Spouses," *American Journal of Public Health* 98 (2008): 2092–2098; F. Elwert and N. A. Christakis, "Widowhood and Race," *American Sociological Review* 71 (2006): 16–41. There is some suggestion in Tesla's accounts that he felt a *romantic* attachment to some of his pigeons.
7. "Pet Industry Market Size and Ownership Statistics," American Pet Products Association, http://www.americanpetproducts.org/press_industrytrends.asp.
8. K. Allen, J. Blascovich, and W. B. Mendes, "Cardiovascular Reactivity and the Presence of Pets, Friends, and Spouses: The Truth About Cats and Dogs," *Psychosomatic Medicine* 64 (2002): 727–739.

9. K. V. A. Johnson and R. I. M. Dunbar, "Pain Tolerance Predicts Human Social Network Size," *Scientific Reports* 6 (2016): 25267. Notably, opioid receptors play a role in both pain relief and social bonding.
10. T. N. Davis et al., "Animal Assisted Interventions for Children with Autism Spectrum Disorder: A Systematic Review," *Education and Training in Autism and Developmental Disabilities* 50 (2015): 316–329; R. A. Johnson et al., "Effects of Therapeutic Horseback Riding on Post-Traumatic Stress Disorder in Military Veterans," *Military Medical Research* 5 (2018): 3.
11. C. Siebert, "What Does a Parrot Know About PTSD?," *New York Times Magazine,* January 28, 2016.
12. E. W. Budge, trans., *The History of Alexander the Great, Being the Syriac Version of the Pseudo-Callisthenes,* vol. 1 (Cambridge, UK: University Press, 1889), pp. 17–18.
13. J. H. Crider, "Fala, Never in the Doghouse," *New York Times,* October 15, 1944. Roosevelt defended Fala, saying, "These Republican leaders have not been content with attacks on me, or my wife, or on my sons. No, not content with that, they now include my little dog, Fala. Well, of course, I don't resent attacks, and my family doesn't resent attacks, but Fala does resent them." For a recording and transcript of Roosevelt's speech, see "Campaign Dinner Address of Franklin Delano Roosevelt (the Fala Speech)" (Washington, DC, September 23, 1944), Wyzant, http://www.wyzant.com/resources/lessons/history/hpol/fdr/fala.
14. "Geese Fly with Man Who Reared Them," BBC News, December 29, 2011, http://www.bbc.com/news/av/science-environment-16301233/geese-fly-with-man-who-reared-them. Bill Lishman was the first person to do something similar. B. Lishman, *C'mon Geese* (Cooper-Lishman Productions, 1989), video recording.
15. J. van Lawick–Goodall, *My Friends the Wild Chimpanzees* (Washington, DC: National Geographic Society, 1967), p. 18.
16. When Goodall relayed this account of primate behavior to Louis Leakey, he replied with the famous statement "We must now redefine man, redefine tool, or accept chimpanzees as humans."
17. Jane Goodall, interview by Bill Moyers, *Bill Moyers Journal,* PBS, November 27, 2009, http://www.pbs.org/moyers/journal/11272009/transcript3.html.
18. Van Lawick–Goodall, *In the Shadow of Man,* p. 76.
19. Goodall, *My Friends,* p. 191.
20. J. Goodall, "Fifi Fights Back," *National Geographic,* April 2003.
21. J. Goodall, *Through a Window: My Thirty Years with the Chimpanzees of Gombe* (Boston: Houghton Mifflin, 1990).
22. Still, as intimately as Goodall came to know the chimpanzees, there were limits. As she explained in an interview in 2014, "It took over a year before I gained the trust of the chimpanzees, but I never was part of their community." J. Shorthouse and A. Gaffney, "Jane Goodall: 80 and Touring Australia," *ABC Sunshine Coast* (Australia), June 4, 2014, http://www.abc.net.au/local/photos/2014/06/03/4017793.htm. Two other young ethologists, Dian Fossey and Birutė Galdikas, who studied gorillas and orangutans, respectively, followed a similar approach, entering primate communities on their terms. Fossey described the way in which her respect for the norms of hierarchy and

submission within gorilla communities put the gorillas at ease. See S. Montgomery, *Walking with the Great Apes* (Boston: Houghton Mifflin, 1991), and D. Fossey, *Gorillas in the Mist* (Boston: Houghton Mifflin, 1983).

23. H. Whitehead, *Analyzing Animal Societies* (Chicago: University of Chicago Press, 2009); K. Faust and J. Skvoretz, "Comparing Networks Across Space, Time, and Species," *Sociological Methodology* 32 (2002): 267–299.
24. R. M. Seyfarth and D. L. Cheney, "The Evolutionary Origins of Friendship," *Annual Review of Psychology* 63 (2012): 153–177; M. Krützen et al., "Contrasting Relatedness Patterns in Bottlenose Dolphins (*Tursiops* sp.) with Different Alliance Strategies," *Proceedings of the Royal Society B* 270 (2003): 497–502; J. C. Mitani, "Cooperation and Competition in Chimpanzees: Current Understanding and Future Challenges," *Evolutionary Anthropology* 18 (2009): 215–227.
25. Mitani, "Cooperation and Competition"; J. B. Silk et al., "Strong and Consistent Social Bonds Enhance the Longevity of Female Baboons," *Current Biology* 20 (2010): 1359–1361.
26. R. C. Connor, "Dolphin Social Intelligence: Complex Alliance Relationships in Bottlenose Dolphins and a Consideration of Selective Environments for Extreme Brain Size Evolution in Mammals," *Philosophical Transactions of the Royal Society B* 362 (2007): 587–602.
27. See, for example, J. E. Tanner, F. G. P. Patterson, G. Francine, and R. W. Byrne, "The Development of Spontaneous Gestures in Zoo-Living Gorillas and Sign-Taught Gorillas: From Action and Location to Object Representation," *Journal of Developmental Processes* 1 (2006): 69–102; J. D. Bonvillian and F. G. P. Patterson, "Early Sign-Language Acquisition: Comparisons Between Children and Gorillas," in S. T. Parker, R. W. Mitchell, and H. L. Miles, eds., *The Mentalities of Gorillas and Orangutans* (New York: Cambridge University Press, 1999), pp. 240–264; and H. S. Terrance, *Nim: A Chimpanzee Who Learned Sign Language* (New York: Columbia University Press, 1987).
28. C. Kasper and B. Voelkl, "A Social Network Analysis of Primate Groups," *Primates* 50 (2009): 343–356.
29. J. C. Mitani, "Male Chimpanzees Form Enduring and Equitable Social Bonds," *Animal Behaviour* 77 (2009): 633–640.
30. Similar findings have been observed in other communities of chimpanzees. I. C. Gilby and R. W. Wrangham, "Association Patterns Among Wild Chimpanzees (*Pan troglodytes schweinfurthii*) Reflect Sex Differences in Cooperation," *Behavioral Ecology and Sociobiology* 62 (2008): 1831–1842.
31. J. Lehmann and C. Boesch, "Sociality of the Dispersing Sex: The Nature of Social Bonds in West African Female Chimpanzees, *Pan troglodytes*," *Animal Behaviour* 77 (2009): 377–387.
32. A. R. Parish, "Female Relationships in Bonobos (*Pan paniscus*)," *Human Nature* 7 (1996): 61–96; D. L. Cheney, "The Acquisition of Rank and the Development of Reciprocal Alliances Among Free-Ranging Baboons," *Behavioral Ecology and Sociobiology* 2 (1977): 303–318. Once again, genetic relatedness is the best predictor of a tie, as measured by proximity and grooming behaviors. Interestingly, baboons appear to have a sort of

hereditary social rank into which they are born (similar to human caste systems), and female social ties are more likely between individuals of the same age and rank.

33. Kasper and Voelkl, "Social Network Analysis of Primate Groups."
34. J. C. Flack, M. Girvan, F. B. M. de Waal, and D. C. Krakauer, "Policing Stabilizes Construction of Social Niches in Primates," *Nature* 439 (2006): 426–429.
35. Something similar happens among wolves. The death of an alpha male causes chaos in the wolf pack and leads younger male wolves to start breeding, increasing the overall number of wolves too. R. B. Wielgus and K. A. Peebles, "Effects of Wolf Mortality on Livestock Depredations," *PLOS ONE* 9 (2014): e113505.
36. Since leaders are much less numerous than others in a group, if an epidemic begins with a random person in the group, it will more often start in, and therefore be confined to, the periphery of the population. Of course, if it happens to begin in a leader, this is worse. But this occurs less often.
37. J. Poole, *Coming of Age with Elephants* (New York: Hyperion, 1996), p. 275.
38. C. J. Moss, H. Croze, and P. C. Lee, *The Amboseli Elephants: A Long-Term Perspective on a Long-Lived Mammal* (Chicago: University of Chicago Press, 2011).
39. S. de Silva, A. D. G. Ranjeewa, and D. Weerakoon, "Demography of Asian Elephants (*Elephas maximus*) at Uda Walawe National Park, Sri Lanka Based on Identified Individuals," *Biological Conservation* 144 (2011): 1742–1752.
40. Poole, *Coming of Age*, pp. 147–148.
41. Ibid., p. 162.
42. Ibid.
43. For some examples, see M. Scully, *Dominion: The Power of Man, the Suffering of Animals, and the Call to Mercy* (New York: St. Martin's, 2002), p. 206. Something similar, involving cross-species helping behavior, may occur with humpback whales that interfere with orca predation of seals, even rescuing seals by lifting them out of the water. See E. Kelsey, "The Power of Compassion: Why Humpback Whales Rescue Seals and Why Volunteering for Beach Cleanups Improves Your Health," *Hakai* (August 17, 2017).
44. F. Bibi, B. Kraatz, N. Craig, M. Beech, M. Schuster, and A. Hill, "Early Evidence for Complex Social Structure in Proboscidea from a Late Miocene Trackway Site in the United Arab Emirates," *Biology Letters* 8 (2012): 670–673.
45. P. Pecnerova et al., "Genome-Based Sexing Provides Clues About Behavior and Social Structure in the Woolly Mammoth," *Current Biology* 27 (2017): 3505–3510. A total of 69 percent of the ninety-eight mammoth specimens was male.
46. Most members of elephant core groups share the same mitochondrial DNA haplotype (meaning that they have a common ancestor in the line of females that preceded them), and ordinarily only about 1 percent of the elephants in a core group join the group as outsiders. E. A. Archie, C. J. Moss, and S. C. Alberts, "The Ties That Bind: Genetic Relatedness Predicts the Fission and Fusion of Social Groups in Wild African Elephants," *Proceedings of the Royal Society B* 273 (2006): 513–522. This applies to the relatively unperturbed population of Amboseli. Of course, males breed with females from many core groups, and the resulting gene flow substantially reduces the genetic differentiation among core groups overall.

47. Poole, *Coming of Age*, pp. 274–275.
48. Archie, Moss, and Alberts, "Ties That Bind"; G. Wittemyer et al., "Where Sociality and Relatedness Diverge: The Genetic Basis for Hierarchical Social Organization in African Elephants," *Proceedings of the Royal Society B* 276 (2009): 3513–3521.
49. K. R. Hill et al., "Co-Residence Patterns in Hunter-Gatherer Societies Show Unique Human Social Structure," *Science* 331 (2011): 1286–1289. See also C. L. Apicella, F. W. Marlowe, J. H. Fowler, and N. A. Christakis, "Social Networks and Cooperation in Hunter-Gatherers," *Nature* 481 (2012): 497–501.
50. P. Fernando and R. Lande, "Molecular Genetic and Behavioral Analysis of Social Organization in the Asian Elephant (*Elephas maximus*)," *Behavioral Ecology and Sociobiology* 48 (2000): 84–91; Archie, Moss, and Alberts, "Ties That Bind"; Wittemyer et al., "Where Sociality and Relatedness Diverge."
51. Wittemyer et al., "Where Sociality and Relatedness Diverge." An apparently similar situation of elephants forming core groups with unrelated individuals has been observed in another high-predation environment in Tanzania. K. Gobush, B. Kerr, and S. Wasser, "Genetic Relatedness and Disrupted Social Structure in a Poached Population of African Elephants," *Molecular Ecology* 18 (2009): 722–734. To be clear, non-kin friendships may exist in elephants (as in other animals), because simply assembling with others provides benefits, such as predator vigilance (especially against lions, who attack newborn elephants) and resource defense. These advantages of social structure are enough to create and maintain it, and evolution can select for it independent of any further benefits that arise from helping one's genetic relatives.
52. C. J. Moss and J. H. Poole, "Relationships and Social Structure of African Elephants," in R. A. Hinde, ed., *Primate Social Relationships: An Integrated Approach* (Oxford: Blackwell, 1983), pp. 315–325. See also C. Moss, *Elephant Memories: Thirteen Years in the Life of an Elephant Family* (New York: William Morrow, 1988).
53. Archie, Moss, and Alberts, "Ties That Bind."
54. S. de Silva and G. Wittemyer, "A Comparison of Social Organization in Asian Elephants and African Savannah Elephants," *International Journal of Primatology* 33 (2012): 1125–1141; G. Wittemyer, I. Douglas-Hamilton, and W. M. Getz, "The Socioecology of Elephants: Analysis of the Processes Creating Multi-Tiered Social Structures," *Animal Behaviour* 69 (2005): 1357–1371. There are also two higher levels, the subpopulation and population levels.
55. Wittemyer, Douglas-Hamilton, and Getz, "Socioecology of Elephants."
56. Ibid.
57. De Silva and Wittemyer, "Comparison of Social Organization." Asian elephants occupy habitats with consistent rainfall, whereas African elephants have to move around more and potentially compete with one another more in order to forage. They also face different predation risks. Elephants in Sri Lanka have no predators (other than humans), whereas African elephant calves are sometimes taken by lions. Gregariousness in African elephants may serve as a defense against predators in three regards: first, animals foraging in open environments may seek one another's company as a form of cover; second, clustering together may dilute the risk faced by

individuals; and third, animals in groups may cooperate in active defense or monitoring of predation risks. Indeed, large-scale phylogenetic analyses show that herbivores in general (not just elephants) that live in more open environments tend to be more social. T. Caro, C. Graham, C. Stoner, and J. Vargas, "Adaptive Significance of Anti-Predator Behaviour in Artiodactyls," *Animal Behaviour* 67 (2004): 205–228.

58. L. Weilgart, H. Whitehead, and K. Payne, "A Colossal Convergence," *American Scientist* 84 (1996): 278–287.
59. L. J. N. Brent, D. W. Franks, E. A. Foster, K. C. Balcomb, M. A. Cant, and D. P. Croft, "Ecological Knowledge, Leadership, and the Evolution of Menopause in Killer Whales," *Current Biology* 25 (2015): 746–750.
60. D. Lusseau, "The Emergent Properties of a Dolphin Social Network," *Proceedings of the Royal Society B* 270 (2003): S186–S188. The transitivity was likely somewhat artificially inflated here for technical reasons having to do with the fact that the mapping of the networks originated in the bipartite graph of 1,292 groups of individual dolphins seen together.
61. D. Lusseau and M. E. J. Newman, "Identifying the Role That Animals Play in Their Social Networks," *Proceedings of the Royal Society B* 271 (2004): S477–S481.
62. J. Wiszniewski, D. Lusseau, and L. M. Moller, "Female Bisexual Kinship Ties Maintain Social Cohesion in a Dolphin Network," *Animal Behaviour* 80 (2010): 895–904.
63. Lusseau, "Emergent Properties." These dolphins appear *not* to manifest degree assortativity.
64. R. Williams and D. Lusseau, "A Killer Whale Social Network Is Vulnerable to Targeted Removals," *Biology Letters* 2 (2006): 497–500; E. A. Foster et al., "Social Network Correlates of Food Availability in an Endangered Population of Killer Whales, *Orcinus orca*," *Animal Behaviour* 83 (2012): 731–736; O. A. Filatova et al., "The Function of Multi-Pod Aggregations of Fish-Eating Killer Whales (*Orcinus orca*) in Kamchatka, Far East Russia," *Journal of Ethology* 27 (2009): 333–341.
65. For an example of a paper taking a dim view of the possibility of primate friendships, see S. P. Henzi and L. Barrett, "Coexistence in Female-Bonded Primate Groups," *Advances in the Study of Behavior* 37 (2007): 43–81.
66. Attributed to David Premack in Seyfarth and Cheney, "Evolutionary Origins."
67. Ibid.
68. Scully, *Dominion,* p. 194.
69. W. C. McGrew and L. Baehren, "'Parting Is Such Sweet Sorrow,' but Only for Humans?," *Human Ethology Bulletin* 31 (2016): 5–14.
70. S. R. de Kort and N. J. Emory, "Corvid Caching: The Role of Cognition," in T. Zentall and E. A. Wasserman, eds., *The Oxford Handbook of Comparative Cognition* (Oxford: Oxford University Press, 2012), pp. 390–408; L. P. Acredolo, "Coordinating Perspectives on Infant Spatial Orientation," in R. Cohen, ed., *The Development of Spatial Cognition* (Hillsdale, NJ: Lawrence Erlbaum, 1985), pp. 115–140; P. Bloom, *Just Babies: The Origins of Good and Evil* (New York: Broadway Books, 2013).
71. S. Perry, C. Barrett, and J. Manson, "White-Faced Capuchin Monkeys Show Triadic Awareness in Their Choice of Allies," *Animal Behaviour* 67 (2004): 165–170. For more examples, see R. W. Wrangham, "Social Relationships in Comparative Perspective,"

in R. A. Hinde, ed., *Primate Social Relationships: An Integrated Approach* (Oxford: Blackwell, 1983), pp. 325–334.

72. Seyfarth and Cheney, "Evolutionary Origins," p. 168.
73. J. B. Silk, "Using the 'F'-Word in Primatology," *Behaviour* 139 (2002): 421–446.
74. A. S. Griffin and S. A. West, "Kin Discrimination and the Benefit of Helping in Cooperatively Breeding Vertebrates," *Science* 302 (2003): 634–636.
75. P. G. Hepper, ed., *Kin Recognition* (Cambridge: Cambridge University Press, 1991).
76. W. D. Hamilton, "The Genetical Evolution of Social Behaviour, Pt. 1," *Journal of Theoretical Biology* 7 (1964): 1–16.
77. To be clear, just because kinship facilitates the emergence of altruism and cooperation does not mean that it is always required for those nice behaviors to appear.
78. Hamilton, "Genetical Evolution, Pt. 1," p. 16.
79. S. A. Frank, "Natural Selection. VII: History and Interpretation of Kin Selection Theory," *Journal of Evolutionary Biology* 26 (2013): 1151–1184; S. A. West, I. Pen, and A. S. Griffin, "Cooperation and Competition Between Relatives," *Science* 296 (2002): 72–75.
80. K. Belson, "Elders Offer Help at Japan's Crippled Reactor," *New York Times,* June 27, 2011. This idea can be accommodated in Hamilton's equation by adding terms for the reproductive value, *v,* for the recipient ($v_r$) and the giver ($v_g$), as in $rBv_r - Cv_g > 0$. What this accomplishes is to modify the relative benefits and costs that accrue according to the reproductive value of who is giving and who is receiving.
81. C. J. Barnard and P. Aldhous, "Kinship, Kin Discrimination, and Mate Choice," in P. G. Hepper, ed., *Kin Recognition* (Cambridge: Cambridge University Press, 1991), pp. 125–147.
82. W. G. Holmes and P. W. Sherman, "Kin Recognition in Animals," *American Scientist* 71 (1983): 46–55.
83. F. W. Peek, E. Franks, D. Case, "Recognition of Nest, Eggs, Nest Site, and Young in Female Red-Winged Blackbirds," *Wilson Bulletin* 84 (1972): 243–249.
84. T. Aubin, P. Jouventin, and C. Hildebrand, "Penguins Use the Two-Voice System to Recognize Each Other," *Proceedings of the Royal Society B* 267 (2000): 1081–1087.
85. J. Mehler, J. Bertoncini, and M. Barriere, "Infant Recognition of Mother's Voice," *Perception* 7 (1978): 491–497. Even fetuses can recognize their mothers' voices. B. S. Kisilevsky et al., "Effects of Experience on Fetal Voice Recognition," *Psychological Science* 14 (2003): 220–224.
86. M. Greenberg and R. Littlewood, "Post-Adoption Incest and Phenotypic Matching: Experience, Personal Meanings and Biosocial Implications," *British Journal of Medical Psychology* 68 (1995): 29–44. For a collection of reports of this phenomenon in the popular press, see M. Bowerman, "Sexual Attraction to a Long-Lost Parent; Is That a Normal Reaction?," *USA Today,* August 10, 2016.
87. Finally, it is conceivable that animals would evolve the capacity to express and recognize signals via what Hamilton called recognition alleles. W. D. Hamilton, "The Genetical Evolution of Social Behaviour," *Journal of Theoretical Biology* 7 (1964): 17–52. A gene could simultaneously enable its possessor to express a trait, empower the possessor to perceive the trait in others, and endow the possessor with the desire to help those with the trait. Richard Dawkins called this the "green-beard effect," imagining

that such genes might give one a green beard and simultaneously make one partial to others with green beards. It's an appealing idea, but it turns out that there are quite a few conceptual and evolutionary problems with it, including the fact that it would be hard for a single gene (or even a set of genes) to have all of these effects. This mechanism may, however, appear in some microorganisms. R. Dawkins, *The Selfish Gene* (Oxford: Oxford University Press, 1976). See also S. A. West and A. Gardner, "Altruism, Spite, and Greenbeards," *Science* 327 (2010): 1341–1344.

88. D. Lieberman, J. Tooby, and L. Cosmides, "The Architecture of Human Kin Detection," *Nature* 445 (2007): 727–731. Regarding recognizing siblings, see M. F. Dal Martello and L. T. Maloney, "Where Are Kin Recognition Signals in the Human Face?," *Journal of Vision* 6 (2006): 1356–1366.
89. G. Palla, A.-L. Barabási, and T. Vicsek, "Quantifying Social Group Evolution," *Nature* 446 (2007): 664–667.
90. This is known as the ship-of-Theseus problem in philosophy, so called because the Athenians allegedly maintained Theseus's original ship in their harbor for centuries after he returned from killing the Minotaur in Crete, but all of its components were replaced at one point or another. A variation of it is the so-called family-knife problem, a reference to an heirloom knife that has had its handle and blade replaced several times over the centuries.

### *Chapter 8: Friends and Networks*

1. C. Gibbons, "The Victims: Real Movie Heroes Saved Their Sweethearts During Colo. Ambush," *New York Post,* July 22, 2012; H. Yan, "Tales of Heroism Abound from Colorado Movie Theater Tragedy," CNN, July 24, 2012; O. Katrandjian, "Colorado Shooting: Victims Who Died While Saving Their Loved Ones," ABC News, July 22, 2012. During the Thousand Oaks shooting near Los Angeles on November 7, 2018, when a gunman entered a bar and started shooting (ultimately killing a dozen people), many men banded together to try to protect others present. According to one eyewitness who appeared in a video interview, "While we were all dogpiled over on the side, there were multiple men that got on their knees and pretty much blocked all of us, with their back towards the shooter, ready to take a bullet for any single one of us"; see https://www.goodmorningamerica.com/news/story/multiple-people-injured-reported-mass-shooting-california-bar-59050130.
2. C. Ng and D. Harris, "Women Who Survived Theater Shooting Grieve for Hero Boyfriends," ABC News, July 24, 2012. See also "Hero Dies Saving Girlfriend in Theater," CNN, July 24, 2012.
3. Yan, "Tales of Heroism."
4. H. Rosin, "In the Aurora Theater the Men Protected the Women. What Does That Mean?," *Slate,* July 23, 2012.
5. For some cases in the news, see M. Wagner, "Buffalo Dad Who Rescued Fiancée, Two Kids from House Fire Dies While Saving Third Child," *New York Daily News,* February 20, 2016; K. French, "Father, 47, Run Over and Killed in Car Crash Saved His Nine-Year-Old Daughter's Life by Shoving Her to Safety," *Daily Mail,* March 17, 2017; D.

Prendergast, K. Sheehan, and P. DeGregory, "Mom Dies After Saving Daughter from Out-of-Control Car," *New York Post,* May 14, 2017; and K. Mettler, "She Dived in the Water to Save Her Son," *Washington Post,* August 29, 2016. See also R. Wright, *The Moral Animal* (New York: Vintage, 1995); W. B. Swann et al., "What Makes a Group Worth Dying For? Identity Fusion Fosters Perception of Familial Ties, Promoting Self-Sacrifice," *Journal of Personality Social Psychology* 106 (2014): 912–926; and R. M. Fields and C. Owens, *Martyrdom: The Psychology, Theology, and Politics of Self-Sacrifice* (Westport, CT: Greenwood, 2004).

6. John 15:13–14 (New International Version).
7. P. Holley, "Zaevion Dobson, High School Football Hero Who Died Shielding Girls from Gunmen, Honored at ESPYS," *Washington Post,* July 13, 2016.
8. S. Goldstein, "Connecticut Teen Is Fatally Hit by Car While Saving Friend, Unwittingly Completes Bucket List," *New York Daily News,* July 13, 2015.
9. Tribune Media Wire, "Teen Completes 'Bucket List' by Sacrificing Her Life to Save Friend," WNEP-TV (Moosic, PA) July 14, 2015.
10. A. Spital, "Public Attitudes Toward Kidney Donation by Friends and Altruistic Strangers in the United States," *Transplantation* 71 (2001): 1061–1064. In the United States, 90 percent of respondents believe friend donation is acceptable, and 80 percent believe stranger donation is acceptable. For friend donations in the news, see C. Watts, "Amy Grant's Daughter Donates Kidney to Best Friend," *USA Today,* January 26, 2017; and A. Wilson, "'Heard Urine Need of a Kidney': Friend Donates Kidney to Man 'Days Away from Failure,'" Global News, July 27, 2016. A particularly haunting set of examples of altruism in concentration camps during the Holocaust is provided in A. B. Shostak, *Stealth Altruism: Forbidden Care as Jewish Resistance in the Holocaust* (London: Routledge, 2017). People also risk their lives for complete strangers, and they often do so intuitively. Regarding a sample of fifty-one winners of the Carnegie Hero Fund Commission medal (average age thirty-six, 82 percent male), given to people who act in this way, see D. G. Rand and Z. G. Epstein, "Risking Your Life Without a Second Thought: Intuitive Decision-Making and Extreme Altruism," *PLOS ONE* 9 (2014): e109687.
11. D. Gilbert, *Stumbling on Happiness* (New York: Knopf, 2006).
12. D. J. Hruschka, *Friendship: Development, Ecology, and Evolution of a Relationship* (Berkeley: University of California Press, 2010), p. 35. A cynical (and incorrect) account of friendship—one not in keeping with the evidence of friendship relationships in nonhuman social species or with the reality that an evolutionary account would always ultimately balance the costs and benefits—would assert that friendship is simply a cultural institution that normalizes and justifies an unequal relationship in which someone is always being exploited for someone else's benefit.
13. Historian Michel Foucault argued that homosexuality might be a way of being friendly, but I see homosexual desire as more similar to heterosexual desire than same-sex friendships. M. Foucault, "Friendship as a Way of Life," in M. Foucault and P. Rabinow, eds., *Essential Works of Foucault, 1954–1984,* vol. 1, *Ethics: Subjectivity and Truth* (New York: New Press, 1997), pp. 135–140.

14. A. Aron, E. N. Aron, and D. Smollan, "Inclusion of Other in the Self Scale and the Structure of Interpersonal Closeness," *Journal of Personality and Social Psychology* 63 (1992): 596–612.
15. Hruschka, *Friendship.* In some societies, people do not necessarily choose their friends; they sometimes inherit friendships from parents or based on clan ties, or they may have friends recommended by elders. Friendships might also be sealed with public or private rituals, just like weddings.
16. F. Kaplan, "The Idealist in the Bluebonnets: What Bush's Meeting with the Saudi Ruler Really Means," *Slate,* April 26, 2005.
17. Hruschka, *Friendship,* p. 17.
18. S. Perry, "Capuchin Traditions Project," UCLA Department of Anthropology, http://www.sscnet.ucla.edu/anthro/faculty/sperry/ctp.html.
19. For instance, in societies characterized by economic and legal unpredictability (such as corrupt states), people are more willing to lie to protect their friends. Hruschka, *Friendship,* p. 186.
20. J. C. Williams, *White Working Class: Overcoming Class Cluelessness in America* (Boston: Harvard Business Review Press, 2017). See also M. Small, *Unanticipated Gains: Origins of Network Inequality in Everyday Life* (Oxford: Oxford University Press, 2009).
21. B. Bigelow, "Children's Friendship Expectations: A Cognitive-Developmental Study," *Child Development* 48 (1977): 246–253.
22. M. Taylor, *Imaginary Companions and the Children Who Create Them* (New York: Oxford University Press, 1999), pp. 30–33. It used to be thought that it was psychologically maladaptive to have imaginary friends, but this is no longer felt to be true. If anything, children with imaginary friends are less shy, more intelligent, and more socially competent than those without.
23. E. A. Madsen, R. J. Tunney, F. Fieldman, and H. C. Plotkin, "Kinship and Altruism: A Cross-Cultural Experimental Study," *British Journal of Psychology* 93 (2007): 339–359.
24. Just because friendships may not necessarily feel like reciprocal relationships *proximately* does not mean that *ultimately* they do not have this origin.
25. J. Tooby and L. Cosmides, "Friendship and the Banker's Paradox: Other Pathways to the Evolution of Adaptations for Altruism," *Proceedings of the British Academy* 88 (1996): 119–143.
26. Ibid., p. 132.
27. Hruschka, *Friendship;* R. M. Seyfarth and D. L. Cheney, "The Evolutionary Origins of Friendship," *Annual Review of Psychology* 63 (2012): 153–177. See also A. Burt, "A Mechanistic Explanation of Popularity: Genes, Rule Breaking, and Evocative Gene-Environment Correlations," *Journal of Personality and Social Psychology* 96 (2009): 783–794; G. Guo, "Genetic Similarity Shared by Best Friends Among Adolescents," *Twin Research and Human Genetics* 9 (2006): 113–121; J. H. Fowler, C. T. Dawes, and N. A. Christakis, "Model of Genetic Variation in Human Social Networks," *PNAS: Proceedings of the National Academy of Sciences* 106 (2009): 1720–1724; J. D. Boardman, B. W. Domingue, and J. M. Fletcher, "How Social and Genetic Factors Predict Friendship Networks," *PNAS: Proceedings of the National Academy of Sciences* 109 (2012): 17377–17381; and M.

Brendgen, "Genetics and Peer Relations: A Review," *Journal of Research on Adolescence* 22 (2012): 419–437.

28. Fowler, Dawes, and Christakis, "Model of Genetic Variation."
29. Tooby and Cosmides, "Friendship and the Banker's Paradox," p. 137.
30. M. McPherson, L. Smith-Lovin, and J. M. Cook, "Birds of a Feather: Homophily in Social Networks," *Annual Review of Sociology* 27 (2001): 415–444.
31. C. Parkinson, A. M. Kleinbaum, and T. Wheatley, "Similar Neural Responses Predict Friendship," *Nature Communications* 9 (2018): 332.
32. N. A. Christakis and J. H. Fowler, "Friendship and Natural Selection," *PNAS: Proceedings of the National Academy of Sciences* 111 (2014): 10796–10801. Regarding homophily in other species, see D. Lusseau and M. E. J. Newman, "Identifying the Role That Animals Play in Their Social Networks," *Proceedings of the Royal Society B* 271 (2004): S477–S481; L. J. H. Brent, J. Lehmann, and G. Ramos-Fernández, "Social Network Analysis in the Study of Nonhuman Primates: A Historical Perspective," *American Journal of Primatology* 73 (2011): 720–730.
33. L. M. Guth and S. M. Roth, "Genetic Influence on Athletic Performance," *Current Opinion in Pediatrics* 25 (2013): 653–658.
34. Y. T. Tan, G. E. McPherson, I. Peretz, S. F. Berkovic, and S. J. Wilson, "The Genetic Basis of Music Ability," *Front Psychology* 5 (2014): 658.
35. Homophily evolves under a much wider variety of conditions than heterophily, even when the fitness advantage to dissimilarity exceeds the fitness advantage to similarity. F. Fu, M. A. Nowak, N. A. Christakis, and J. H. Fowler, "The Evolution of Homophily," *Scientific Reports* 2 (2012): 845.
36. Christakis and Fowler, "Friendship and Natural Selection." See also B. W. Domingue, D. W. Belsky, J. M. Fletcher, D. Conley, J. D. Boardman, and K. M. Harris, "The Social Genome of Friends and Schoolmates in the National Longitudinal Study of Adolescent to Adult Health," *PNAS: Proceedings of the National Academy of Sciences* 115 (2018): 702–707; and J. H. Fowler, J. E. Settle, and N. A. Christakis, "Correlated Genotypes in Friendship Networks," *PNAS: Proceedings of the National Academy of Sciences* 108 (2011): 1993–1997.
37. A one-standard-deviation change in the friendship score can explain approximately 1.4 percent of the variance in the existence of friendship ties. Christakis and Fowler, "Friendship and Natural Selection." This is similar to the variance explained using the best currently available genetic scores for schizophrenia and bipolar disorder (0.4 percent to 3.2 percent) and body mass index (1.5 percent). For comparison, see S. M. Purcell et al., "Common Polygenic Variation Contributes to Risk of Schizophrenia and Bipolar Disorder," *Nature* 460 (2009): 748–752; and E. K. Speliotes et al., "Association Analyses of 249,796 Individuals Reveal 18 New Loci Associated with Body Mass Index," *Nature Genetics* 42 (2010): 937–948.
38. D. Lieberman, J. Tooby, and L. Cosmides, "The Architecture of Human Kin Detection," *Nature* 445 (2007): 727–731.
39. In fact, the customs related to fictive kin seen in so many societies comport with this idea; humans pick compadres, have godparents and "aunties," fight side by side with brothers-in-arms, and call their friends by kin terms such as "bro" or "sis."

40. E. Herrmann et al., "Humans Have Evolved Specialized Skills of Social Cognition: The Cultural Intelligence Hypothesis," *Science* 317 (2007): 1360–1366.
41. Such an effect would especially speed up the evolution of phenotypes that are intrinsically synergistic, and this may help shed light on the observation that evolution in humans is accelerating. J. Hawks, E. T. Wang, G. M. Cochran, H. C. Harpending, and R. K. Moyzis, "Recent Acceleration of Human Adaptive Evolution," *PNAS: Proceedings of the National Academy of Sciences* 104 (2007): 20753–20758.
42. W. D. Hamilton, "Innate Social Aptitudes of Man: An Approach from Evolutionary Genetics," in R. Fox, ed., *Biosocial Anthropology* (London: Malaby Press, 1975), pp. 133–153; J. M. Smith, "Group Selection," *Quarterly Review of Biology* 51 (1976): 277–283.
43. H. B. Shakya, N. A. Christakis, and J. H. Fowler, "An Exploratory Comparison of Name Generator Content: Data from Rural India," *Social Networks* 48 (2017): 157–168. There are, obviously, many other questions one can use to identify all sorts of specialized or minor connections, such as "Who do you play sports with?" and "Who do you ask for health advice?"
44. A. J. O'Malley, S. Arbesman, D. M. Steiger, J. H. Fowler, and N. A. Christakis, "Egocentric Social Network Structure, Health, and Pro-Social Behaviors in a National Panel Study of Americans," *PLOS ONE* 7 (2012): e36250. This corresponds with prior work; see P. V. Marsden, "Core Discussion Networks of Americans," *American Sociological Review* 52 (1987): 122–131; M. McPherson, L. Smith-Lovin, and M. E. Brashears, "Social Isolation in America: Changes in Core Discussion Networks over Two Decades," *American Sociological Review* 71 (2006): 353–375. People understandably include spouses and siblings when answering these questions, so if we want to identify strictly unrelated friends, it helps to remove those individuals from the list.
45. The panels in this figure are as follows: (a) A gift network among 91 men in the Nyangatom people of Sudan (the ties indicate who would give an anonymous gift to whom); 34 ties are family ties (to siblings) and 239 are friendship connections. (b) Another gift network among 96 men residing in a village in Uganda; 35 ties are family ties, and 151 are friendship ties. (c) A network of 103 women among the Hadza people of Tanzania, based on whom the women said they would ideally like to share camp with in the future; 179 ties are family ties, and 183 are friendship ties. (d) A rural village in Honduras with 216 people (78 men and 138 women); 235 ties are family ties, and 505 are friendship ties. Men and women seem well mixed in this village (the blue and red dots, respectively, are interspersed). Also note that if a line was drawn from approximately where one o'clock would be on a clock face to about where seven o'clock would be, the village would seem to be divisible into two social communities (with more ties within each community than between them). (e) A rural village in Uganda with 261 people (121 men and 140 women); 173 ties are family ties, and 657 are friendship ties. Note the relative separation of men and women in this village (the blue and red dots are not as uniformly interspersed, meaning that men tend to socialize with men, and women with women). (f) A rural village in India with 214 people (95 men and 119 women); 107 ties are family ties, and 569 are friendship ties. Men and women are socially segregated here too, into clusters of blue and red dots. All of these data are from our own data-collection efforts and published papers,

except for the Indian village network (for which the raw data were collected by others). See L. Glowacki, A. Isakov, R. W. Wrangham, R. McDermott, J. H. Fowler, and N. A. Christakis, "Formation of Raiding Parties for Intergroup Violence Is Mediated by Social Network Structure," *PNAS: Proceedings of the National Academy of Sciences* 113 (2016): 12114–12119; J. M. Perkins et al., "Food Insecurity, Social Networks and Symptoms of Depression Among Men and Women in Rural Uganda: A Cross-Sectional, Population-Based Study," *Public Health Nutrition* 21 (2018): 838–848; C. L. Apicella, F. W. Marlowe, J. H. Fowler, and N. A. Christakis, "Social Networks and Cooperation in Hunter-Gatherers," *Nature* 481 (2012): 497–501; H. N. Shakya et al., "Exploiting Social Influence to Magnify Population-Level Behaviour Change in Maternal and Child Health: Study Protocol for a Randomised Controlled Trial of Network Targeting Algorithms in Rural Honduras," *BMJ Open* 7 (2017): e012996; H. B. Shakya, N. A. Christakis, and J. H. Fowler, "Social Network Predictors of Latrine Ownership," *Social Science and Medicine* 125 (2015): 129–138. Slightly different name generators to identify important social ties were used for the villages in Honduras, Uganda, and India, but in general, the ties were defined based on whom people said they would get social support from or spend time with. Sometimes, such people were also family members (shown in orange ties between the nodes), but most of the time they were not close family (shown in gray ties).

46. J. Perkins, S. Subramanian, and N. A. Christakis, "A Systematic Review of Sociocentric Network Studies on Health Issues in Low- and Middle-Income Countries," *Social Science and Medicine* 125 (2015): 60–78.
47. Apicella et al., "Social Networks and Cooperation."
48. C. M. Rawlings and N. E. Friedkin, "The Structural Balance Theory of Sentiment Networks: Elaboration and Test," *American Journal of Sociology* 123 (2017): 510–548.
49. S. Sampson, "Crisis in a Cloister" (PhD diss., Cornell University, 1969).
50. For studies of bullying involving sociocentric network mapping, see C. Salmivalli, A. Huttunen, and K. M. J. Lagerspetz, "Peer Networks and Bullying in Schools," *Scandinavian Journal of Psychology* 38 (1997): 305–312; and G. Huitsing and R. Veenstra, "Bullying in Classrooms: Participant Roles from a Social Network Perspective," *Aggressive Behavior* 38 (2012): 494–509. Regarding workplaces, see, for example, L. Xia, Y. C. Yuan, and G. Gay, "Exploring Negative Group Dynamics: Adversarial Network, Personality, and Performance in Project Groups," *Management Communication Quarterly* 23 (2009): 32–62; A. Gerbasi, C. L. Porath, A. Parker, G. Spreitzer, and R. Cross, "Destructive De-Energizing Relationships: How Thriving Buffers Their Effect on Performance," *Journal of Applied Psychology* 100 (2015): 1423–1433; and G. Labianca and D. J. Brass, "Exploring the Social Ledger: Negative Relationships and Negative Asymmetry in Social Networks in Organizations," *Academy of Management Review* 31 (2006): 596–614.
51. An analysis of 18,819 players over a period of 445 days documented different sorts of positive ties (e.g., sending a direct message) and negative ties (e.g., placing bounties on enemies). Positive ties outnumbered negative ties by a factor of about 10 to 1. Positive ties were reciprocated roughly 60 to 80 percent of the time, whereas negative ties were reciprocated roughly only 10 to 20 percent of the time. M. Szell, R. Lambiotte, and S. Thurner, "Multirelational Organization of Large-Scale Social Networks in an

Online World," *PNAS: Proceedings of the National Academy of Sciences* 107 (2010): 13636–13641. For another example of negative ties in an online network, see G. Facchetti, G. Iacono, and C. Altafini, "Computing Global Structural Balance in Large-Scale Signed Social Networks" *PNAS: Proceedings of the National Academy of Sciences* 108 (2011): 20953–20958.

52. Shakya et al., "Exploiting Social Influence." Our work on antagonistic ties is described in A. Isakov, J. H. Fowler, E. M. Airoldi, and N. A. Christakis, "The Structure of Negative Ties in Human Social Networks" (unpublished manuscript, 2018).
53. According to the World Bank, the murder rate in Honduras peaked at 93.2 per 100,000 in 2011, declining to 74.6 per 100,000 in 2014. For comparison, other country statistics are as follows: United States, 3.9 per 100,000 in 2013; United Kingdom, 0.9 per 100,000 in 2013; and Russian Federation, 9.5 per 100,000 in 2014. "Intentional Homicides (per 100,000 People)," World Bank, https://data.worldbank.org/indicator/VC.IHR.PSRC.P5?year_high_desc=false.
54. The standard deviation of this measure was 2.6.
55. The standard deviation was 1.2.
56. The standard deviation was 1.3.
57. G. Simmel, *The Sociology of Georg Simmel* (New York: Simon and Schuster, 1950); F. Heider, "Attitudes and Cognitive Organization," *Journal of Psychology* 21 (1946): 107–112; D. Cartwright and F. Harary, "Structural Balance: A Generalization of Heider's Theory," *Psychology Review* 63 (1956): 277–293. The earliest known exposition of the claim that the "enemy of my enemy is my friend" appears to be from the fourth century BCE. L. N. Rangarajan, *The Arthashastra* (New Delhi: Penguin Books India, 1992), p. 520.
58. A. Rapoport, "Mathematical Models of Social Interaction," in R. A. Galanter, R. R. Lace, and E. Bush, eds., *Handbook of Mathematical Sociology*, vol. 2 (New York: John Wiley and Sons, 1963), 493–580.
59. H. Tajfel, M. Billig, R. Bundy, and C. Flament, "Social Categorization in Intergroup Behaviour," *European Journal of Social Psychology* 1 (1971): 149–178.
60. "Paul Klee and Wassily Kandinsky," Wassily Kandinsky: Biography, Paintings, and Quotes, Wassily-Kandinsky.org, 2011, http://www.wassily-kandinsky.org/kandinsky-and-paul-klee.jsp.
61. M. Billig and H. Tajfel, "Social Categorization and Similarity in Intergroup Behaviour," *European Journal of Social Psychology* 3 (1973): 27–55.
62. Tajfel et al., "Social Categorization."
63. T. Yamagishi, N. Jin, and T. Kiyonari, "Bounded Generalized Reciprocity: Ingroup Boasting and Ingroup Favoritism," *Advances in Group Processes* 16 (1999): 161–197.
64. Experiments show that if rewards do *not* depend on the behaviors of others in the group, people do not show in-group favoritism. J. M. Rabbie and H. F. M. Lodewijkx, "Conflict and Aggression: An Individual-Group Continuum," *Advances in Group Processes* 11 (1994): 139–174.
65. Cooperation is costly, and numerous explanations have been advanced for how it can arise or be sustained in human groups. These explanations include a central authority who can coordinate or sanction; inclusive fitness based on kinship (as described by

Hamilton); market interactions; reciprocity based on serial interactions; decentralized enforcement, including via social norms; reputation effects; and group selection.

66. M. B. Brewer, "The Psychology of Prejudice: Ingroup Love or Outgroup Hatred?," *Journal of Social Issues* 55 (1999): 429–444.
67. M. Sherif, O. J. Harvey, B. J. White, W. R. Hood, and C. W. Sherif, *Intergroup Conflict and Cooperation: The Robbers Cave Experiment* (Norman: Institute of Group Relations, University of Oklahoma, 1961). It appears that the experiments reported in this study were not the first ones Sherif conducted; another group of boys did not perform as expected, and Sherif rejected the results. G. Perry, *The Lost Boys: Inside Muzafer Sherif's Robbers Cave Experiment* (Melbourne: Scribe, 2018).
68. Sherif et al., *Intergroup Conflict*, p. 98.
69. Ibid., p. 151.
70. Ibid.
71. Something similar happened on the evening of September 11, 2001, when one hundred and fifty members of Congress from both parties gathered on the steps of the Capitol and sang "God Bless America." "The Singing of 'God Bless America' on September 11, 2001," History, Art and Archives, U.S. House of Representatives, http://history.house.gov/HistoricalHighlight/Detail/36778.
72. W. G. Sumner, *Folkways: A Study of the Sociological Importance of Usages, Manners, Customs, Mores, and Morals* (Boston: Ginn, 1906), pp. 12–13.
73. The killing of conspecifics is also rare in animals. See J. M. Gomez, M. Verdo, A. Gonzalez-Negras, and M. Mendez, "The Phylogenetic Roots of Human Lethal Violence," *Nature* 538 (2016): 233–237.
74. J. K. Choi and S. Bowles, "The Coevolution of Parochial Altruism and War," *Science* 318 (2007): 636–640. See also M. R. Jordan, J. J. Jordan, and D. G. Rand, "No Unique Effect of Intergroup Competition on Cooperation: Non-Competitive Thresholds Are as Effective as Competitions Between Groups for Increasing Human Cooperative Behavior," *Evolution and Human Behavior* 38 (2017): 102–108.
75. R. A. Hammond and R. Axelrod, "The Evolution of Ethnocentrism," *Journal of Conflict Resolution* 50 (2006): 1–11.
76. F. Fu, C. E. Tarnita, N. A. Christakis, L. Wang, D. G. Rand, and M. A. Nowak, "Evolution of Ingroup Favoritism," *Scientific Reports* 2 (2012): 460.
77. This story is told in R. M. Sapolsky, *Behave: The Biology of Humans at Our Best and Worst* (New York: Penguin, 2017), p. 409. Armistead died at the hospital, alas.
78. Y. Dunham, E. E. Chen, and M. R. Banaji, "Two Signatures of Implicit Intergroup Attitudes: Developmental Invariance and Early Enculturation," *Psychological Science* 24 (2013): 860–868; Y. Dunham, A. S. Baron, and M. R. Banaji, "The Development of Implicit Intergroup Cognition," *Trends in Cognitive Sciences* 12 (2008): 248–253. The role of enculturation is also clearly important.
79. A. V. Shkurko, "Is Social Categorization Based on Relational Ingroup/Outgroup Opposition? A Meta-Analysis," *Social Cognitive and Affective Neuroscience* 8 (2013): 870–877.
80. M. B. Brewer, "The Psychology of Prejudice: Ingroup Love or Outgroup Hatred?," *Journal of Social Issues* 55 (1999): 429–444.

81. Yamagishi, Jin, and Kiyonari, "Bounded Generalized Reciprocity," p. 173.
82. G. Allport, *The Nature of Prejudice* (Reading, MA: Addison-Wesley, 1954), p. 42.
83. In-group bias can even subvert the impact of sharing superordinate goals. Even if people do share superordinate goals with an out-group, it does not necessarily lead to holding positive views of them, let alone offering them kindness, because the in-group might still project in-group expectations onto this larger group. All that might be achieved is that the in-group would judge out-group individuals as simply bad members of the superordinate group to which everyone now belongs.
84. H. C. Triandis, *Individualism and Collectivism* (Boulder, CO: Westview Press, 1995).
85. C. Lévi-Strauss, *Structural Anthropology,* trans. C. Jacobson and B. G. Schoepf (New York: Basic Books, 1967). Incidentally, the way humans see social opposition as binary ("us" versus "them") helps explain why we are entertained and confounded by more complex competitive interactions in fiction, such as the three-way gunfight in *The Good, the Bad, and the Ugly* and the five-way contest among the armies in *The Hobbit: The Battle of the Five Armies.*
86. R. W. Emerson, *Essays and English Traits by Ralph Waldo Emerson* (1841; New York: Cosimo Classics, 1909), pp. 109–124.
87. Hruschka, *Friendship.*

### *Chapter 9: One Way to Be Social*

1. A. Starr and M. L. Edwards, "Mitral Replacement: Clinical Experience with a Ball-Valve Prosthesis," *Annals of Surgery* 154 (1961): 726–740.
2. J. P. Binet, A. Carpentier, J. Langlois, C. Duran, and P. Colvez, "Implantation de valves hétérogènes dans le traitement des cardiopathies aortiques," *Comptes rendus des séances de l'Académie des sciences. Série D, Sciences naturelles* 261 (1965): 5733–5734.
3. Artists and storytellers have explored the unnerving nature of human-animal hybrids for a very long time, at least as far back as the Chimera of Greek mythology. But modern science has had its own fixation with hybrids: animal-to-human blood transfusions in the seventeenth century, the first xenotransplantation of a pig cornea to a human in 1838, the insertion of a baboon heart into the newborn Baby Fae in 1984, and the successful integration of human stem cells into a growing pig embryo in 2017. D. K. C. Cooper, "A Brief History of Cross-Species Organ Transplantation," *Baylor University Medical Center Proceedings* 25 (2012): 49–57; K. Reemtsma, "Xenotransplantation: A Historical Perspective," *Institute for Laboratory Animal Research Journal* 37 (1995): 9–12.
4. B. Hölldobler and E. O. Wilson, *The Ants* (Cambridge, MA: Harvard University Press, 1990).
5. Several members of the family Bathyergidae, which includes the naked mole rat, the Damaraland mole rat, and others, can be considered eusocial. It is possible that certain social voles may also exhibit eusocial behaviors. H. Burda, R. L. Honeycutt, S. Begall, O. Locker-Grütjen, and A. Scharff, "Are Naked and Common Mole-Rats Eusocial and If So, Why?," *Behavioral Ecology and Sociobiology* 47 (2000): 293–303.
6. An additional technical criterion for eusociality is that there are overlapping generations in the colony. Eusociality has evolved independently in several insect taxa, including bees, wasps, ants, and termites; in coral-reef shrimp (where it has evolved

several times!); and in mammals such as mole rats (which is a fascinating exception). *Eusocial* is defined as a form of social organization that involves three properties: cooperative care of offspring, overlapping generations, and a division of labor into reproductive and nonreproductive subgroups of the population.

7. M. dos Reis, J. Inoue, M. Hasegawa, R. J. Asher, P. C. J. Donoghue, and Z. Yang, "Phylogenomic Datasets Provide Both Precision and Accuracy in Estimating the Timescale of Placental Mammal Phylogeny," *Proceedings of the Royal Society B* 279 (2012): 3491–3500. These estimates are necessarily imprecise; for instance, the point of divergence between humans and chimpanzees is thought to have occurred sometime between thirteen and four million years ago.
8. J. Parker, G. Tsagkogeorga, J. A. Cotton, Y. Liu, P. Provero, E. Stupka, and S. J. Rossiter, "Genome-Wide Signatures of Convergent Evolution in Echolocating Mammals," *Nature* 502 (2013): 228–231. In the case of echolocation, we know that genes used for the convergent phenotypes may be similar across widely divergent taxa, such as bats and dolphins.
9. S. C. Morris, *Life's Solution: Inevitable Humans in a Lonely Universe* (Cambridge: Cambridge University Press, 2003), p. 128.
10. Ibid.
11. Ibid., p. 248.
12. S. Gould, *Wonderful Life: The Burgess Shale and the Nature of History* (New York: W. W. Norton, 1990).
13. One interesting question about personality is why the personality of individuals varies little over time and yet there is so much variation between individuals. See, for example, M. Wolf, G. S. van Doorn, O. Leimar, and F. J. Weissing, "Life-History Trade-Offs Favour the Evolution of Animal Personalities," *Nature* 447 (2007): 581–584; and M. Wolf and F. J. Weissing, "An Explanatory Framework for Adaptive Personality Differences," *Philosophical Transactions of the Royal Society B* 365 (2010): 3959–3968.
14. M. J. Sheehan and M. W. Nachman, "Morphological and Population Genomic Evidence That Human Faces Have Evolved to Signal Individual Identity," *Nature Communications* 5 (2014): 4800. See also G. Yovel and W. A. Freiwald, "Face Recognition Systems in Monkey and Human: Are They the Same Thing?," *F1000Prime Reports* 5 (2013): 10.
15. C. Schlitz et al., "Impaired Face Discrimination in Acquired Prosopagnosia Is Associated with Abnormal Response to Individual Faces in the Right Middle Fusiform Gyrus," *Cerebral Cortex* 16 (2006): 574–586; P. Shah, "Identification, Diagnosis and Treatment of Prosopagnosia," *British Journal of Psychiatry* 208 (2016): 94–95.
16. E. Prichard, "Prosopagnosia: How Face Blindness Means I Can't Recognize My Mum," *BBC News Magazine*, July 1, 2016.
17. Sheehan and Nachman, "Morphological and Population Genomic Evidence."
18. In principle, such cues could also be acquired, like barnacles on a whale's tail or rips in an elephant's ears that come to be unique and identifiable.
19. Examples of individual recognition (especially beyond mating pairs) are exceedingly rare outside of mammals and birds. R. W. Wrangham, "Social Relationships in Com-

parative Perspective," in R. A. Hinde, ed., *Primate Social Relationships: An Integrated Approach* (Oxford: Blackwell, 1983), pp. 325–334; P. d'Ettorre, "Multiple Levels of Recognition in Ants: A Feature of Complex Societies," *Biological Theory* 3 (2008): 108–113. Regarding cooperation, see J. M. McNamara, Z. Barta, and A. I. Houston, "Variation in Behaviour Promotes Cooperation in the Prisoner's Dilemma Game," *Nature* 428 (2004): 745–748; and S. F. Brosnan, L. Salwiczek, and R. Bshary, "The Interplay of Cognition and Cooperation," *Philosophical Transactions of the Royal Society B* 365 (2010): 2699–2710.

20. Travelers to Xian, China, can find startling evidence of human facial diversity; scientists believe that each one of the thousands of famous terra-cotta warriors there features a unique set of ears, likely evidence that the statues were modeled on individual human soldiers. E. Quill, "Were the Terracotta Warriors Based on Actual People?," *Smithsonian*, March 2015.
21. Interestingly, faces generally vary only slightly from side to side, and facial symmetry is often viewed as a marker of beauty. B. C. Jones et al., "Facial Symmetry and Judgements of Apparent Health," *Evolution and Human Behavior* 22 (2001): 417–429; K. Grammer and R. Thornhill, "Human (*Homo sapiens*) Facial Attractiveness and Sexual Selection: The Role of Symmetry and Averageness," *Journal of Comparative Psychology* 108 (1994): 233–242; J. E. Scheib, S. W. Gangestad, and R. Thornhill, "Facial Attractiveness, Symmetry and Cues of Good Genes," *Proceedings of the Royal Society B* 266 (1999): 1913–1917.
22. Sheehan and Nachman, "Morphological and Population Genomic Evidence."
23. J. Freund et al., "Emergence of Individuality in Genetically Identical Mice," *Science* 340 (2013): 756–759.
24. One such advantage might be to communicate how cooperative someone is, and there is evidence that adult humans can make instantaneous, intuitive, and accurate decisions about how likely other people are to reciprocate generosity based solely on their faces. See J. F. Bonnefon, A. Hopfensitz, and W. De Neys, "Can We Detect Cooperators by Looking at Their Face?," *Current Directions in Psychological Science* 26 (2017): 276–281.
25. R. A. Hinde, "Interactions, Relationships and Social Structure," *Man* 11 (1976): 1–17.
26. J. van Lawick-Goodall, *In the Shadow of Man* (Boston: Houghton Mifflin, 1971).
27. L. A. Parr, "The Evolution of Face Processing in Primates," *Philosophical Transactions of the Royal Society B* 366 (2011): 1764–1777.
28. L. A. Parr, J. T. Winslow, W. D. Hopkins, and F. B. de Waal, "Recognizing Facial Cues: Individual Discrimination by Chimpanzees (*Pan troglodytes*) and Rhesus Monkeys (*Macaca mulatta*)," *Journal of Comparative Psychology* 114 (2000): 47–60. See also S. A. Rosenfeld and G. W. Van Hoesen, "Face Recognition in the Rhesus Monkey," *Neuropsychologia* 17 (1979): 503–509.
29. J. A. Pineda, G. Sebestyen, and C. Nava, "Face Recognition as a Function of Social Attention in Non-Human Primates: An ERP Study," *Cognitive Brain Research* 2 (1994): 1–12.
30. L. A. Parr, M. Heintz, E. Lonsdorf, and E. Wroblewski, "Visual Kin Recognition in Nonhuman Primates (*Pan troglodytes* and *Macaca mulatta*): Inbreeding Avoidance or Male Distinctiveness?," *Journal of Comparative Psychology* 124 (2010): 343–350; C. Almstrom

and M. Knight, "Using a Paired-Associate Learning Task to Assess Parent-Child Phenotypic Similarity," *Psychology Reports* 97 (2005): 129–137.

31. K. McComb, C. Moss, S. M. Durant, L. Baker, and S. Sayialel, "Matriarchs as Repositories of Social Knowledge in African Elephants," *Science* 292 (2001): 491–494. This is useful knowledge because elephants will often group themselves in what is known as a bunching response when they come near unfamiliar individuals, and the ability to correctly distinguish friend from foe is important.
32. S. L. King and V. M. Janik, "Bottlenose Dolphins Can Use Learned Vocal Labels to Address Each Other," *PNAS: Proceedings of the National Academy of Sciences* 110 (2013): 13216–13221.
33. S. L. King, L. S. Sayigh, R. S. Wells, W. Fellner, and V. M. Janik, "Vocal Copying of Individually Distinctive Signature Whistles in Bottlenose Dolphins," *Proceedings of the Royal Society B* 280 (2013): 20130053. Parrots may also have names for one another, possibly ones that their mothers gave them while they were still in the nest. K. S. Berg et al., "Vertical Transmission of Learned Signatures in a Wild Parrot," *Proceedings of the Royal Society B* 279 (2012): 585–591.
34. B. Amsterdam, "Mirror Self-Image Reactions Before Age Two," *Developmental Psychobiology* 5 (1972): 297–305. Mirror self-recognition is a cultural universal (even in places that lack mirrors), but the precise age of onset can vary a bit according to cultural circumstance. See J. Kartner, H. Keller, N. Chaudhary, and R. D. Yovsi, "The Development of Mirror Self-Recognition in Different Sociocultural Contexts," *Monographs of the Society for Research in Child Development* 77 (2012).
35. J. R. Anderson and G. G. Gallup Jr., "Which Primates Recognize Themselves in Mirrors?," *PLOS Biology* 9 (2011): e1001024; J. M. Plotnik, F. B. M. de Waal, and D. Reiss, "Self-Recognition in an Asian Elephant," *PNAS: Proceedings of the National Academy of Sciences* 103 (2006): 17053–17057; D. Reiss and L. Marino, "Mirror Self-Recognition in the Bottlenose Dolphin: A Case of Cognitive Convergence," *PNAS: Proceedings of the National Academy of Sciences* 98 (2001): 5937–5942. The list of animals that have been shown to have mirror self-recognition is short; beyond certain primates, elephants, and cetaceans, it includes magpies and possibly ants and manta rays. H. Prior, A. Schwarz, and O. Gunturkun, "Mirror-Induced Behavior in the Magpie (*Pica pica*): Evidence of Self-Recognition," *PLOS Biology* 6 (2008): e202; M. Cammaerts and R. Cammaerts, "Are Ants (*Hymenoptera, Formicidae*) Capable of Self-Recognition?," *Journal of Science* 5 (2015): 521–532; C. Ari and D. P. D'Agostino, "Contingency Checking and Self-Directed Behaviors in Giant Manta Rays: Do Elasmobranchs Have Self-Awareness?," *Journal of Ethology* 34 (2016): 167–174.
36. G. G. Gallup Jr., "Chimpanzees: Self-Recognition," *Science* 167 (1970): 86–87.
37. J. R. Anderson and G. G. Gallup, "Mirror Self-Recognition: A Review and Critique of Attempts to Promote and Engineer Self-Recognition in Primates," *Primates* 56 (2015): 317–326. See also C. W. Hyatt and W. D. Hopkins, "Self-Awareness in Bonobos and Chimpanzees: A Comparative Perspective," in S. T. Parker, R. W. Mitchell, and M. L. Boccia, eds., *Self-Awareness in Animals and Humans* (Cambridge: Cambridge University Press, 1994), pp. 248–253; S. D. Suarez and G. G. Gallup Jr., "Self-Recognition in Chim-

panzees and Orangutans, but Not Gorillas," *Journal of Human Evolution* 10 (1981): 175–188; J. Riopelle, R. Nos, and A. Jonch, "Situational Determinants of Dominance in Captive Gorillas," in J. Biegert and W. Leutenegger, eds., *Proceedings of the Third International Congress on Primatology, Zurich, 1970* (Basel: Karger, 1971), pp. 86–91.

38. G. G. Gallup, M. K. McClure, S. D. Hill, and R. A. Bundy, "Capacity for Self-Recognition in Differentially Reared Chimpanzees," *Psychological Record* 21 (1971): 69–74.
39. It's even possible that something similar happens within our species; one study found that the nature and timing of self-recognition behaviors in toddlers varied somewhat according to cross-cultural differences in child-rearing that emphasized individuality and autonomy. Kartner et al., "Development of Mirror Self-Recognition."
40. F. G. P. Patterson and R. H. Cohn, "Self-Recognition and Self-Awareness in Lowland Gorillas," in S. T. Parker, R. Mitchell, and M. L. Boccia, eds., *Self-Awareness in Animals and Humans* (Cambridge: Cambridge University Press, 1994), pp. 273–290.
41. R. Cohn, *Michael's Story, Where He Signs About His Family* (KokoFlix, March 23, 2008), video recording, https://www.youtube.com/watch?v=DXKsPqQ0Ycc. See also R. Morin, "A Conversation with Koko the Gorilla," *Atlantic*, August 28, 2015.
42. K. Gold and B. Scassellati, "A Bayesian Robot That Distinguishes 'Self' from 'Other,'" *Proceedings of the Annual Meeting of the Cognitive Science Society* 29 (2007): 1037–1042.
43. Cats and dogs fail the mirror self-recognition test and generally treat mirror images as strangers. Dogs might actually be more self-aware than we realize but simply not sufficiently visually oriented to pass the mirror test. In one study, dogs were able to distinguish their own urine from other dogs' urine. M. Bekoff, "Observations of Scent-Marking and Discriminating Self from Others by a Domestic Dog (*Canis familiaris*): Tales of Displaced Yellow Snow," *Behavioural Processes* 55 (2001): 75–79. See also A. Horowitz, "Smelling Themselves: Dogs Investigate Their Own Odours Longer When Modified in an 'Olfactory Mirror' Test," *Behavioural Processes* 143 (2017): 17–24.
44. Plotnik, de Waal, and Reiss, "Self-Recognition in an Asian Elephant."
45. Ibid. The fact that only Happy out of the three elephants passed the mark test is not necessarily concerning, because similar percentages are observed in primates and because of the other observational evidence that Maxine and Patty understood that it was their images in the mirror.
46. Reiss and Marino, "Mirror Self-Recognition in the Bottlenose Dolphin." Interestingly, while other species (such as chimpanzees) pay attention to the marks on *others*, dolphins cared only about the marks on *themselves*—perhaps because they do not have natural grooming behaviors.
47. F. Delfour and K. Marten, "Mirror Image Processing in Three Marine Mammal Species: Killer Whales (*Orcinus orca*), False Killer Whales (*Pseudorca crassidens*) and California Sea Lions (*Zalophus californianus*)," *Behavioural Processes* 53 (2001): 181–190.
48. While grief is universal, mourning rituals vary cross-culturally. J. Archer, *The Nature of Grief: The Evolution and Psychology of Reactions to Loss* (London: Routledge, 1999); P. C. Rosenblatt, R. P. Walsh, and D. A. Jackson, *Grief and Mourning in Cross-Cultural Perspective* (New Haven, CT: Human Relations Area File Press, 1976); W. Stroebe and M. S. Stroebe, *Bereavement and Health: The Psychological and Physical Consequences of Partner*

*Loss* (Cambridge: Cambridge University Press, 1987). On the physical nature of emotional pain, see D. N. DeWall et al., "Acetaminophen Reduces Social Pain: Behavioral and Neural Evidence," *Psychological Science* 21 (2010): 931–937.

49. H. Williams, *Historical and Archaeological Aspects of Egyptian Funerary Culture* (Leiden: Brill, 2014); J. Toynbee, *Death and Burial in the Roman World* (Ithaca, NY: Cornell University Press, 1971); B. Effros, *Merovingian Mortuary Archaeology and the Making of the Early Middle Ages* (Berkeley: University of California Press, 2003); A. Reynolds, *Anglo-Saxon Deviant Burial Customs* (Oxford: Oxford University Press, 2009).
50. M. F. Oxenham et al., "Paralysis and Severe Disability Requiring Intensive Care in Neolithic Asia," *Anthropological Science* 117 (2009): 107–112. See also L. Tilley and M. F. Oxenham, "Survival Against the Odds: Modeling the Social Implications of Care Provision to Seriously Disabled Individuals," *International Journal of Paleopathology* 1 (2011): 35–42.
51. E. Crubezy and E. Trinkaus, "Shanidar 1: A Case of Hyperostotic Disease (DISH) in the Middle Paleolithic," *American Journal of Physical Anthropology* 89 (1992): 411–420. This individual may also have had hearing loss, providing further indirect evidence of social support and cooperation among these hominids. E. Trinkaus and S. Villotee, "External Auditory Exostoses and Hearing Loss in the Shanidar 1 Neanderthal," *PLOS ONE* 12 (2017): e0186684.
52. D. W. Frayer, W. E. Horton, R. Macchiarelli, and M. Mussi, "Dwarfism in an Adolescent from the Italian Late Upper Paleolithic," *Nature* 330 (1987): 60–62. Judging from the remains, he was probably about three and a half feet tall and had severely restricted elbow movement, handicaps that would have been serious impediments to the nomadic foraging of his group. When he died, he was buried with special funerary treatment in an "important cave," suggestive of his high status.
53. D. N. Dickel and G. H. Doran, "Severe Neural Tube Defect Syndrome from the Early Archaic of Florida," *American Journal of Physical Anthropology* 80 (1989): 325–334. This boy suffered from spina bifida and had paralysis and severely limited locomotion. There is also evidence of sensory loss and infection in the lower extremities (which can arise from paralysis of this type even in modern times).
54. J. Goodall, *Through a Window: My Thirty Years with the Chimpanzees of Gombe* (Boston: Houghton Mifflin, 1990).
55. J. Anderson, A. Gillies, and L. Lock, "*Pan* Thanatology," *Current Biology* 20 (2010): R349–R351.
56. A. L. Engh et al., "Behavioural and Hormonal Responses to Predation in Female Chacma Baboons (*Papio hamadryas ursinus*)," *Proceedings of the Royal Society B* 273 (2006): 707–712.
57. A. J. Willingham, "A Mourning Orca Mom Carried Her Dead Baby for Days Through the Ocean," CNN, July 27, 2018, https://www.cnn.com/2018/07/27/us/killer-whale-mother-dead-baby-trnd/index.html?no-st=1532790132. There are reports of similar behavior by dolphins "guarding" the bodies of dead peers. K. M. Dudzinski, M. Sakai, M. Masaki, K. Kogi, T. Hishii, and M. Kurimoto, "Behavioural Observations of Bottlenose Dolphins Towards Two Dead Conspecifics," *Aquatic Mammals* 29 (2003): 108–

116; F. Ritter, "Behavioral Responses of Rough-Toothed Dolphins to a Dead Newborn Calf," *Marine Mammal Science* 23 (2007): 429–433.

58. Y. Warren and E. A. Williamson, "Transport of Dead Infant Mountain Gorillas by Mothers and Unrelated Females," *Zoo Biology* 23 (2004): 375–378; D. Biro, T. Humle, K. Koops, C. Sousa, M. Hayashi, and T. Matsuzawa, "Chimpanzee Mothers at Bossou, Guinea Carry the Mummified Remains of Their Dead Infants," *Current Biology* 20 (2010): R351–R352; T. Li, B. Ren, D. Li, Y. Zhang, and M. Li, "Maternal Responses to Dead Infants in Yunnan Snub-Nosed Monkey (*Rhinopithecus bieti*) in the Baimaxueshan Nature Reserve, Yunnan, China," *Primates* 53 (2012): 127–132.
59. E. J. C. van Leeuwen, K. A. Cronin, and D. B. M. Haun, "Tool Use for Corpse Cleaning in Chimpanzees," *Scientific Reports* 7 (2017): 44091. However, the cleaning behavior might simply have been *social* cleaning and not *corpse* cleaning. Possibly, it was maternal instinct, not grief, driving this behavior. W. C. McGrew and D. E. G. Tutin, "Chimpanzee Tool Use in Dental Grooming," *Nature* 241 (1973): 477–478. Incidentally, it is noteworthy here that the tools are being used *socially*, not just to advance an individual's actions.
60. C. Moss, *Echo of the Elephants: The Story of an Elephant Family* (New York: William Morrow, 1992), p. 60. Moss also notes that elephants do not show such respect for the bones of any other animal except possibly a human who has been killed by elephants (p. 61).
61. J. Poole, *Coming of Age with Elephants: A Memoir* (New York: Hyperion, 1996), p. 95. Examples of such death rituals pervade both the scientific and the popular literature. See, for example, M. Meredith, *Elephant Destiny: Biography of an Endangered Species in Africa* (New York: PublicAffairs, 2004).
62. Poole, *Coming of Age*, p. 165.
63. Ibid., p. 161.
64. "World: South Asia Elephant Dies of Grief," BBC News, May 6, 1999, http://news.bbc.co.uk/2/hi/south_asia/337356.stm.
65. G. A. Bradshaw, A. N. Schore, J. L. Brown, J. H. Poole, and C. J. Moss, "Elephant Breakdown," *Nature* 433 (2005): 807.
66. See, for example, O. Karasapan, "Syria's Mental Health Crisis," Brookings Institution, April 25, 2016, https://www.brookings.edu/blog/future-development/2016/04/25/syrias-mental-health-crisis/.
67. M. P. Crawford, "The Cooperative Solving of Problems by Young Chimpanzees," *Comparative Psychology Monographs* 14 (1937).
68. K. A. Mendres and F. B. M. de Waal, "Capuchins Do Cooperate: The Advantage of an Intuitive Task," *Animal Behaviour* 60 (2000): 523–529. Unrelated cotton-top tamarin monkeys proved highly cooperative as well, successfully working together in a similar test 97 percent of the time. K. A. Cronin, A. V. Kurian, and C. T. Snowdon, "Cooperative Problem Solving in a Cooperatively Breeding Primate (*Saguinus oedipus*)," *Animal Behaviour* 69 (2005): 133–142.
69. J. M. Plotnik, R. Lair, W. Suphachoksahakun, and F. B. M. de Waal, "Elephants Know When They Need a Helping Trunk in a Cooperative Task," *PNAS: Proceedings of the*

*National Academy of Sciences* 108 (2011): 5116–5121. Similar experiments have been done with dolphins; see K. Jaakkola, E. Guarino, K. Donegan, and S. L. King, "Bottlenose Dolphins Can Understand Their Partner's Role in a Cooperative Task," *Proceedings of the Royal Society B* 285 (2018): 20180948.

70. M. A. Nowak, "Five Rules for the Evolution of Cooperation," *Science* 314 (2006): 1560–1563.
71. M. Lynn and A. Grassman, "Restaurant Tipping: An Examination of Three 'Rational' Explanations," *Journal of Economic Psychology* 11, no. 2 (1990): 169–181; O. H. Azar, "What Sustains Social Norms and How They Evolve? The Case of Tipping," *Journal of Economic Behavior and Organization* 54 (2004): 49–64.
72. National Volunteer Fire Council, "Final Report: The Role of Volunteer Fire Service in the September 11, 2001, Terrorist Attacks" (Washington, DC, August 1, 2002). In New York City alone, for example, an estimated 2,600 emergency personnel from 285 volunteer fire-rescue departments provided more than 43,700 hours of service in the response to and recovery from the attacks on the World Trade Center. After Hurricane Katrina in 2005, an estimated 575,554 Americans volunteered in the Gulf Coast region, and eighteen million people donated money to relief efforts. "The Power of Help and Hope After Katrina by the Numbers: Volunteers in the Gulf," Corporation for National and Community Service, September 18, 2006, https://www.nationalservice.gov/pdf/katrina_volunteers_respond.pdf.
73. Regarding cooperation and non-kin ties among the Hadza, see C. L. Apicella, F. W. Marlowe, J. H. Fowler, and N. A. Christakis, "Social Networks and Cooperation in Hunter-Gatherers," *Nature* 481 (2012): 497–501, and K. M. Smith, T. Larroucau, I. A. Mabulla, and C. L. Apicella, "Hunter-Gatherers Maintain Assortativity in Cooperation Despite High Levels of Residential Change and Mixing," *Current Biology* 28 (2018): 1–6.
74. R. Axelrod and W. D. Hamilton, "The Evolution of Cooperation," *Science* 211 (1981): 1390–1396.
75. M. A. Nowak and K. Sigmund, "Evolution of Indirect Reciprocity," *Nature* 437 (2005): 1291–1298.
76. G. Hardin, "The Tragedy of the Commons," *Science* 162 (1968): 1243–1248.
77. V. Capraro and H. Barcelo, "Group Size Effect on Cooperation in One-Shot Social Dilemmas. II: Curvilinear Effect," *PLOS ONE* 10 (2015): e0138744; R. M. Isaac and J. M. Walker, "Group Size Effects in Public Goods Provision: The Voluntary Contributions Mechanism," *Quarterly Journal of Economics* 103 (1988): 179–199.
78. B. Allen et al., "Evolutionary Dynamics on Any Population Structure," *Nature* 544 (2017): 227–230.
79. C. Boehm, *Hierarchy in the Forest: The Evolution of Egalitarian Behavior* (Cambridge, MA: Harvard University Press, 2001).
80. J. Henrich et al., "Costly Punishment Across Human Societies," *Science* 312 (2006): 1767–1770; J. Henrich et al., " 'Economic Man' in Cross-Cultural Perspective: Ethnography and Experiments from 15 Small-Scale Societies," *Behavioral and Brain Sciences* 28 (2005): 795–855.
81. W. Güth, R. Schmittberger, and B. Schwarze, "An Experimental Analysis of Ultimatum Bargaining," *Journal of Economic Behavior and Organization* 3 (1982): 367–388; M.

A. Nowak, K. M. Page, and K. Sigmund, "Fairness Versus Reason in the Ultimatum Game," *Science* 289 (2000): 1773–1775.

82. Still another variant is the trust game. Here player 1 decides how much to give to player 2, and the researcher triples that amount. Then player 2 gets to decide how much money to give back to player 1. Player 1 must trust player 2 if giving a lot, and player 2 can be seen as very trustworthy if he gives a lot back to player 1.
83. Henrich et al., "Costly Punishment."
84. J. Henrich, "Does Culture Matter in Economic Behavior? Ultimatum Game Bargaining Among the Machiguenga of the Peruvian Amazon," *American Economic Review* 90 (2000): 973–979.
85. Henrich et al., "Costly Punishment."
86. J. Henrich et al., "Overview and Synthesis," in J. Henrich, R. Boyd, S. Bowles, C. Camerer, W. Fehr, and H. Gintis, eds., *Foundations of Human Sociality: Economic Experiments and Ethnographic Evidence from Fifteen Small-Scale Societies* (Oxford: Oxford University Press, 2004), pp. 8–54.
87. When third-party punishers were added to typical dictator-game situations worldwide, approximately 60 percent of the punishers chose to punish dictators who were too selfish (for instance, those who donated less than half their endowment). E. Fehr and U. Fischbacher, "Third-Party Punishment and Social Norms," *Evolution and Human Behavior* 25 (2004): 63–87.
88. Henrich et al., "Costly Punishment." These authors also argue that there might be an effect of these cultural norms on the *genetic* basis for altruistic behaviors *across* populations, reflecting gene-culture coevolution (see chapter 11). Local punishment behaviors can create social environments favoring the genetic evolution of psychological traits that predispose people to administer, anticipate, and avoid punishment.
89. S. Lotz, T. G. Okimoto, T. Schlösser, and D. Fetchenhauer, "Punitive Versus Compensatory Reactions to Injustice: Emotional Antecedents to Third-Party Interventions," *Journal of Experimental Social Psychology* 47 (2011): 477–480.
90. E. Fehr and S. Gächter, "Altruistic Punishment in Humans," *Nature* 415 (2002): 137–140.
91. R. Boyd, H. Gintis, S. Bowles, and P. J. Richerson, "The Evolution of Altruistic Punishment," *PNAS: Proceedings of the National Academy of Sciences* 100 (2003): 3531–3535.
92. C. Hauert, S. De Monte, J. Hofbauer, and K. Sigmund, "Volunteering as Red Queen Mechanism for Cooperation in Public Goods Games," *Science* 296 (2002): 1129–1132.
93. This cyclical pattern has been called Red Queen dynamics, a reference to a character in Lewis Carroll's *Through the Looking-Glass.* When Alice remarks that where she's from, if you ran very fast for a long time, you'd get somewhere, the Red Queen responds, "A slow sort of country!... Now, *here,* you see, it takes all the running you can do, to keep in the same place."
94. These mathematical models of evolution, therefore, yield an important insight about how punishment could have arisen. The existence of the option of being a loner would drive everyone but cooperators and punishers to extinction, and then, with no defectors left requiring costly punishment, punishers could emerge and survive (and reap the same rewards as the cooperators). J. H. Fowler, "Altruistic Punishment and

the Origin of Cooperation," *PNAS: Proceedings of the National Academy of Sciences* 102 (2005): 7047–7049. Subsequent work built on this. Simple evolutionary models assume that changes in the population are *deterministic*—if one type does better than another, then it will definitely increase in the population. But there is no reason to think that the fittest organism will *always* survive. Evolution is stochastic. C. Hauert, A. Traulsen, H. Brandt, M. A. Nowak, and K. Sigmund, "Via Freedom to Coercion: The Emergence of Costly Punishment," *Science* 316 (2007): 1905–1907.

95. B. Wallace, D. Cesarini, P. Lichtenstein, and M. Johannesson, "Heritability of Ultimatum Game Responder Behavior," *PNAS: Proceedings of the National Academy of Sciences* 104 (2007): 15631–15634; D. Cesarini, C. Dawes, J. H. Fowler, M. Johannesson, P. Lichtenstein, and B. Wallace, "Heritability of Cooperative Behavior in the Trust Game," *PNAS: Proceedings of the National Academy of Sciences* 105 (2008): 3271–3276; D. Cesarini, C. T. Dawes, M. Johannesson, P. Lichtenstein, and B. Wallace, "Genetic Variation in Preferences for Giving and Risk Taking," *Quarterly Journal of Economics* 124 (2009): 809–842.
96. There are benefits to a species from merely living in a group without any other aspects of the social suite. These benefits can include increased feeding efficiency, better exploitation of a variable environment (as when herding animals trample the grasslands in ways that benefit them), and reduction in predation risk (as with schooling fish). Not all of these are seen in every group-living animal, of course.
97. T. M. Caro and M. D. Hauser, "Is There Teaching in Non-Human Animals?," *Quarterly Review of Biology* 67 (1992): 151–174.
98. B. S. Hewlett and C. J. Roulette, "Teaching in Hunter-Gatherer Infancy," *Royal Society Open Science* 3 (2015): 150403.
99. For experimental evidence of teaching in ants, meerkats, and babblers, see N. R. Franks and T. Richardson, "Teaching in Tandem Running Ants," *Nature* 439 (2006): 153; A. Thornton and K. McAuliffe, "Teaching in Wild Meerkats," *Science* 313 (2006): 227–229; and N. J. Raihani and A. R. Ridley, "Experimental Evidence for Teaching in Wild Pied Babblers," *Animal Behaviour* 75 (2008): 3–11.
100. An interesting set of examples involves primates teaching one another things originally learned from humans. For instance, a mother chimpanzee well versed in sign language was spotted showing her infant how to sign. R. S. Fouts, A. D. Hirsch, and D. H. Fouts, "Cultural Transmission of a Human Language in a Chimpanzee Mother-Infant Relationship," in H. E. Fitzgerald, J. A. Mullins, and P. Gage, eds., *Child Nurturance* (New York: Plenum Press, 1982), pp. 159–193. Along the same lines, Koko the gorilla attempted to teach a human to sign by taking the human's hand and molding it into the appropriate shape. F. Patterson and E. Linden, *The Education of Koko* (New York: Holt, Rinehart and Winston, 1981).
101. C. Boesch, "Teaching Among Wild Chimpanzees," *Animal Behaviour* 41 (1991): 530–532.
102. S. Yamamoto, T. Humle, and M. Tanaka, "Basis for Cumulative Cultural Evolution in Chimpanzees: Social Learning of a More Efficient Tool-Use Technique," *PLOS ONE* 8 (2013): e55768.

103. T. Humle and T. Matsuzawa, "Ant-Dipping Among the Chimpanzees of Bossou, Guinea, and Some Comparisons with Other Sites," *American Journal of Primatology* 58 (2002): 133–148.
104. A. Whiten et al., "Cultures in Chimpanzees," *Nature* 399 (1999): 682–685.
105. F. Brotcorne et al., "Intergroup Variation in Robbing and Bartering by Long-Tailed Macaques at Uluwatu Temple (Bali, Indonesia)," *Primates* 58 (2017): 505–516.
106. B. Owens, "Monkey Mafia Steal Your Stuff, Then Sell It Back for a Cracker," *New Scientist*, May 25, 2017.
107. P. I. Chiyo, C. J. Moss, and S. C. Alberts, "The Influence of Life History Milestone and Association Networks on Crop-Raiding Behavior in Male African Elephants," *PLOS ONE* 7 (2012): e31382. For a related example of tool use in dolphins, see J. Mann, M. A. Stanton, E. M. Patterson, E. J. Bienenstock, and L. O. Singh, "Social Networks Reveal Cultural Behaviour in Tool-Using Dolphins," *Nature Communications* 3 (2012): 980.
108. To get a feel for network density, imagine a network of ten individuals. There are (10 x 9)/ 2 = 45 possible ties among them, some fraction or all of which might be present. This fraction is the density.
109. Chiyo, Moss, and Alberts, "Influence of Life History."
110. C. Foley, N. Pettorelli, and L. Foley, "Severe Drought and Calf Survival in Elephants," *Biology Letters* 4 (2008): 541–544.
111. Ironically, while El Niño leads to food shortages for whales, it leads to food surpluses for elephants. G. Wittemyer, I. Douglas-Hamilton, and W. M. Getz, "The Socioecology of Elephants: Analysis of the Processes Creating Multi-Tiered Social Structures," *Animal Behaviour* 69 (2005): 1357–1371. It is interesting that exogenous events such as weather can have an impact on social structures in animals, here via the mechanism of synchronizing reproduction and aligning birth cohorts.
112. L. Weilgart, H. Whitehead, and K. Payne, "A Colossal Convergence," *American Scientist* 84 (1996): 278–287. Elephants and whales also share the qualities of large size, longevity, low birth rates, and communal care of offspring.
113. Whiten et al., "Cultures in Chimpanzees."
114. B. Kenward et al., "Behavioural Ecology: Tool Manufacture by Naïve Juvenile Crows," *Nature* 433 (2005): 121. Regional variation in the shape of crow tools may also reflect cumulative cultural evolution.
115. K. N. Laland and V. M. Janik, "The Animal Culture Debate," *Trends in Ecology and Evolution* 21 (2006): 542–547.
116. S. Mineka and M. Cook, "Social Learning and the Acquisition of Snake Fear in Monkeys," in T. R. Zentall and E. G. Galef Jr., eds., *Social Learning: Psychological and Biological Perspective* (Hillsdale, NJ: Lawrence Erlbaum, 1988), pp. 51–74.
117. B. Sznajder, M. W. Sabelis, and M. Egas, "How Adaptive Learning Affects Evolution: Reviewing Theory on the Baldwin Effect," *Evolutionary Biology* 39 (2012): 301–310.
118. A. Whiten, "The Second Inheritance System of Chimpanzees and Humans," *Nature* 437 (2005): 52–55.v
119. D. P. Schofield, W. C. McGrew, A. Takahashi, and S. Hirata, "Cumulative Culture in Nonhumans: Overlooked Findings from Japanese Monkeys?," *Primates* 59 (2017): 113–122.

120. Boesch, "Teaching Among Wild Chimpanzees."
121. Across six populations, orangutans showed variation in the making of leaf dolls, the building of a sun cover over their nest, and the use of tools for masturbation. A similar analysis was completed for capuchin monkeys. C. P. van Schaik et al., "Orangutan Cultures and the Evolution of Material Culture," *Science* 299 (2003): 102–105; S. Perry et al., "Social Conventions in Wild Capuchin Monkeys: Evidence for Behavioral Traditions in a Neotropical Primate," *Current Anthropology* 44 (2003): 241–268.
122. C. Hobaiter, T. Poisot, K. Zuberbuhler, W. Hoppitt, and T. Gruber, "Social Network Analysis Shows Direct Evidence for Social Transmission of Tool Use in Wild Chimpanzees," *PLOS Biology* 12 (2014): e1001960.
123. J. Allen, M. Weinrich, W. Hoppitt, and L. Rendell, "Network-Based Diffusion Analysis Reveals Cultural Transmission of Lobtail Feeding in Humpback Whales," *Science* 340 (2013): 485–488.
124. D. Wroclavsky, "Killer Whales Bring the Hunt onto Land," Reuters, April 17, 2008, https://www.reuters.com/article/us-argentina-orcas-feature-idUSMAR719014 20080417?src=RSS-SCI.
125. H. Whitehead and L. Rendell, *The Cultural Lives of Whales and Dolphins* (Chicago: University of Chicago Press, 2014).
126. E. J. C. van Leewen, K. A. Cronin, and D. B. M. Haun, "A Group-Specific Arbitrary Tradition in Chimpanzees (*Pan troglodytes*)," *Animal Cognition* 17 (2014): 1421–1425.
127. D. Kim et al., "Social Network Targeting to Maximise Population Behaviour Change: A Cluster Randomised Controlled Trial," *Lancet* 386 (2015): 145–153.
128. To be clear, there are other ways that useless or even harmful practices could be maintained in a population. For instance, if punishment is possible, even fitness-lowering behaviors can be sustained. See R. Boyd and P. J. Richerson, "Punishment Allows the Evolution of Cooperation (or Anything Else) in Sizable Groups," *Ethology and Sociobiology* 13 (1992): 171–195.

### *Chapter 10: Remote Control*

1. D. Attenborough, *Animal Behavior of the Australian Bowerbird,* BBC Studios, February 9, 2007, https://www.youtube.com/watch?v=GPbWJPsBPdA. In the clip, Attenborough notes that he is in New Guinea, not Australia.
2. R. O. Prum, *The Evolution of Beauty: How Darwin's Forgotten Theory of Mate Choice Shapes the Animal World—and Us* (New York: Doubleday, 2017), p. 188.
3. J. Diamond, "Animal Art: Variation in Bower Decorating Style Among Male Bowerbirds *Amblyornis inornatus,*" *PNAS: Proceedings of the National Academy of Sciences* 83 (1986): 3042–3046.
4. L. A. Kelly and J. A. Ender, "Male Great Bowerbirds Create Forced Perspective Illusions with Consistently Different Individual Quality," *PNAS Proceedings of the National Academy of Sciences* 109 (2012): 20980–20985.
5. Prum, *Evolution of Beauty,* p. 199.
6. In 2005, James Fowler and I were exploring how depression might spread among interconnected people in a social network, which we eventually published as J. N. Rosenquist, J. H. Fowler, and N. A. Christakis, "Social Network Determinants of

Depression," *Molecular Psychiatry* 16 (2011): 273–281. In the course of doing that work, we became familiar with the way that psychiatrists think about phenotypes, because, for psychiatrists, phenotypes are not necessarily visible and can be quite subtle. Psychiatrists use the term *endophenotype* to refer to an intermediate, internal phenotype that is on the causal path to a more apparent one. For, example, people who manifest bipolar disorder may have difficulty with face perception, which may be traced to a disorder in the function of at least one gene; that difficulty is the endophenotype. So, in thinking about that topic, it occurred to us that there could also be exophenotypes. The term *endophenotype* was originally coined by Bernard John and Kenneth Lewis with respect to grasshoppers who looked morphologically the same but who manifested different behaviors. B. John and K. R. Lewis, "Chromosome Variability and Geographic Distribution in Insects," *Science* 152 (1966): 711–721.

7. R. Dawkins, *The Extended Phenotype: The Long Reach of the Gene* (Oxford: W. H. Freeman, 1982), p. vi.
8. Writing in 1982, Dawkins noted that there was scant evidence for this argument and that his book should be seen as a work of "advocacy." Ibid., p. vii. One of the most humbling but also reassuring experiences you can have as a scientist is to realize that you are not the first one to have a particular idea.
9. Ibid. Dawkins returned to this topic twenty years later. R. Dawkins, "Extended Phenotype—But Not Too Extended: A Reply to Laland, Turner, and Jablonka," *Biology and Philosophy* 19 (2004): 377–396.
10. H. Eiberg et al., "Blue Eye Color in Humans May Be Caused by a Perfectly Associated Founder Mutation in a Regulatory Element Located Within the *HERC2* Gene Inhibiting *OCA2* Expression," *Human Genetics* 123 (2008): 177–187. This blue color is not ascribable to a blue pigment but to the way the substances in the eye are physically structured (the same way that peacock feather pigments are all actually brown, but they scatter light in ways that make them appear blue or green).
11. Ibid. See also J. J. Negro, M. C. Blázquez, and I. Galván, "Intraspecific Eye Color Variability in Birds and Mammals: A Recent Evolutionary Event Exclusive to Humans and Domestic Animals," *Frontiers in Zoology* 14 (2017): 53.
12. See, for example, R. N. Frank, J. E. Puklin, C. Stock, and L. A. Canter, "Race, Iris Color, and Age-Related Macular Degeneration," *Transactions of the American Ophthalmological Society* 98 (2000): 109–117; and R. Ferguson et al., "Genetic Markers of Pigmentation Are Novel Risk Loci for Uveal Melanoma," *Scientific Reports* 6 (2016): 31191.
13. Anthropologist John Hawks notes that, ten thousand years ago, nobody had blue eyes. Considering the large population of blue-eyed individuals that exist now, it raises the question of why blue-eyed people have a perhaps 5 percent advantage in reproducing compared with non-blue-eyed people. J. Hawks et al., "Recent Acceleration of Human Adaptive Evolution," *PNAS: Proceedings of the National Academy of Sciences* 104 (2007): 20753–20758.
14. D. Peshek, N. Semmaknejad, D. Hoffman, and P. Foley, "Preliminary Evidence That the Limbal Ring Influences Facial Attractiveness," *Evolutionary Psychology* 9 (2011): 137–146.

15. However, brown-eyed people were perceived as more trustworthy in one study. K. Kleisner, L. Priplatova, P. Frost, and J. Flegr, "Trustworthy-Looking Face Meets Brown Eyes," *PLOS ONE* 8 (2013): e53285.
16. Dawkins, "Extended Phenotype."
17. Housing architecture belongs to the realm of culture; however, architecture might affect our evolution by other means, a process known as gene-culture coevolution, discussed in chapter 11.
18. I. Arndt and J. Tautz, *Animal Architecture* (New York: Harry N. Abrams, 2014); M. Hansell, *Built by Animals: The Natural History of Animal Architecture* (Oxford: Oxford University Press, 2007).
19. T. A. Blackledge, N. Scharff, J. A. Coddington, T. Szüts, J. W. Wenzel, C. Y. Hayashi, and I. Agnarssona, "Reconstructing Web Evolution and Spider Diversification in the Molecular Era," *PNAS: Proceedings of the National Academy of Sciences* 106 (2009): 5229–5234. The evolution and origin of webs across Araneae is still in dispute. J. E. Garb, T. DiMauro, V. Vo, and C. Y. Hayashi, "Silk Genes Support the Single Origin of Orb Webs," *Science* 312 (2006): 1762. Webs are differentiated not only by their architecture but also by features such as ultraviolet reflection (less reflection of ultraviolet light makes the webs harder for prey to see), stickiness, fiber strength, and tension-maintaining mechanisms.
20. The concept of an adaptive radiation was most famously illustrated by Charles Darwin, who, during his voyage to the Galápagos Islands in 1835, observed that finches had different sorts of beaks depending on the food sources available at their locations in the islands—thick beaks to break the coverings of seeds, thin ones to get nectar from cactuses. A gene known as *Alx1* varies in its type across the species, affecting the shape of their beaks. S. Lamichhaney et al., "Evolution of Darwin's Finches and Their Beaks Revealed by Genome Sequencing," *Nature* 518 (2015): 371–375. Incidentally, this same gene is known to affect facial features in mice and humans. In the case of spiderwebs, of course, it's not a body part that varies but an external artifact.
21. Species with drab plumage often have more elaborate bowers, while species with colorful plumage tend to build less impressive bowers, as if some species over time have shifted from emphasizing physical phenotypes to emphasizing behavioral exophenotypes. Dawkins, *Extended Phenotype,* p. 199.
22. J. N. Weber, B. K. Peterson, and H. E. Hoekstra, "Discrete Genetic Modules Are Responsible for Complex Burrow Evolution in *Peromyscus* Mice," *Nature* 493 (2013): 402–405.
23. D. P. Hughes, "On the Origins of Parasite Extended Phenotypes," *Integrative and Comparative Biology* 54 (2014): 210–217.
24. W. M. Ingram, L. M. Goodrich, E. A. Robey, and M. B. Eisen, "Mice Infected with Low-Virulence Strains of *Toxoplasma gondii* Lose Their Innate Aversion to Cat Urine, Even After Extensive Parasite Clearance," *PLOS ONE* 8 (2013): e75246.
25. D. G. Biron, F. Ponton, C. Joly, A. Menigoz, B. Hanelt, and F. Thomas, "Water-Seeking Behavior in Insects Harboring Hairworms: Should the Host Collaborate?," *Behavioral Ecology* 16 (2005): 656–660.

26. S. A. Adamo, "The Strings of the Puppet Master: How Parasites Change Host Behavior," in D. P. Hughes, J. Brodeur, and F. Thomas, eds., *Host Manipulation by Parasites* (Oxford: Oxford University Press, 2012), pp. 36–53.
27. M. A. Fredericksen et al., "Three-Dimensional Visualization and a Deep-Learning Model Reveal Complex Fungal Parasite Networks in Behaviorally Manipulated Ants," *PNAS: Proceedings of the National Academy of Sciences* 114 (2017): 12590–12595.
28. D. P. Hughes, T. Wappler, and C. C. Labandeira, "Ancient Death-Grip Leaf Scars Reveal Ant-Fungal Parasitism," *Biology Letters* 7 (2011): 67–70.
29. T. R. Sampson and S. K. Mazmanian, "Control of Brain Development, Function, and Behavior by the Microbiome," *Cell Host and Microbe* 17 (2015): 565–576.
30. A. D. Blackwell et al., "Helminth Infection, Fecundity, and Age of First Pregnancy in Women," *Science* 350 (2015): 970–972.
31. A. Y. Panchin, A. I. Tuzhikov, and Y. V. Panchin, "Midichlorians—the Biomeme Hypothesis: Is There a Microbial Component to Religious Rituals?," *Biology Direct* 9 (2014): 14. See also S. K. Johnson et al., "Risky Business: Linking *Toxoplasma gondii* Infection and Entrepreneurship Behaviours Across Individuals and Countries," *Proceedings of the Royal Society B* 285 (2018): 20180822.
32. L. T. Morran et al., "Running with the Red Queen: Host-Parasite Coevolution Selects for Biparental Sex," *Science* 333 (2011): 216–218. This "running in place" phenomenon may explain the origins of sexual reproduction. See M. Ridley, *The Red Queen: Sex and the Evolution of Human Nature* (New York: Macmillan, 1993).
33. J. W. Bradbury and S. L. Vehrencamp, *Principles of Animal Communication,* 2nd ed. (Oxford: Oxford University Press, 2011).
34. N. Demandt, B. Saus, R. H. J. M. Kurvers, J. Krause, J. Kurtz, and J. P. Scharsack, "Parasite-Infected Sticklebacks Increase the Risk-Taking Behavior of Uninfected Group Members," *Proceedings of the Royal Society B* 285 (2018): 20180956.
35. Even this example could get complicated. If the manner of deposition of feces resulted in the ground becoming more fertile in a way that was beneficial to the organism, then, yes, this would be seen as an exophenotypic effect.
36. Dawkins, *Extended Phenotype,* pp. 206–207.
37. L. Glowacki, A. Isakov, R. W. Wrangham, R. McDermott, J. H. Fowler, and N. A. Christakis, "Formation of Raiding Parties for Intergroup Violence Is Mediated by Social Network Structure," *PNAS: Proceedings of the National Academy of Sciences* 113 (2016): 12114–12119.
38. F. Biscarini, H. Bovenhuis, J. van der Poel, T. B. Rodenburg, A. P. Jungerius, and J. A. M. van Arendonk, "Across-Line SNP Association Study for Direct and Associative Effect on Feather Damage in Laying Hens," *Behavior Genetics* 40 (2010): 715–727.
39. Dawkins, *Extended Phenotype,* p. 230.
40. P. Lieberman, "The Evolution of Human Speech," *Current Anthropology* 48 (2007): 39–66; D. Ploog, "The Neural Basis of Vocalization," in T. J. Crow, ed., *The Speciation of Modern Homo Sapiens* (Oxford: Oxford University Press, 2002), pp. 121–135; W. Enard et al., "Molecular Evolution of *FOXP2,* a Gene Involved in Speech and Language," *Nature* 418 (2002): 869–872; F. Vargha-Khadem, D. G. Gadian, A. Copp, and M. Mishkin,

"*FOXP2* and the Neuroanatomy of Speech and Language," *Nature Reviews Neuroscience* 6 (2005): 131–138; E. G. Atkinson, "No Evidence for Recent Selection at *FOXP2* Among Diverse Human Populations," *Cell* 174 (2018): 1424–1435 (which casts doubt on a crucial role of *FOXP2* in human speech).

41. J. H. Fowler, C. T. Dawes, and N. A. Christakis, "Model of Genetic Variation in Human Social Networks," *PNAS: Proceedings of the National Academy of Sciences* 106 (2009): 1720–1724.
42. L. N. Trut, "Early Canid Domestication: The Farm-Fox Experiment," *American Scientist* 87 (1999): 160–169.
43. Ibid., p. 163.
44. E. Ratliff, "Taming the Wild," *National Geographic,* March 2011.
45. B. Hare, V. Wobber, and R. Wrangham, "The Self-Domestication Hypothesis: Evolution of Bonobo Psychology Is Due to Selection Against Aggression," *Animal Behaviour* 83 (2012): 573–585.
46. K. Prufer et al., "The Bonobo Genome Compared with the Chimpanzee and Human Genomes," *Nature* 486 (2012): 527–531.
47. B. Hare and S. Kwetuenda, "Bonobos Voluntarily Share Their Own Food with Others," *Current Biology* 20 (2010): 230–231.
48. Hare, Wobber, and Wrangham, "Self-Domestication Hypothesis."
49. C. Theofanopoulou et al., "Self-Domestication in *Homo sapiens:* Insights from Comparative Genomics," *PLOS ONE* 12 (2017): e0185306.
50. S. Pinker, *Better Angels of Our Nature: Why Violence Has Declined* (New York: Viking, 2011).
51. "Intentional Homicides (per 100,000 People)," World Bank, https://data.worldbank.org/indicator/VC.IHR.PSRC.P5?year_high_desc=false.

### *Chapter 11: Genes and Culture*

1. A. D. Carlson, "The Wheat Farmer's Dilemma: Notes from Tractor Land," *Harper's,* July 1931, pp. 209–210. Some punctuation changed for clarity. See also R. C. Williams, *Fordson, Farmall, and Poppin' Johnny: A History of the Farm Tractor and Its Impact on America* (Champaign: University of Illinois Press, 1987).
2. Agricultural historian Bruce Gardner estimates that a tractor did the work of five horses and that the number of horses and mules peaked in about 1920 (at perhaps twenty-five million animals) whereas the number of tractors peaked around 1960 (at perhaps five million), with a crossover point in about 1945 when the two sources provided roughly equal power for farms. B. L. Gardner, *American Agriculture in the Twentieth Century: How It Flourished and What It Cost* (Cambridge, MA: Harvard University Press, 2006).
3. "Mechanization on the Farm in the Early Twentieth Century," excerpt from *The People in the Pictures: Stories from the Wettach Farm Photos* (Iowa Public Television, 2003), Iowa Pathways, http://www.iptv.org/iowapathways/artifact/mechanization-farm-early-20th-century.
4. Some potentially disadvantageous implications also affected everyone. For instance, tractors made farmers more independent, so social customs regarding the exchange of labor died out, and neighbors no longer needed to rely on one another as much.

5. D. Thompson, "How America Spends Money: 100 Years in the Life of the Family Budget," *Atlantic,* April 5, 2012. See also: US Department of Agriculture Economic Research Service, "Food Expenditures," data available at https://www.ers.usda.gov/data-products/food-expenditures/. In 1900, 41 percent of the workforce was employed in agriculture; in 2000, only 1.9 percent was. See C. Dimitri, A. Effland, and N. Conklin, "The 20th Century Transformation of U.S. Agriculture and Farm Policy," United States Department of Agriculture, *Economic Information Bulletin* 3 (2005).
6. Where you were born also mattered, as technological innovation proceeded at different speeds on different continents.
7. D. Tuzin, *The Cassowary's Revenge: The Life and Death of Masculinity in a New Guinea Society* (Chicago: University of Chicago Press, 1997), p. 102.
8. One estimate offered that "to till an acre of land with a spade required 96 hours (5,760 minutes); to plow an acre with a yoke of oxen and a crude wooden plow took twenty-four hours (1,440 minutes); with a steel plow such as John Deere developed took five to eight hours (300 to 480 minutes); but in 1998, a 425-horsepower John Deere 9400 four-wheel-drive pulling a fifteen-bottom plow tilled an acre every 3.2 minutes.... Everyone who enjoys the abundant supply of inexpensive food should be grateful for John Deere and his plow." H. M. Drache, "The Impact of John Deere's Plow," *Illinois History Teacher* 8, no. 1 (2001): 2–13, http://www.lib.niu.edu/2001/iht810102.html. For one exploration of the growth of knowledge, see C. Hidalgo, *Why Information Grows: The Evolution of Order, from Atoms to Economies* (New York: Basic Books, 2015).
9. M. Fackler, "Tsunami Warnings, Written in Stone," *New York Times,* April 20, 2011. There is also the related phenomenon of designating the low-water marks in European rivers. The Elbe River in the Czech Republic is dotted with "hunger stones" commemorating historic droughts, with inscriptions like IF YOU SEE ME, WEEP, going back five hundred years. See C. Domonoske, "Drought in Central Europe Reveals Cautionary 'Hunger Stones' in Czech Republic," NPR, August 24, 2018.
10. S. Bhuamik, "Tsunami Folklore 'Saved Islanders,'" BBC News, January 20, 2005.
11. There are some exceptions, such as the invention of brush heads on insect fishing rods used by chimps and the birdsongs that are specific to locations and that can get ever more complex with passing generations. J. Henrich and C. Tennie, "Cultural Evolution in Chimpanzees and Humans," in M. Muller, R. Wrangham, and D. Pilbeam, eds., *Chimpanzees and Human Evolution* (Cambridge, MA: Harvard University Press, 2017), pp. 645–702.
12. P. J. Richerson and R. Boyd, *Not by Genes Alone: How Culture Transformed Human Evolution* (Chicago: University of Chicago Press, 2005), p. 5.
13. J. Henrich, *The Secret of Our Success: How Culture Is Driving Human Evolution, Domesticating Our Species, and Making Us Smarter* (Princeton, NJ: Princeton University Press, 2016). See also K. N. Laland, J. Odling-Smee, and S. Myles, "How Culture Shaped the Human Genome: Bringing Genetics and the Human Sciences Together," *Nature Reviews Genetics* 11 (2010): 137–148; and P. J. Richerson, R. Boyd, and J. Henrich, "Gene-Culture Coevolution in the Age of Genomics," *PNAS: Proceedings of the National Academy of Sciences* 107 (2010): 8985–8992.

14. J. Henrich and J. Broesch, "On the Nature of Cultural Transmission Networks: Evidence from Fijian Village for Adaptive Learning Biases," *Philosophical Transactions of the Royal Society B* 366 (2011): 1139–1148; M. Chudek, S. Heller, S. Birch, and J. Henrich, "Prestige-Biased Cultural Learning: Bystanders' Differential Attention to Potential Models Influences Children's Learning," *Evolution and Human Behavior* 38 (2012): 46–56.
15. M. Nielsen and K. Tomaselli, "Overimitation in Kalahari Bushman Children and the Origins of Human Cultural Cognition," *Psychological Science* 21 (2010): 729–736.
16. B. G. Galef, "Strategies for Social Learning: Testing Predictions from Formal Theory," *Advances in the Study of Behavior* 39 (2009): 117–151; W. Hoppitt and K. N. Laland, "Social Processes Influencing Learning in Animals: A Review of the Evidence," *Advances in the Study of Behavior* 38 (2008): 105–165.
17. J. Henrich and F. J. Gil-White, "The Evolution of Prestige: Freely Conferred Deference as a Mechanism for Enhancing the Benefits of Cultural Transmission," *Evolution and Human Behavior* 22 (2001): 165–196.
18. I. G. Kulanci, A. A. Ghazanfar, and D. I. Rubenstein, "Knowledgeable Lemurs Become More Central in Social Networks," *Current Biology* 28 (2018): 1306–1310.
19. M. Chudek et al., "Prestige-Biased Cultural Learning"; P. L. Harris and K. H. Corriveau, "Young Children's Selective Trust in Informants," *Philosophical Transactions of the Royal Society B* 366 (2011): 1179–1187.
20. Henrich and Broesch, "Nature of Cultural Transmission Networks."
21. The possible role of genes and heritability in social hierarchy is still being worked out. See M. A. Vanderkooij and C. Sandi, "The Genetics of Social Hierarchies," *Current Opinions in Behavioral Sciences* 2 (2015): 52–57. Regardless of whether dominance (in males) arises from genetic or other sources, this attribute does not seem to be transmitted across many generations in the male line, however. See J. S. Lansing et al., "Male Dominance Rank Skews the Frequency Distribution of Y Chromosome Haplotypes in Human Populations," *PNAS: Proceedings of the National Academy of Sciences* 105 (2008): 11645–11650.
22. J. L. Martin, "Is Power Sexy?," *American Journal of Sociology* 111 (2005): 408–446.
23. J. Snyder, L. Kirkpatrick, and C. Barrett, "The Dominance Dilemma: Do Women Really Prefer Dominant Men as Mates?," *Personal Relations* 15 (2008): 425–444.
24. C. von Ruden, M. Gurven, and H. Kaplan, "Why Do Men Seek? Fitness Payoffs to Dominance and Prestige," *Proceedings of the Royal Society B* 278 (2011): 2223–2232. Much of what counted as prestige here was simply the number of friends the men had. Dominance status peaks about a decade before prestige status in the Tsimané (and likely in other populations). See C. C. von Ruden, M. Gurven, and H. Kaplan, "The Multiple Dimensions of Male Social Status in an Amazonian Society," *Evolution and Human Behavior* 29 (2008): 402–415.
25. This expression was coined by Robert Boyd. Henrich, *Secret of Our Success,* p. 26.
26. Ibid., p. 27.
27. R. E. Schultes, "Ethnopharmacological Conservation: A Key to Progress in Medicine," *Acta Amazonica* 18 (1988): 393–406.
28. Some Amazonian tribes have even developed ways of assessing the strength of their poisons by counting the number of times a captive test frog can jump after being pricked.

Another metric involves counting how many trees a monkey can leap to after being struck with a dart. A one-tree concentration is very potent and deadly, but the weaker three-tree curare is used to subdue live animals to keep as pets. In general, blowgun darts kill small animals quickly, but it might take twenty minutes and many darts for large monkeys or tapirs to die. Ethnobotanist Steve Beyer was told by a member of the formerly headhunting Shapra who live on the border between Peru and Ecuador that it can take up to twenty darts to bring down a human. S. Beyer, "Arrow Poisons," *Singing to the Plants: Steve Beyer's Blog on Ayahuasca and the Amazon,* http://www.singingtotheplants.com/2008/01/arrow-poisons/; S. V. Beyer, *Singing to the Plants: A Guide to Mestizo Shamanism in the Upper Amazon* (Albuquerque: University of New Mexico Press, 2009).

29. Preparation procedures in other tribes can be simpler and faster. L. Rival, "Blowpipes and Spears: The Social Significance of Huaorani Technological Choices," in P. Descola and G. Palsson, eds., *Nature and Society: Anthropological Perspectives* (London: Routledge, 1996), pp. 145–164.
30. C. R. Darwin, *The Descent of Man, and Selection in Relation to Sex* (London: John Murray, 1871). This idea traces through important work by geneticists Marcus Feldman and Luigi Luca Cavalli-Sforza in the 1970s, to work by Robert Boyd and Peter Richerson in the 1980s, through ongoing work by Joe Henrich, Ken Laland, and others more recently. See M. Feldman and L. Cavalli-Sforza, "Cultural and Biological Evolutionary Processes, Selection for a Trait Under Complex Transmission," *Theoretical Population Biology* 9 (1976): 238–259.
31. M. T. Pfeffer, "Implications of New Studies of Hawaiian Fishhook Variability for Our Understanding of Polynesian Settlement History," in G. Rakita and T. Hurt, eds., *Style and Function: Conceptual Issues in Evolutionary Archaeology* (Westport, CT: Bergin and Garvey, 2001), pp. 165–181.
32. The spread and simultaneity of fishhook invention is a challenging topic because we do not yet know the complete story. And, of course, natural selection, unlike an inventor, does not have a purpose in advance. F. Jacob, "Evolution and Tinkering," *Science* 196 (1977): 1161–1166. See also P. V. Kirch, *Feathered Gods and Fishhooks: An Introduction to Hawaiian Archaeology and Prehistory* (Honolulu: University of Hawaii Press, 1997).
33. R. C. Dunnell, "Style and Function: A Fundamental Dichotomy," *American Antiquity* 43 (1978): 192–202.
34. Scientists have long taken note of the possible existence of common traits (e.g., biological, technological, cultural) across disparate exemplars (e.g., organisms, machines, societies). Beginning in the nineteenth century, anatomists (such as Richard Owen) started conceptually distinguishing between different classes of biological structures with shared functions and forms, designating these structures as either "homologous" or "analogous." Whereas homologous structures arise from a common evolutionary origin and transmission, analogous structures evolve independently as a common solution to a similar set of environmental challenges. Analogous traits, therefore, emerge through the process of convergent evolution. It is, of course, possible that homologues with common cultural lineages are functional. Many different processes—from convergence to diffusion—can ultimately perpetuate both functional and stylistic traits.

35. F. M. Reinman, "Fishing: An Aspect of Oceanic Economy; An Archaeological Approach," *Fieldiana: Anthropology* 56 (1967): 95–208.
36. S. O'Connor, R. Ono, and C. Clarkson, "Pelagic Fishing at 42,000 Years Before the Present and the Maritime Skills of Modern Humans," *Science* 334 (2011): 1117–1121.
37. B. Gramsch, J. Beran, S. Hanik, and R. S. Sommer, "A Palaeolithic Fishhook Made of Ivory and the Earliest Fishhook Tradition in Europe," *Journal of Archaeological Science* 40 (2013): 2458–2463.
38. D. Sahrhage and J. Lundbeck, *A History of Fishing* (Berlin: Springer-Verlag, 1992). These hooks appear to have incorporated a harpoon-type barb approximately eleven thousand years ago, and it seems that thereafter, examples of typical barbed, line-attached hooks spread from northern Eurasia, first to northeastern (but not western) Europe and China and later to Japan, Polynesia, and the northwestern American coast. By the Bronze Age, roughly forty-three hundred years ago, fishhooks made of metal, as well as more traditional materials (such as bone and flint), were virtually ubiquitous in coastal populations.
39. R. F. Heizer, "Artifact Transport by Migratory Animals and Other Means," *American Antiquity* 9 (1944): 395–400. See also L. C. W. Landberg, "Tuna Tagging and the Extra-Oceanic Distribution of Curved, Single-Piece Shell Fishhooks in the Pacific," *American Antiquity* 31 (1966): 485–493; and F. M. Reinman, "Tuna Tagging and Shell Fishhooks: A Comment from Oceania," *American Antiquity* 33 (1968): 95–100.
40. Reinman, "Tuna Tagging"; F. M. Reinman, "Fishhook Variability: Implications for the History and Distribution of Fishing Gear in Oceania," in R. C. Green and M. Kelly, eds., *Studies in Oceanic Culture History*, vol. 1 (Honolulu: Bernice Pauahi Bishop Museum, 1970) pp. 47–59; P. V. Kirch, "The Archaeological Study of Adaptation: Theoretical and Methodological Issues," *Advances in Archaeological Method and Theory* 3 (1980): 101–156.
41. Knowledge can be lost entirely. A complex astronomical device with toothed gears known as the Antikythera mechanism (which was found underwater in Greece in 1902) was not re-created for over one thousand years. T. Freeth et al., "Decoding the Ancient Greek Astronomical Calculator Known as the Antikythera Mechanism," *Nature* 444 (2006): 587–591. In another example, as of 2017, there was one last surviving sea-silk seamstress, in Italy; when she dies, the secrets of the craft, which have been in the same matrilineal clan for more than a thousand years, will be lost. E. Stein, "The Last Surviving Sea Silk Seamstress," BBC, September 6, 2017, http://www.bbc.com/travel/story/20170906-the-last-surviving-sea-silk-seamstress. Incidentally, this observation means that cognitively identical populations, such as ancestral hominin species, may leave different archaeological evidence not because of their brains but because of their population size. Henrich, *Secret of Our Success*, chapter 13.
42. N. Casey, "Thousands Spoke His Language in the Amazon. Now, He's the Only One," *New York Times*, December 26, 2017.
43. L. Bromham, X. Hua, T. G. Fitzpatrick, and S. J. Greenhill, "Rate of Language Evolution Is Affected by Population Size," *PNAS: Proceedings of the National Academy of Sciences* 112 (2015): 2097–2102.

44. M. A. Kline and R. Boyd, "Population Size Predicts Technological Complexity in Oceania," *Proceedings of the Royal Society B* 277 (2010): 2559–2564. But see the following two papers, which examined populations drawn from the northwestern coastal regions of North America, for conflicting evidence regarding the relationship of population size and tool-kit size: M. Collard, M. Kemery, and S. Banks, "Causes of Tool Kit Variation Among Hunter-Gatherers: A Test of Four Competing Hypotheses," *Canadian Journal of Archeology* 29 (2005): 1–19; D. Read, "An Interaction Model for Resource Implement Complexity Based on Risk and Number of Annual Moves," *American Antiquity* 73 (2008): 599–625.
45. W. Oswalt, *An Anthropological Analysis of Food-Getting Technology* (New York: John Wiley and Sons, 1976); R. Torrence, "Hunter-Gatherer Technology: Macro and Microscale Approaches," in C. Panter-Brick, R. H. Layton, and P. Rowley-Conwy, eds., *Hunter-Gatherers: An Interdisciplinary Perspective* (Cambridge: Cambridge University Press, 2000), pp. 99–143.
46. Collard, Kemery, and Banks, "Causes of Tool Kit Variation."
47. M. Derex, M.-P. Beugin, B. Godelle, and M. Raymond, "Experimental Evidence for the Influence of Group Size on Cultural Complexity," *Nature* 503 (2013): 389–391.
48. For an early and simplified model of this, see J. Henrich, "The Evolution of Innovation-Enhancing Institutions," in M. J. O'Brien and S. J. Shennan, eds., *Innovation in Cultural Systems: Contributions from Evolutionary Anthropology* (Cambridge, MA: MIT Press, 2010), pp. 99–120.
49. There was likely culture before our species, *Homo sapiens*. Humans have surely had some aspect of it as long as we have been engaged in complex foraging, which dates back at least to *Homo erectus* (who lived from about 1.9 million years ago to about one hundred and forty-three thousand years ago). Hence, culture probably began to shape our genes even earlier than one million years ago.
50. Moreover, by about four hundred thousand years ago, we find regional variation in stone tools in the archaeological record, which is consistent with variation from place to place in local culture.
51. At the other extreme, if the environment is sufficiently stable, individual learning may be sufficient, and social learning may offer little, if any, gain in efficiency or sophistication, so why bother with evolving this capacity?
52. R. Wrangham, *Catching Fire: How Cooking Made Us Human* (New York: Basic Books, 2009).
53. On running barefoot, see D. E. Lieberman et al., "Foot Strike Patterns and Collision Forces in Habitually Barefoot Versus Shod Runners," *Nature* 463 (2010): 531–535.
54. D. E. Lieberman, *The Story of the Human Body: Evolution, Health, and Disease* (New York: Pantheon, 2013).
55. M. W. Feldman and L. L. Cavalli-Sforza, "On the Theory of Evolution Under Genetic and Cultural Transmission, with Application to the Lactose Absorption Problem," in M. W. Feldman, ed., *Mathematical Evolutionary Theory* (Princeton, NJ: Princeton University Press, 1989), pp. 145–173; K. Aoki, "A Stochastic Model of Gene-Culture Coevolution Suggested by the 'Culture Historical Hypothesis' for the Evolution of

Adult Lactose Absorption in Humans," *PNAS: Proceedings of the National Academy of Sciences* 83 (1986): 2929–2933.

56. Y. Itan, B. L. Jones, C. J. E. Ingram, D. M. Swallow, and M. G. Thomas, "A Worldwide Correlation of Lactase Persistence Phenotype and Genotype," *BMC Evolutionary Biology* 10 (2019): 36.
57. S. A. Tishkoff et al., "Convergent Adaptation of Human Lactase Persistence in Africa and Europe," *Nature Genetics* 39 (2007): 31–40. Other studies have documented that the relevant gene variants were not present in ancestral humans. J. Burger, M. Kirchner, B. Bramanti, W. Haak, and M. G. Thomas, "Absence of Lactase-Persistence-Associated Alleles in Early Neolithic Europeans," *PNAS: Proceedings of the National Academy of Sciences* 104 (2007): 3736–3741. Regarding the likely independent evolution of lactase persistence in response to camel domestication, see N. S. Enattah et al., "Independent Introduction of Two Lactase-Persistence Alleles into Human Populations Reflects Different History of Adaptation to Milk Culture," *American Journal of Human Genetics* 82 (2008): 57–72.
58. C. Sather, *The Bajau Laut: Adaptations, History, and Fate in a Maritime Fishing Society of South-Eastern Sabah* (Oxford: Oxford University Press, 1997).
59. E. Schagatay, A. Lodin-Sundstrom, and E. Abrahamsson, "Underwater Working Times in Two Groups of Traditional Apnea Divers in Asia: The Ama and the Bajau," *Diving and Hyperbaric Medicine* 41 (2011): 27–30.
60. M. A. Ilardo et al., "Physiological and Genetic Adaptations to Diving in Sea Nomads," *Cell* 173 (2018): 569–580.
61. S. Myles et al., "Identification of a Candidate Genetic Variant for the High Prevalence of Type Two Diabetes in Polynesians," *European Journal of Human Genetics* 15 (2007): 584–589; J. R. Binden and P. T. Baker, "Bergmann's Rule and the Thrifty Genotype," *American Journal of Physical Anthropology* 104 (1997): 201–210; P. Houghton, "The Adaptive Significance of Polynesian Body Form," *Annals of Human Biology* 17 (1990): 19–32. See also R. L. Minster et al., "A Thrifty Variant in *CREBRF* Strongly Influences Body Mass Index in Samoans," *Nature Genetics* 48 (2016): 1049–1054.
62. D. Dediu and D. R. Ladd, "Linguistic Tone Is Related to the Population Frequency of the Adaptive Haplogroups of Two Brain Size Genes, *ASP* and *Microcephalin*," *PNAS: Proceedings of the National Academy of Sciences* 104 (2007): 10944–10949.
63. O. Galor and Ö. Özak, "The Agricultural Origins of Time Preference," *American Economic Review* 106 (2016): 3064–3103.
64. In the case of starchy foods and amylase, the specific genetic mechanism here may be different, involving something known as copy-number variation. G. H. Perry, "Diet and the Evolution of Human Amylase Gene Copy Number Variation," *Nature Genetics* 39 (2007): 1256–1260.
65. W. H. Durham, *Coevolution: Genes, Culture, and Human Diversity* (Stanford, CA: Stanford University Press, 1991), pp. 103–109. See also M. J. O'Brien and K. N. Laland, "Genes, Culture, and Agriculture: An Example of Human Niche Construction," *Current Anthropology* 53 (2012): 434–470.

66. Alas, these same mutations also increase the risk of cystic fibrosis. E. van de Vosse et al., "Susceptibility to Typhoid Fever Is Associated with a Polymorphism in the Cystic Fibrosis Transmembrane Conductance Regulator (CFTR)," *Human Genetics* 118 (2005): 138–140; E. M. Poolman and A. P. Galvani, "Evaluating Candidate Agents of Selective Pressure for Cystic Fibrosis," *Journal of the Royal Society Interface* 4 (2007): 91–98. See also J. Hawks, E. T. Wang, G. M. Cochran, H. C. Harpending, and R. K. Moyzis, "Recent Acceleration of Human Adaptive Evolution," *PNAS: Proceedings of the National Academy of Sciences* 104 (2007): 20753–20758; and W. McNeill, *Plagues and Peoples* (Garden City, NY: Doubleday, 1976).
67. C. L. Apicella, "High Levels of Rule-Bending in a Minimally Religious and Largely Egalitarian Forager Population," *Religion, Brain and Behavior* 8 (2018): 133–148. See also A. Norenzayan et al., "The Cultural Evolution of Prosocial Religions," *Behavioral and Brain Sciences* 39 (2016): 1–65.
68. C. Apicella, personal communication, November 1, 2017.
69. For a review, see P. B. Gray and B. C. Campbell, "Human Male Testosterone, Pair-Bonding, and Fatherhood," in P. T. Ellison and P. B. Gray, eds., *Endocrinology of Social Relationships* (Cambridge, MA: Harvard University Press, 2009), pp. 270–293. See also P. B. Gray, S. M. Kahlenberg, E. S. Barrett, S. F. Lipson, and P. T. Ellison, "Marriage and Fatherhood Are Associated with Lower Testosterone in Males," *Evolution and Human Behavior* 23 (2002): 193–201; A. E. Storey, C. J. Walsh, R. L. Quinton, and K. E. Wynne-Edwards, "Hormonal Correlates of Paternal Responsiveness in New and Expectant Fathers," *Evolution and Human Behavior* 21 (2000): 79–95; and S. M. van Anders and N. V. Watson, "Relationship Status and Testosterone in North American Heterosexual and Non-Heterosexual Men and Women: Cross-Sectional and Longitudinal Data," *Psychoneuroendocrinology* 31 (2006): 715–723. Entry into marriage and fatherhood may not reduce testosterone in polygynous societies, however, and the reason is likely that married men are still seeking reproductive partners. P. B. Gray, "Marriage, Parenting, and Testosterone Variation Among Kenyan Swahili Men," *American Journal of Physical Anthropology* 122 (2003): 279–286. Regarding the necessity of face-to-face interaction with one's children, see M. N. Muller, F. W. Marlowe, R. Bugumba, and P. T. Ellison, "Testosterone and Paternal Care in East African Foragers and Pastoralists," *Proceedings of the Royal Society B* 276 (2009): 347–354.
70. J. F. Schulz, "The Churches' Ban on Consanguineous Marriages, Kin-Networks, and Democracy" (paper, June 12, 2017), https://ssrn.com/abstract=2877828.
71. A. H. Bittles and M. L. Black, "Consanguinity, Human Evolution, and Complex Diseases," *PNAS: Proceedings of the National Academy of Sciences* 107 (2010): 1779–1786. Mortality among first-cousin progeny is about 3.5 percent higher than it is among nonconsanguineous offspring. While there are therefore such inbreeding costs, those costs might be compensated by the *social* benefits of tight families, at least in some environments. The impact of consanguineous marriage may also vary between modern environments and older or more traditional environments. And women in societies with cousin marriage have much higher fertility (that is, more babies), which might compensate for their increased infant mortality. The tight family bonds in

consanguineous marriages might also supply lots of alloparents to help the mother. Hence, overall, the selection pressure might actually go the other way, in favor of such consanguineous marriages, in some circumstances.

72. R. Boyd and P. J. Richerson, "Cultural Transmission and the Evolution of Cooperative Behavior," *Human Ecology* 10 (1982): 325–351; R. Boyd and P. J. Richerson, "The Evolution of Reciprocity in Sizeable Groups," *Journal of Theoretical Biology* 132 (1988): 337–356; M. Chudek and J. Henrich, "Culture-Gene Coevolution, Norm-Psychology and the Emergence of Human Prosociality," *Trends in Cognitive Sciences* 15 (2011): 218–226; R. Boyd, H. Gintis, S. Bowles, and P. J. Richerson, "The Evolution of Altruistic Punishment," *PNAS: Proceedings of the National Academy of Sciences* 100 (2003): 3531–3535; J. Henrich et al., "Costly Punishment Across Human Societies," *Science* 312 (2006): 1767–1770; H. Gintis, "The Hitchhiker's Guide to Altruism: Gene Culture Coevolution and the Internalization of Norms," *Journal of Theoretical Biology* 220 (2003): 407–418; H. Gintis, "The Genetic Side of Gene-Culture Coevolution: Internalization of Norms and Prosocial Emotions," *Journal of Economic Behavior and Organization* 53 (2004): 57–67. The same types of mathematical analyses can help explain the emergence of prosocial emotions (that warm glow you feel in the company of your friends).
73. Laland, Odling-Smee, and Myles, "How Culture Shaped the Human Genome"; Hawks, et al., "Recent Acceleration of Human Adaptive Evolution." Of course, any acceleration in human evolution may be due to factors other than the emergence of cultural selection pressures. Another issue, for example, is the rise in the number of humans on the planet. With larger populations of an animal, beneficial mutations are more likely to occur somewhere in the population simply by chance; these larger populations of our species may have been facilitated by the invention of agriculture. However, it is also the case that cultural impacts can cease or reverse, which would mean that the genetic changes would be incomplete (a "partial genetic sweep" that did not reach "fixation" in the population).
74. X. Yi et al., "Sequencing of Fifty Human Exomes Reveals Adaptation to High Altitude," *Science* 329 (2010): 75–78.
75. I imagine, however, that the settlers could have created cultural or religious rules that "required" them to make periodic trips or pilgrimages to the lowlands (which might have helped to reduce the stress of being at high altitudes the rest of the time). In modern times, I suppose they could import bottled oxygen.
76. Some scientists have speculated that as cultural adaptation supplanted genetic adaptation, the pace of human genetic evolution should have *slowed*. But this seems unlikely not only because of evidence of a quickening pace to evolution in the past forty thousand years, but also because the migration into new environments and the creation of new cultural niches could themselves serve to increase the selective pressures on new allelic variants in human populations (e.g., to cope with new infections made likely by dense settlements or to cope with new foodstuffs).
77. To be clear, the recent increase in myopia is not primarily due to the weakening of selection pressure but rather arises because of changes in lifestyle that bring people indoors and affect what distance they are focusing on while their eyes are developing in childhood. Focusing on items far away, in bright light, as when people spend more time

outside and less time indoors, may be necessary to avoid myopia. East Asia, in particular, is seeing an epidemic of myopia. Sixty years ago, about 10 to 20 percent of the Chinese population was nearsighted; now up to 90 percent of teenagers and young adults are. For a good review, see E. Dolgin, "The Myopia Boom," *Nature* 519 (2015): 276–278.

78. O. S. Platt et al., "Mortality in Sickle Cell Disease—Life Expectancy and Risk Factors for Early Death," *New England Journal of Medicine* 330 (1994): 1639–1644.
79. Plus, for female patients with sickle-cell disease, pregnancy used to be so dangerous to the mother and fetus that doctors would recommend birth control if patients made it to sexual maturity, hence obviating reproduction.
80. D. G. Finniss, T. J. Kaptchuk, F. Miller, and F. Benedetti, "Biological, Clinical, and Ethical Advances of Placebo Effects," *Lancet* 375 (2010): 686–695.
81. There is some evidence that, in a modern British population, intelligence and educational achievement are being selected *against* (i.e., such populations have lower fecundity). J. S. Sanjak, J. Sidorenko, M. R. Robinson, K. R. Thornton, and P. M. Visscher, "Evidence of Directional and Stabilizing Selection in Contemporary Humans," *PNAS: Proceedings of the National Academy of Sciences* 115 (2017): 151–156.
82. J. Tooby and L. Cosmides, "Evolutionary Psychology and the Generation of Culture. I: Theoretical Considerations," *Ethology and Sociobiology* 10 (1989): 29–49; J. H. Barkow, L. Cosmides, and J. Tooby, eds., *The Adapted Mind: Evolutionary Psychology and the Generation of Culture* (Oxford: Oxford University Press, 1992).
83. G. Cochran and H. Harpending, *The 10,000 Year Explosion: How Civilization Accelerated Human Evolution* (New York: Basic Books, 2009).

### *Chapter 12: Natural and Social Laws*

1. Sometimes, the analogy is applied to explain the workings of the body, for example when the immune system is seen as an "army." But usually, the metaphor of the body is used the other way around, to illuminate the function of society. See: E. Martin, *Flexible Bodies* (Boston: Beacon Press, 1995). Rudolf Virchow, one of the leading physicians of the nineteenth century, described the functioning of the human body in terms of a society, with each cell a citizen. He submitted that living organisms represented a kind of "cellular democracy," a "republic of cells," and a "cellular state." R. Porter, *The Greatest Benefit to Mankind: A Medical History of Humanity* (New York: W. W. Norton, 1999), p. 331. Conversely, Herbert Spencer, a nineteenth-century Englishman who dabbled in a wide array of disciplines, saw human society as a scaled-up analogue of the human body. H. Spencer, *The Principles of Biology* (London: Williams and Norgate, 1864).
2. T. L. Patavinus, *History of Rome,* trans. C. Roberts, bk. 2 (London: J. M. Dent and Sons, 1905).
3. One of Aesop's fables, "The Belly and the Members," closely parallels this speech by Menenius Agrippa. J. Jacobs, *The Fables of Aesop* (London: Macmillan, 1902), pp. 72–73. Indeed, fables of the body and the belly occur in an assortment of other texts, according to Jacobs, including a 1250 BCE Egyptian allegory located in the Upanishads; a Buddhist fable tucked within the Chinese Avadanas; and Jewish and Christian varieties scattered throughout the Bible.

4. A. F. Jensen, *India: Its Culture and People* (New York: Longman, 1991), p. 32. This order of castes is typically as follows: the Brahman, or priestly, caste represents society's head; the Kshatriya, or warrior, caste is analogized to its arms; the Vaishya caste (traders and landowners) are the legs; and the Sudra caste (servants) are the feet.
5. Plato, *Republic,* bk. 4, 436b.
6. 1 Cor. 12:15–26 (English Standard Version).
7. See, for example, A. D. Harvey, *Body Politic: Political Metaphor and Political Violence* (Newcastle, UK: Cambridge Scholars, 2007). Writing in 1518, statesman Thomas More provided a typical characterization: "A kingdom in all its parts is like a man.... The king is the head; the people form the other parts." Ibid., p. 23. See also C. W. Mills, "Body Politic, Bodies Impolitic," *Social Research* 78 (2011): 583–606.
8. T. Hobbes, *Leviathan* (Whitefish, MT: Kessinger, 2004), p. 1.
9. The Leviathan wears a crown and holds a sword in his right hand and a crook in his left—symbols, perhaps, of force and justice. These symbols also relate to Hobbes's vision of a commonwealth that is both civil (sword/king) and ecclesiastical (crook/bishop). The two sides of the frontispiece, below the giant Leviathan, follow this dichotomy. Each side element reflects corresponding powers: castle to church, crown to miter, canon to excommunication, weapons to logic, and battlefield to religious courts. Hobbes sees the Leviathan as having humanlike qualities, including being prone to use violence for self-preservation. L. Ostman, "The Frontispiece of Leviathan—Hobbes' Bible Use," *Akademeia* 2 (2012): ea0112.
10. This is usually given as Job 41:33 in modern English translations. Hobbes was surely familiar with the biblical notion of a fearsome, seagoing leviathan, which is mentioned in Job. And, indeed, the people on the sovereign's body appear like scales.
11. Hobbes, *Leviathan,* p. 16.
12. Regarding animal rights, see P. Singer, *Animal Liberation: A New Ethics for Our Treatment of Animals* (New York: Random House, 1975); and M. Scully, *Dominion: The Power of Man, the Suffering of Animals, and the Call to Mercy* (New York: St. Martin's, 2002).
13. N. K. Sanders, *The Epic of Gilgamesh* (Assyrian International News Agency Books Online, n.d.), tablet 1, p. 4, http://www.aina.org/books/eog/eog.pdf. At the beginning of the epic, these two beings—one untamed, another civilized—are in opposition. They eventually come together to form a harmonious friendship, but only after Enkidu is domesticated.
14. R. N. Bellah, *Religion in Human Evolution* (Cambridge, MA: Harvard University Press, 2011).
15. God commands humans to "be fruitful and multiply and fill the earth and subdue it, and have dominion over the fish of the sea and over the birds of the heavens and over every living thing that moves on the earth." Genesis 1:28 (English Standard Version). Humankind's ranking above, and separation from, nature is further cemented when God commands Adam to name all the creatures of the earth. Throughout the Bible, the wilderness is often portrayed as the cradle of evil and danger, the crucible where Jesus was tested by the Devil. Philosopher John Passmore contends that the Bible does not prescribe a single unified principle concerning how humans ought to

behave in relation to nature. J. A. Passmore, *Man's Responsibility for Nature: Ecological Problems and Western Traditions* (London: Duckworth, 1974).

16. Aristotle, *Politics,* trans. C. Lord (Chicago: University of Chicago Press, 2013). To be clear, Aristotle saw human minds as apart from nature, but he believed that human bodies, including their desires, were as animalistic as any part of nature and needed to be mastered.
17. L. White Jr., "The Historical Roots of Our Ecological Crisis," *Science* 155 (1967): 1203–1207.
18. T. Aquinas, *Summa Contra Gentiles,* trans. V. J. Bourke, bk. 3 (Notre Dame, IN: Notre Dame Press, 1975). Beyond underscoring human ascendancy and clarifying the anthropocentric purpose of nature, Aquinas also considered and then dismissed the possibility that humans imbued with a soul and reason might have certain ethical obligations to other creatures. T. Aquinas, *Summa Theologica,* trans. Fathers of the English Dominica Province (Cincinnati: Benziger Brothers, 1974). The teachings of Saint Francis are often highlighted by those seeking to demonstrate that early Christianity did not necessarily create a schism between humans and the natural world. In stark opposition to Aquinas, Francis asserted that one's treatment of the natural world should be directly modeled on one's treatment of one's own kind. But in fact, Francis's views were so discordant with Catholic Church dogma that historian Lynn White Jr. concluded, "The prime miracle of Saint Francis is the fact that he did not end at the stake." L. White, "The Historical Roots of Our Ecological Crisis," *Science* 155 (1967): 1203–1207.
19. In some sense, therefore, the seemingly secular Scientific Revolution fostered the idea that humans had a right and a duty to subdue nature in alignment with God's divine will. See J. Agassi, *The Very Idea of Modern Science: Francis Bacon and Robert Boyle* (New York: Springer Dordrecht Heidelberg, 2012). See also C. Merchant, *The Death of Nature: Women, Ecology, and the Scientific Revolution* (New York: HarperCollins, 1980).
20. R. Descartes, *Meditations on First Philosophy,* trans. A. Anderson and L. Anderson (Baltimore: Agora, 2012).
21. I. Kant, *Groundwork of the Metaphysics of Morals,* ed. M. Gregor and J. Timmermann, rev. ed. (Cambridge: Cambridge University Press, 2012).
22. R. W. Emerson, *Nature, Addresses, and Lectures,* ed. A. R. Ferguson (Cambridge, MA: Belknap Press, 1971), pp. 13–28.
23. C. R. Darwin, *The Descent of Man, and Selection in Relation to Sex* (London: John Murray, 1871).
24. However, to attribute this surging interest in the untamed and sublime natural world entirely to the Industrial Revolution would be a gross oversimplification; these ideas first came to fruition before the Industrial Revolution, and many transcendentalists thought well of the progress induced by steam and coal power.
25. The social sciences—spanning sociology, economics, anthropology, political science, and psychology—are composed of a number of different disciplinary traditions with varied topical, methodological, and philosophical underpinnings. L. McDonald, *Early Origins of the Social Sciences* (Montreal: McGill-Queen's University

Press, 1993). The fields of psychology and, to a lesser extent, anthropology have always had a strong biological component.

26. J. Searle, *The Construction of Social Reality* (New York: Free Press, 1995).
27. S. M. Lindberg, J. S. Hyde, and J. L. Petersen, "New Trends in Gender and Mathematics Performance: A Meta-Analysis," *Psychological Bulletin* 136 (2010): 1123–1135.
28. S. A. Cartwright, "Diseases and Peculiarities of the Negro Race," *DeBow's Review, Southern and Western States*, vol. 9, (New Orleans: n.p., 1851).
29. S. Arbesman, *The Half-Life of Facts: Why Everything We Know Has an Expiration Date* (New York: Current, 2012).
30. A. Comte, *A General View of Positivism*, trans. J. H. Bridges (London: Trubner, 1865).
31. E. Durkheim, *The Rules of Sociological Method*, trans. W. D. Halls (New York: Free Press, 1982). Durkheim argued that a set of specifically social facts existed that was not reducible to individual humans, that there was a kind of holistic social reality that transcended the thoughts and actions of individuals, and that this reality required a special scientific method to appreciate.
32. Plato, *The Republic*, trans. T. Griffith (Cambridge: Cambridge University Press, 2000); T. Gould, *The Ancient Quarrel Between Poetry and Philosophy* (Princeton, NJ: Princeton University Press, 1990).
33. B. F. Skinner, *About Behaviorism* (New York: Knopf, 1974).
34. V. Reppert, *C. S. Lewis's Dangerous Idea: In Defense of the Argument from Reason* (Downers Grove, IL: InterVarsity Press, 2003).
35. Regarding beauty, see R. O. Prum, *The Evolution of Beauty: How Darwin's Forgotten Theory of Mate Choice Shapes the Animal World—and Us* (New York: Doubleday, 2017).
36. As eloquently encapsulated by physicist Werner Heisenberg: "The positivists have a simple solution: the world must be divided into that which we can say clearly and the rest, which we had better pass over in silence. But can anyone conceive of a more pointless philosophy, seeing that what we can say clearly amounts to next to nothing? If we omitted all that is unclear, we would probably be left with completely uninteresting and trivial tautologies." W. Heisenberg, *Physics and Beyond: Memories of a Life in Science* (London: George Allen and Unwin, 1971), p. 213. By emphasizing only observable scientific facts, positivism misses much of the larger picture, on "the great ocean of truth." Heisenberg also argued that positivists could actually undermine their own program and illustrated this with an example from the history of science in which claims of meteorites in the eighteenth century were "dismissed as rank superstition," whereas, of course, they do exist.
37. D. Kevles, *In the Name of Eugenics: Genetics and the Uses of Human Heredity* (New York: Knopf, 1985); R. Merton, *The Sociology of Science: Theoretical and Empirical Investigations* (Chicago: University of Chicago Press, 1973). We should also be cognizant of the "replication crisis" afflicting so many branches of science in the 2010s, including psychology, economics, physics, biology, epidemiology, and oncology.
38. P. W. Anderson, "More Is Different," *Science* 177 (1972): 393–396.
39. Interestingly, humans are natural-born essentialists. From an early age, we categorize objects according to fundamental commonalities, discriminate between these categories, and assign each category a basic essence. P. Bloom, *How Pleasure Works: The*

*New Science of Why We Like What We Like* (New York: W. W. Norton, 2010); S. A. Gelman, *The Essential Child: Origins of Essentialism in Everyday Thought* (New York: Oxford University Press, 2010).

40. In the end, we would get Laplace's Demon. For "such an intellect," French mathematician Pierre-Simon Laplace argued in 1814, "nothing would be uncertain and the future just like the past would be present before its eyes." P. S. Laplace, *A Philosophical Essay on Probabilities,* 6th ed., trans. F. W. Truscott and F. L. Emory (New York: Dover, 1951), p. 4.
41. There is actually much debate about whether the idea that the world obeys natural laws and is predictable can be reconciled with the idea that humans can truly have free will. Since both ideas—determinism and free will—seem so sensible and have so much philosophical and empirical support, people who hold the view that the two are compatible actually have a special name: *compatibilists.* The reason this is important is that determinism subverts the entire basis for moral assessment. How can we hold someone responsible for his or her choices and actions if they are determined by past events outside his or her control or, indeed, by the choices and actions of others?
42. R. Lewontin, *Biology as Ideology: The Doctrine of DNA* (Concord, ON: House of Anansi Press, 1991); S. J. Gould, *The Mismeasure of Man* (New York: W. W. Norton, 1981).
43. D. Nelkin, "Biology Is Not Destiny," *New York Times,* September 28, 1995. For two examples of such studies, see G. Guo, M. E. Roettger, and T. Cai, "The Integration of Genetic Propensities into Social-Control Models of Delinquency and Violence Among Male Youths," *American Sociological Review* 73 (2008): 543–568; and A. Feder, E. J. Nestler, and D. S. Charney, "Psychobiology and Molecular Genetics of Resilience," *Nature Reviews Neuroscience* 10 (2009): 446–457.
44. J. Hibbing, "Ten Misconceptions Concerning Neurobiology and Politics," *Perspectives on Politics* 11 (2013): 475–489.
45. For an early history, see M. H. Haller, *Eugenics: Hereditarian Attitudes in American Thought* (New Brunswick, NJ: Rutgers University Press, 1963).
46. Such extreme adherence to the doctrine of a blank slate is very common in sociology. See M. Horowitz, A. Haynor, and K. Kickham, "Sociology's Sacred Victims and the Politics of Knowledge: Moral Foundations Theory and Disciplinary Controversies," *American Sociologist* (2018). For crime statistics, see E. A. Carson and D. Golinelli, "Prisoners in 2012—Advance Counts" (report no. NCJ 242467, Bureau of Justice Statistics, July 2013). Regarding the overwhelmingly male violence in chimpanzees in similar proportions as humans (in terms of both perpetrators and victims), see M. L. Wilson et al., "Lethal Aggression in *Pan* Is Better Explained by Adaptive Strategies Than Human Impact," *Nature* 513 (2014): 414–417; and J. M. Gomez, M. Verdo, A. Gonzalez-Negras, and M. Mendez, "The Phylogenetic Roots of Human Lethal Violence," *Nature* 538 (2016): 233–237.
47. Some argue that even if knowledge of scientific reality led inexorably to eugenics and racial discrimination, it would not justify ignorance of that reality, because the supreme value is recognizing the truth, whatever it may be and whatever the costs of that knowledge.

48. E. Fromm, *Man for Himself: An Inquiry into the Psychology of Ethics* (New York: Rinehart, 1947), p. 20.
49. Ibid., pp. 20–21.
50. B. D. Earp, A. Sandeberg, and J. Savulescu, "Brave New Love: The Threat of High-Tech 'Conversion' Therapy and the Bio-Oppression of Sexual Minorities," *AJOB Neuroscience* 5 (2014): 4–12.
51. Historicus, "Stalin on Revolution," *Foreign Affairs*, January 1949, p. 196.
52. R. C. Tucker, "Stalin and the Uses of Psychology," *World Politics* 8 (1956): 455–483. See also E. van Ree, *The Political Thought of Joseph Stalin: A Study in Twentieth-Century Revolutionary Patriotism* (London: Routledge Curzon, 2002), p. 290.
53. J. A. Getty, G. T. Rittersporn, and V. N. Zemskov, "Victims of the Soviet Penal System in the Pre-War Years," *American Historical Review* 98 (1993): 1017–1049; S. G. Wheatcroft, "The Scale and Nature of German and Soviet Repression and Mass Killings, 1930–45," *Europe-Asia Studies* 48 (1996): 1319–1353; S. G. Wheatcroft, "More Light on the Scale of Repression and Excess Mortality in the Soviet Union in the 1930s," *Soviet Studies* 42 (1990): 355–367; S. G. Wheatcroft, "Victims of Stalinism and the Soviet Secret Police: The Comparability and Reliability of the Archival Data. Not the Last Word," *Europe-Asia Studies* 51 (1999): 315–345.
54. A. G. Walder, "Marxism, Maoism, and Social Change: A Re-Examination of the 'Voluntarism' in Mao's Strategy and Thought," *Modern China* 3 (1977): 125–160.
55. In a well-known speech at the Yenan forum, for instance, Mao argued that " 'the theory of human nature' which some people in Yenan advocate as the basis of their so-called theory of literature and art puts the matter in just this way and is wholly wrong." M. Tse-Tung, *Selected Works of Mao Tse-Tung*, vol. 3 (Peking: People's Publishing House, 1960), p. 90.
56. My friend psychologist Dan Gilbert argues that to ask whether morality is outside us is a riddle that traps the mind; it's like asking, "Is it longer to New York City or by bus?" It's a riddle that has proved to be quite provocative, however.
57. D. C. Lahti and B. S. Weinstein, "The Better Angels of Our Nature: Group Stability and the Evolution of Moral Tension," *Evolution and Human Behavior* 26 (205): 47–63.
58. D. Hume, "Concerning Moral Sentiment," appendix 1 in *An Enquiry Concerning the Principles of Morals* (London: A. Millar, 1751), p. 289.
59. Taken from a documentary featuring a conversation between Iris Murdoch and David Pears that was part of the TV series *Logic Lane* (1972), directed by Michael Chanan; cited in N. Krishna, "Is Goodness Natural?," *Aeon*, November 28, 2017, https://aeon.co/essays/how-philippa-foot-set-her-mind-against-prevailing-moral-philosophy.
60. R. M. Hare, *The Language of Morals* (Oxford: Clarendon Press, 1952).
61. D. Gilbert, *Stumbling on Happiness* (New York: Knopf, 2016).
62. P. Foot, cited in R. Hursthouse, *On Virtue Ethics* (Oxford: Oxford University Press, 2002), p. 196. See also P. Foot, "Does Moral Subjectivism Rest on a Mistake?," *Oxford Journal of Legal Studies* 15 (1995): 1–14.
63. Foot, "Does Moral Subjectivism Rest on a Mistake?"
64. A. H. Maslow, "A Theory of Human Motivation," *Psychological Review* 50 (1943): 370–396.

65. A. H. Maslow, *The Farther Reaches of Human Nature* (New York: Viking, 1971), p. 279.
66. H. Shirado and N. A. Christakis, "Locally Noisy Autonomous Agents Improve Global Human Coordination in Network Experiments," *Nature* 545 (2017): 370–375. In this experiment, involving four thousand people in 230 groups working together to solve a coordination problem, we evaluated what happened when we secretly placed some artificial-intelligence bots within the groups. How do people behave when they are part of a hybrid system made up of humans and machines? We found that if the bots acted in certain ways (paradoxically, if we deliberately made them just a little bit imperfect in their decision-making), the real humans did not mind their presence and actually performed better.
67. M. L. Traeger, S. S. Sebo, M. Jung, B. Scassellati, and N. A. Christakis, "Vulnerable Robots Positively Shape Human Conversational Dynamics in a Human-Robot Team" (unpublished manuscript, 2018). Other work by roboticist Brian Scassellati's research group has shown that embedding a robot in a group can affect how an autistic child communicates not only with the robot but also with other humans. E. S. Kim et al., "Social Robots as Embedded Reinforcers of Social Behavior in Children with Autism," *Journal of Autism and Developmental Disorders* 43 (2013): 1038–1049.
68. E. Awad et al., "The Moral Machine Experiment," *Nature* 563 (2018): 59–64.
69. Or perhaps, after it becomes more common to have sex with humanoid robots, people will modify how they have sex with each other.
70. D. Silver et al., "Mastering the Game of Go with Deep Neural Networks and Tree Search," *Nature* 529 (2016): 484–489; D. Silver et al., "Mastering the Game of Go Without Human Knowledge," *Nature* 550 (2017): 354–359.
71. J. D. Sander and J. K. Joung, "CRISPR-Cas Systems for Editing, Regulating, and Targeting Genomes," *Nature Biotechnology* 32 (2014): 347–355. More generally, see J. Enriquez and S. Gullans, *Evolving Ourselves: How Unnatural Selection and Nonrandom Mutation Are Changing Life on Earth* (New York: Current, 2015).
72. J. Hughes, *Citizen Cyborg: Why Democratic Societies Must Respond to the Redesigned Human of the Future* (New York: Basic Books, 2004); J. Harris, *Enhancing Evolution: The Ethical Case for Making Better People* (Princeton, NJ: Princeton University Press, 2007). In November 2018, a Chinese scientist announced that he had used CRISPR technology to edit the genomes of a pair of twin girls so as to disable a pathway that HIV uses to infect cells. Though the girls appear to have been born healthy, this first use of CRISPR to edit the human germline provoked an international outcry. See D. Cyranoski and H. Ledford, "Genome-Edited Baby Claim Provokes International Outcry," *Nature* 563 (2018): 607–608.
73. This word is very rarely used, and prior uses are not quite like mine. For instance, Daniel Bell uses the term to describe the act of explaining the evolution of the meaning of sociological concepts in D. Bell, "Sociodicy: A Guide to Modern Usage," *American Scholar* 35 (1966): 696–714. Pierre Bourdieu appears to use it to explain how ideology works to justify a then-current state of affairs in P. Bourdieu, "Symbolic Power," *Critique of Anthropology* 4 (1979): 77–85.
74. S. A. Pinker, *Enlightenment Now: The Case for Reason, Science, Humanism, and Progress* (New York: Viking, 2018).

# Index

Note: Italic page numbers refer to figures and tables.